Metalworking Processes: Theoretical and Experimental Study

Metalworking Processes: Theoretical and Experimental Study

Guest Editors

Konrad Laber
Janusz Tomczak
Beata Leszczyńska-Madej
Grażyna Mrówka-Nowotnik
Magdalena Barbara Jabłońska

Basel • Beijing • Wuhan • Barcelona • Belgrade • Novi Sad • Cluj • Manchester

Guest Editors

Konrad Laber
Department of Metallurgy
and Metal Technology
Czestochowa University
of Technology
Czestochowa
Poland

Janusz Tomczak
Faculty of Mechanical
Engineering
Lublin University
of Technology
Lublin
Poland

Beata Leszczyńska-Madej
Department of Materials
Science and Non-Ferrous
Metals Engineering
AGH University of Science
and Technology
Cracow
Poland

Grażyna Mrówka-Nowotnik
Department of Material
Science
Rzeszów University
of Technology
Rzeszów
Poland

Magdalena Barbara Jabłońska
Department of Technology
of Materials
Silesian University
of Technology
Katowice
Poland

Editorial Office
MDPI AG
Grosspeteranlage 5
4052 Basel, Switzerland

This is a reprint of the Special Issue, published open access by the journal *Materials* (ISSN 1996-1944), freely accessible at: www.mdpi.com/journal/materials/special_issues/6SSIIQGQ4A.

For citation purposes, cite each article independently as indicated on the article page online and using the guide below:

Lastname, A.A.; Lastname, B.B. Article Title. *Journal Name* **Year**, *Volume Number*, Page Range.

ISBN 978-3-7258-3370-2 (Hbk)
ISBN 978-3-7258-3369-6 (PDF)
https://doi.org/10.3390/books978-3-7258-3369-6

© 2025 by the authors. Articles in this book are Open Access and distributed under the Creative Commons Attribution (CC BY) license. The book as a whole is distributed by MDPI under the terms and conditions of the Creative Commons Attribution-NonCommercial-NoDerivs (CC BY-NC-ND) license (https://creativecommons.org/licenses/by-nc-nd/4.0/).

Contents

Preface vii

Bożena Gzik-Zroska, Kamil Joszko, Agata Piatek, Wojciech Wolański, Edyta Kawlewska and Arkadiusz Szarek et al.
The Influence of Hot Isostatic Pressing on the Mechanical Properties of Ti-6Al-4V Samples Printed Using the LENS Method
Reprinted from: *Materials* **2025**, *18*, 612, https://doi.org/10.3390/ma18030612 1

Hui Li, Yingxia Zhu, Wei Chen, Chen Yuan and Lei Wang
Advanced Bending and Forming Technologies for Bimetallic Composite Pipes
Reprinted from: *Materials* **2024**, *18*, 111, https://doi.org/10.3390/ma18010111 21

Wenting Wei, Zheng Liu, Qinglong Liu, Guanghua Zhou, Guocheng Liu and Yanxiong Liu et al.
Flow Behavior Analysis of the Cold Rolling Deformation of an M50 Bearing Ring Based on the Multiscale Finite Element Model
Reprinted from: *Materials* **2024**, *18*, 77, https://doi.org/10.3390/ma18010077 45

Maciej Suliga, Piotr Szota, Monika Gwoździk, Joanna Kulasa and Anna Brudny
The Influence of Temperature in the Wire Drawing Process on the Wear of Drawing Dies
Reprinted from: *Materials* **2024**, *17*, 4949, https://doi.org/10.3390/ma17204949 67

Dariusz Leśniak, Józef Zasadziński, Wojciech Libura, Beata Leszczyńska-Madej, Marek Bogusz and Tomasz Latos et al.
Structure and Mechanical Properties of AlMgSi(Cu) Extrudates Straightened with Dynamic Deformation
Reprinted from: *Materials* **2024**, *17*, 3983, https://doi.org/10.3390/ma17163983 90

Xinpan Yu, Yong Wang, Huibin Wu and Na Gong
Ultra-Fine Bainite in Medium-Carbon High-Silicon Bainitic Steel
Reprinted from: *Materials* **2024**, *17*, 2225, https://doi.org/10.3390/ma17102225 114

Konrad Błażej Laber
Analysis of the Uniformity of Mechanical Properties along the Length of Wire Rod Designed for Further Cold Plastic Working Processes for Selected Parameters of Thermoplastic Processing
Reprinted from: *Materials* **2024**, *17*, 905, https://doi.org/10.3390/ma17040905 131

Krzysztof Remsak, Sonia Boczkal, Kamila Limanówka, Bartłomiej Płonka, Konrad Żyłka and Mateusz Wegrzyn et al.
Effects of Zn, Mg, and Cu Content on the Properties and Microstructure of Extrusion-Welded Al–Zn–Mg–Cu Alloys
Reprinted from: *Materials* **2023**, *16*, 6429, https://doi.org/10.3390/ma16196429 153

Rong Wang, Size Peng, Bowen Zhou, Xiaoyang Jiang, Maojun Li and Pan Gong
Experimental Investigation on Microstructure Alteration and Surface Morphology While Grinding 20Cr2Ni4A Gears with Different Grinding Allowance Allocation
Reprinted from: *Materials* **2023**, *16*, 6111, https://doi.org/10.3390/ma16186111 169

Grzegorz Banaszek, Kirill Ozhmegov, Anna Kawałek, Sylwester Sawicki, Alexandr Arbuz and Abdrakhman Naizabekov
Modeling of Closure of Metallurgical Discontinuities in the Process of Forging Zirconium Alloy
Reprinted from: *Materials* **2023**, *16*, 5431, https://doi.org/10.3390/ma16155431 187

Runze Zhang, Jinshan He, Shiguang Xu, Fucheng Zhang and Xitao Wang
The Optimized Homogenization Process of Cast 7Mo Super Austenitic Stainless Steel
Reprinted from: *Materials* **2023**, *16*, 3438, https://doi.org/10.3390/ma16093438 . . . . **216**

Konrad Błażej Laber
Innovative Methodology for Physical Modelling of Multi-Pass Wire Rod Rolling with the Use of a Variable Strain Scheme
Reprinted from: *Materials* **2023**, *16*, 578, https://doi.org/10.3390/ma16020578 . . . . **230**

Dariusz Leśniak, Józef Zasadziński, Wojciech Libura, Krzysztof Żaba, Sandra Puchlerska and Jacek Madura et al.
FEM Numerical and Experimental Study on Dimensional Accuracy of Tubes Extruded from 6082 and 7021 Aluminium Alloys
Reprinted from: *Materials* **2023**, *16*, 556, https://doi.org/10.3390/ma16020556 . . . . **254**

Preface

The currently used technologies of plastic working processes and thermomechanical processes of metals and alloys, in addition to shaping the manufactured products and minimizing the energy consumption of the production process, must firstly ensure the required microstructure and related mechanical and technological properties. This Special Issue covers new groundbreaking trends in the plastic working and thermomechanical treatment processes of metals and alloys. The editors would like to thank all the authors for preparing their manuscripts.

Konrad Laber, Janusz Tomczak, Beata Leszczyńska-Madej, Grażyna Mrówka-Nowotnik, and
Magdalena Barbara Jabłońska
Guest Editors

Article

The Influence of Hot Isostatic Pressing on the Mechanical Properties of Ti-6Al-4V Samples Printed Using the LENS Method

Bożena Gzik-Zroska [1,*], Kamil Joszko [2], Agata Piątek [3], Wojciech Wolański [2], Edyta Kawlewska [2], Arkadiusz Szarek [4], Wojciech Kajzer [1] and Grzegorz Stradomski [5]

1. Department of Biomaterials and Medical Devices Engineering, Faculty of Biomedical Engineering, Silesian University of Technology, 41-800 Zabrze, Poland; wojciech.kajzer@polsl.pl
2. Department of Biomechatronics, Faculty of Biomedical Engineering, Silesian University of Technology, 41-800 Zabrze, Poland; kamil.joszko@polsl.pl (K.J.); wwolanski@polsl.pl (W.W.); edyta.kawlewska@polsl.pl (E.K.)
3. SKN "Biokreatywni", Faculty of Biomedical Engineering, Silesian University of Technology, 41-800 Zabrze, Poland; agatpia347@student.polsl.pl
4. Department of Technology and Automation, Faculty of Mechanical Engineering and Computer Science, Czestochowa University of Technology, 42-200 Częstochowa, Poland; arkadiusz.szarek@pcz.pl
5. Faculty of Production Engineering and Materials Technology, Czestochowa University of Technology, 42-201 Częstochowa, Poland; grzegorz.stradomski@pcz.pl
* Correspondence: bozena.gzikzroska@polsl.pl

Abstract: The aim of this work was to assess the influence of the parameters of the hot isostatic pressing (HIP) process and the direction of printing of Ti-6Al-4V samples made using the laser-engineered net shaping (LENS) method on strength properties. The tests were carried out using a static testing machine and a digital image correlation system. Samples before and after the HIP process were tested. The HIP process was carried out at a temperature of 1150 °C, a heating time of 240 min and various pressure values of 500, 1000 and 1500 bar. Based on the comparative analysis of the test results, it has been shown that the ability to adjust the parameters of the HIP process has a significant impact on the final mechanical properties of the samples.

Keywords: titanium alloy; static tensile test; digital image correlation; 3D printing parameters

1. Introduction

Additive technologies, which have been developing very dynamically in recent years, have their origins in rapid prototyping techniques. These technologies date back to the 1980s. Various types of rapid prototyping have been introduced since the Cold War for the development of advanced military industry and space technology. Depending on their purpose, they would be likely to be implemented; choosing of materials has become a sensitive issue with a lot of challenges. Many improvements have been made since then, but the idea itself is still based on the same assumptions. The early application of these solutions was very closely linked to foundry engineering, and the main task was to obtain a 3D model to assess manufacturability. Since then, the development of laser technology, electronics, computer science and also material engineering has allowed the creation of fully fledged products using 3D printing technology. Figure 1 shows such an example. However, 3D printing technology itself has its drawbacks, in addition to its huge advantages. Therefore, for some applications, especially those where the printed element is to be used, additional finishing processing is necessary. One such technique, the task of which is to change the functional properties of the finished product, is heat pressure processing [1–3].

Figure 1. High-temperature isostatic press HIP AIP8-30H-PED.

The isostatic pressing process is an advanced material processing method used in industry to produce metal parts with high durability. This method involves compressing the material in a chamber using the appropriate temperature and pressure [4,5].

Based on the appropriately selected temperature and pressure, the isostatic pressing process is divided into a high-temperature process, using high temperatures in the range of 1200–2200 °C and high isostatic pressure for processing [6,7], and a low-temperature process, using high hydrostatic pressure and an ambient temperature.

Hot isostatic pressing (HIP) is a technology that, in general, improves the mechanical properties of the processed material; however, this technique can also degrade mechanical properties [7,8]. It should be remembered that the interaction of pressure and temperature is very complex. The HIP process closes the porosity, but may also lead to grain growth, which in most cases will negatively affect some mechanical properties. Many works by previously un-published authors indicated the possibility of a decrease in the tensile strength of the material after the HIP process, but at the same time, an increase in fatigue or impact resistance was observed. This is why it is so important to correctly define the scope that is required and what functional properties are crucial.

Its main advantage is the ability to remove internal porosity in castings, elements printed using additive methods or obtained by metal injection. The hot isostatic pressing process takes place in a special, tightly closed vessel, which is placed in an appropriate furnace. The vessel with the sample inside is subjected to high temperature and high isostatic pressure. The pressure is constant throughout the vessel and affects the material from every direction. In a material that is subjected to elevated temperature and high pressure, phenomena such as plastic deformation, creep and diffusion occur. Both factors are kept constant, which allows the material to form evenly. High pressure affects changes in the material's phases and interphase reactions, while increased temperature allows it to become flexible. Changes in the structure of the processed material are irreversible. The HIP process also aims to reduce internal stresses to avoid micro-cracks that could lead to serious damage to the material [4–6]. The isostatic pressing process is becoming increasingly popular among engineers. Research is being carried out to determine changes in the properties of samples subjected to heat treatment, such as isostatic pressing. The research methodology used differs in terms of the materials tested, the method of preparing samples for testing and the parameters of the isostatic pressing process [7].

The process of isostatic pressing and shaping the geometry of surfaces by 3D printing can also be used in biomedical engineering. Currently, medical implants are mainly manufactured from titanium alloys, which have a favorable set of mechanical properties (apart from insufficient resistance to abrasive wear), a high degree of biocompatibility and positive bioactivity in relation to the surrounding cellular structures. Biomaterials of this type are characterized by high specific strength and relatively low (favorable) stiffness, and the favorable surface properties in terms of biotolerance are obtained thanks to a naturally formed, compact oxide layer. The phenomena of biological integration at the implant-plant–tissue interface, apart from medical factors, depends on so-called primary stabilization. The best results in this respect are obtained by appropriately shaping the implant in terms of geometry and surface topography in such a way that it is possible to mechanically anchor the implant in the surrounding tissues. The quality and condition of the implant's surface layer has a direct impact on the ability to structurally and functionally connect the "used" implant with living tissue [9,10].

The presented arguments encourage a deeper analysis of the problem of modification in terms of the geometry and topography of the surface layer. This possibility is provided by, among others, advanced additive techniques, which not only create new opportunities for the surface modification of existing medical implants, but also enable the concurrent shaping of geometry (adapted to the patient's individual anatomical features) and structures with increased biofunctionality for medical implants [11–14]. One of the more promising, innovative manufacturing technologies in this respect is the laser-engineered net shaping (LENS) method. This enables, unlike techniques from the selective laser melting (SLM) and electron beam melting (EBM) groups, the implementation of hybrid manufacturing processes, e.g., shaping the final geometry on a core with a simplified shape, or applying a coating to functionalize the surface of a fully formed element with specific geometry.

LENS technology has many advantages compared to other 3D printing methods. The basic one is the lack of restrictions in the powder feeding process. Some LENS models can realize five-axis control. Additionally, the typical build volume is comparatively large and can be 900 × 1500 × 900 mm (LENS 1500, 2018). However, the truly defining feature of the LENS process is the ability to fabricate with multiple materials and create functionally graded materials, owing to the multiple powder feed lines. Due to the high temperatures generated in the printing process, it is possible to produce products from materials with a high melting point. This method also makes it possible to produce complex porous structures. The method also makes it possible to reprint used parts, as there is no need to start printing on a flat platform. The LENS method also has disadvantages. Due to the possibility of deposition of partially melted powder particles on the printing surface, a high surface roughness is achieved. In the case of the Ti-6Al-4V material, increasing the surface roughness reduces the fatigue strength [15]. The technology is used to produce parts in industries such as medicine and aviation [16–18].

The authors undertook research on the Ti-6Al-4V alloy because, due to its properties, it is more often used in the production of implants than pure Ti.

Many scientific works have been devoted to assessing the influence of hot isostatic pressing on mechanical properties and microstructures [15–29]. In these papers the impact of the hot isostatic pressing process on samples produced, for example, by the selective laser melting (SLM) method [11] or powder metallurgy for orthopedic and dental applications was evaluated. The results of tests on samples made using the SLM method revealed an increase in the density of the tested material. The best results were obtained for samples subjected to the HIP process at a temperature of 910 °C and a pressure of 130 MPa [9].

Research conducted at the German University of Applied Sciences Aschaffenburg on implants made of the titanium alloy Ti-6Al-4V provided interesting results. It turned out that heat treatment in the form of hot isostatic pressing increased the tensile strength of the tested samples by 15% and increased the plasticity of the samples by an impressive 53% [25]. An increase in the fatigue strength of samples printed after the HIP process was also demonstrated [30].

The cited works did not assess the mechanical properties of samples produced using the LENS method and did not check whether changing the direction of printing the samples would also have a significant impact.

In additive technology using the LENS method, an important aspect is the possibility of printing in two directions. This gives great possibilities as to the shape of printed elements. However, the question arises whether the products will have the same strength regardless of the printing direction. This issue is poorly analyzed in the literature reports compared to, for example, various heat treatment methods [31,32]. According to these hesitations, the authors decided to check whether the printing direction of Ti-6Al-4V samples affects the change in mechanical properties. Do the mechanical properties of samples subjected to the HIP process solely depend on the adopted pressure values (500, 1000, 1500 bar) or also on the printing direction? Providing an answer to the above question will provide a significant contribution to the better understanding of the factors influencing the change in the properties of the Ti-6Al-4V alloy.

2. Materials and Methods

Within the study the used samples were made of the Ti-6Al-4V titanium alloy, whose chemical composition is shown in Table 1, using the LENS method. The samples were originally in the form of powder with a grain size ranging from 20 μm to 63 μm.

Table 1. Chemical composition of the used material.

	Ti	Al	V	Fe	C	N	H	O_{limit}	$O_{typical}$
Grade 5	Bal.	5.5–6.75	3.5–4.5	≤0.40	≤0.08	≤0.05	≤0.015	≤0.20	0.11

The choice of material was dictated by its medical applications. This alloy is biocompatible and the tested technique allows for the creation of spatial structures adapted to specific individual needs. However, before it is possible to make ready-made correct prints that can be implanted, the assumptions must be examined. Therefore, durability tests of the prints were performed, the aim of which was to determine the directions of changes for individual parameters.

The LENS method melted the powder and connected the material using a laser beam. Ultimately, the samples took the form of paddles measuring 65 × 10 × 2 mm. The samples were printed in two different orientations on the build plate: vertical (samples marked Y) and horizontal (samples marked X). After the printing stage, the samples were annealed at a temperature of 920 °C. This process lasted 4 h and was aimed at removing impurities such as inert gases (Ar) and inclusions, and removing internal stresses. After annealing, the samples were rinsed twice in an ultrasonic bath using waves with a frequency of 37 kHz and then dried. Both the rinsing and washing processes took 60 min. The final stage was the mechanical grinding and polishing of each sample surface. In the next stage, the samples were divided into two groups. One group of samples was subjected to hot isostatic pressing, while in the second group, this step was omitted (X1 ÷ X6, Y1 ÷ Y6). The samples subjected to the HIP process were further divided into three groups. Each group differed in the set pressure that prevailed in the chamber during heat treatment. Table 1 contains detailed parameters of the HIP process for individual groups.

The hot isostatic pressing process took place in a special chamber located in an appropriate furnace (Figure 1). The vessel with the sample inside was subjected to a high temperature and high isostatic pressure (Table 2). The pressure in the entire chamber was constant and affected the material in the same way from every direction.

Table 2. Parameters of the hot isostatic pressing process.

Sample Number	Temperature [°C]	Pressure [Bar]	Time [min]
X1.1 ÷ X1.4 Y1.1 ÷ Y1.3		500	
X2.1 ÷ X2.4 Y2.1 ÷ Y2.4	1150	1000	240
X3.1 ÷ X3.2 Y3.1 ÷ Y3.3		1500	

Figures 2–4 show the course of the entire HIP process for the three groups of tested samples.

The evaluation of the microstructure in the unetched and etched state of the samples after printing and after the HIP process is presented in Figures 5–12. The analysis was performed using a Keyence (Osaka, Japan) VHX 7000 series optical microscope. The studies were carried out in a bright field and the sample preparation technique was standard, using diamond pastes. In order to reveal the microstructure of the material, the samples were etched with titan 2 reagent (HNO$_3$ 2 mL, HF—3 mL, H$_2$O—96 mL).

a)

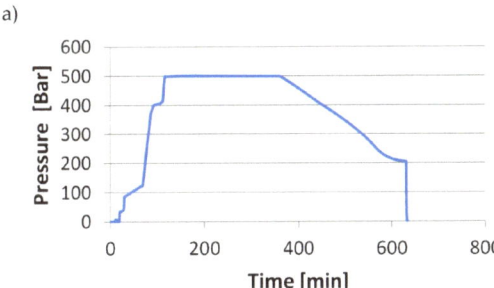

b)

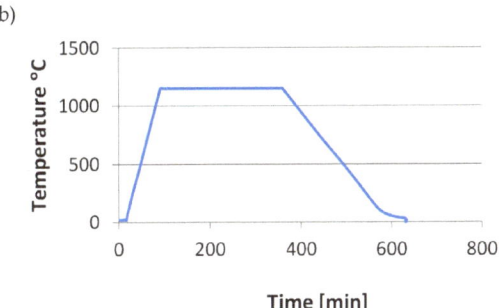

Figure 2. Process curves for the samples (X1.1 ÷ X1.4, Y1.1 ÷ Y1.3): (a) pressure changes over time, (b) temperature changes over time.

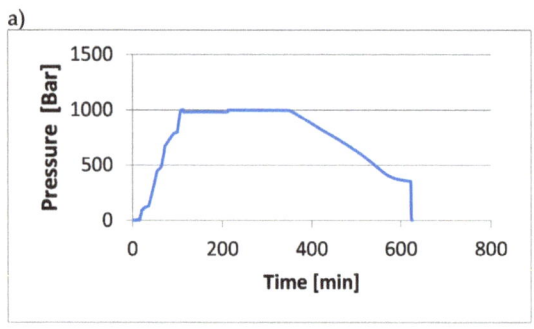
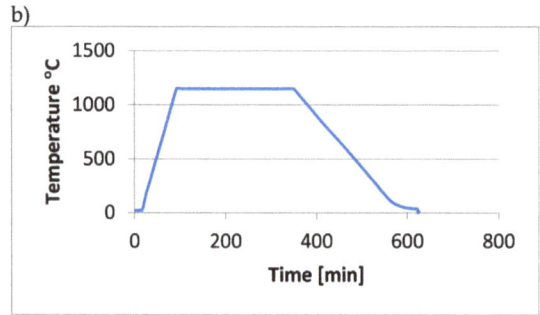

Figure 3. Process curves for the samples (X2.1 ÷ X2.4, Y2.1 ÷ Y2.4): (**a**) pressure changes over time, (**b**) temperature changes over time.

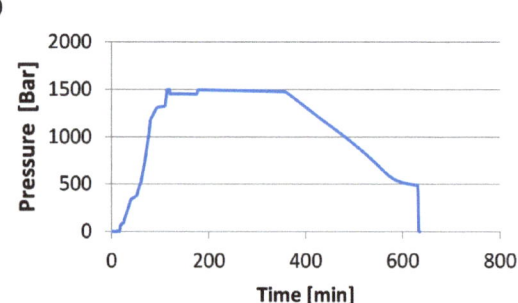
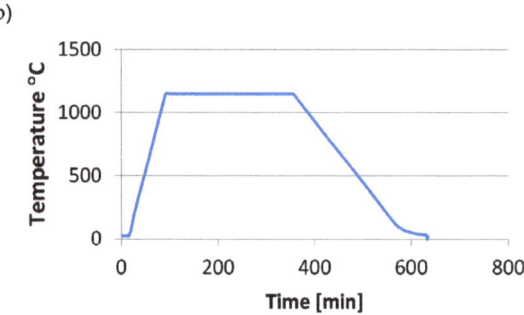

Figure 4. Process curves for the samples (X3.1 ÷ X3.2, Y3.1 ÷ Y3.3): (**a**) pressure changes over time, (**b**) temperature changes over time.

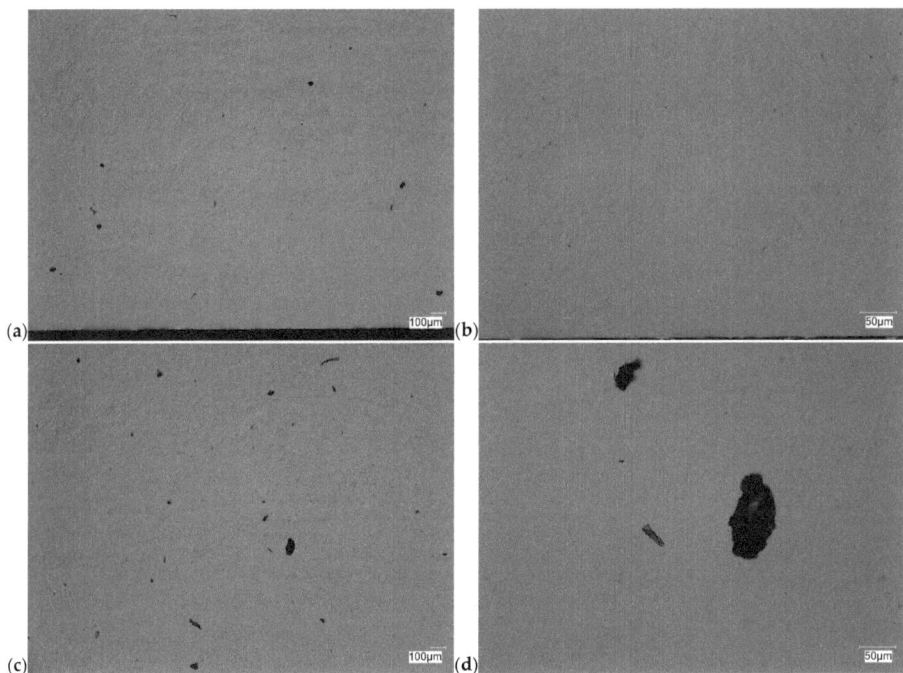

Figure 5. Microscopic view of the surface of sample X, unetched state: (**a**) 100× edge, (**b**) 500× edge, (**c**) 100× central part, (**d**) 500× central part.

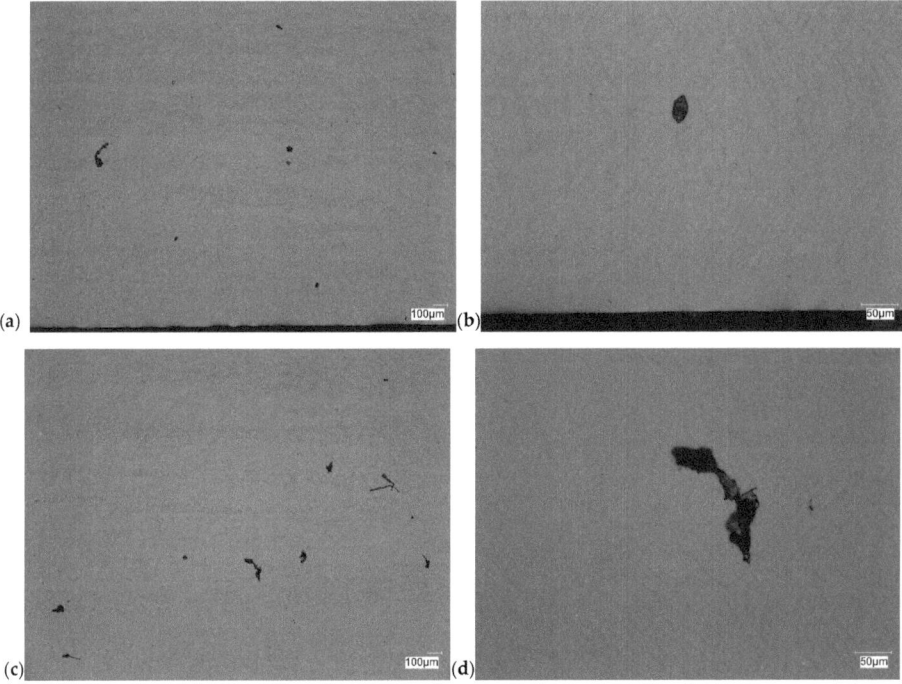

Figure 6. Microscopic view of the surface of sample Y, unetched state: (**a**) 100× edge, (**b**) 500× edge, (**c**) 100× central part, (**d**) 500× central part.

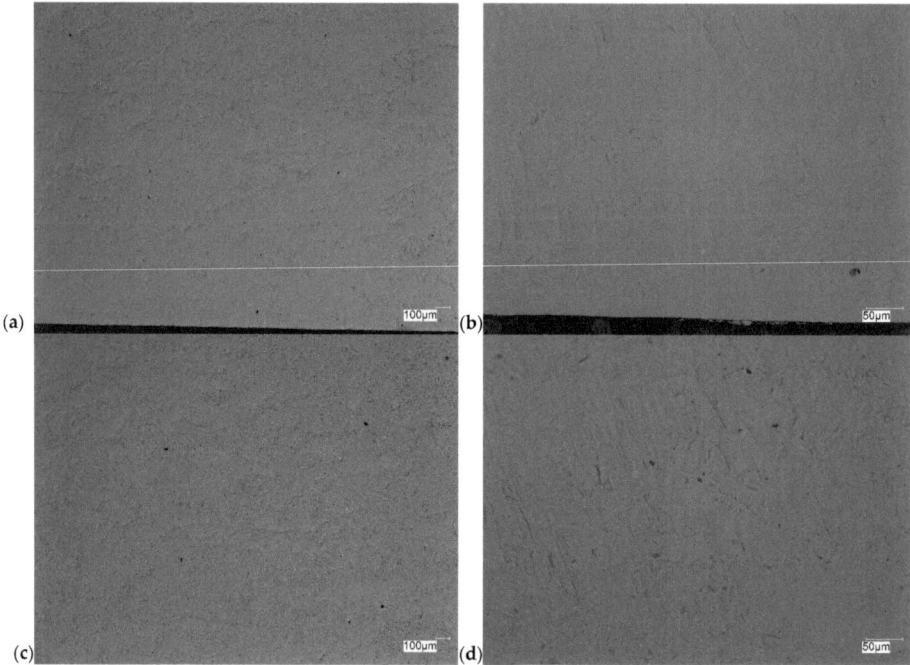

Figure 7. Microscopic view of the surface of sample XX, unetched state: (**a**) magnification 100× edge, (**b**) 500× edge, (**c**) 100× central part, (**d**) 500× central part.

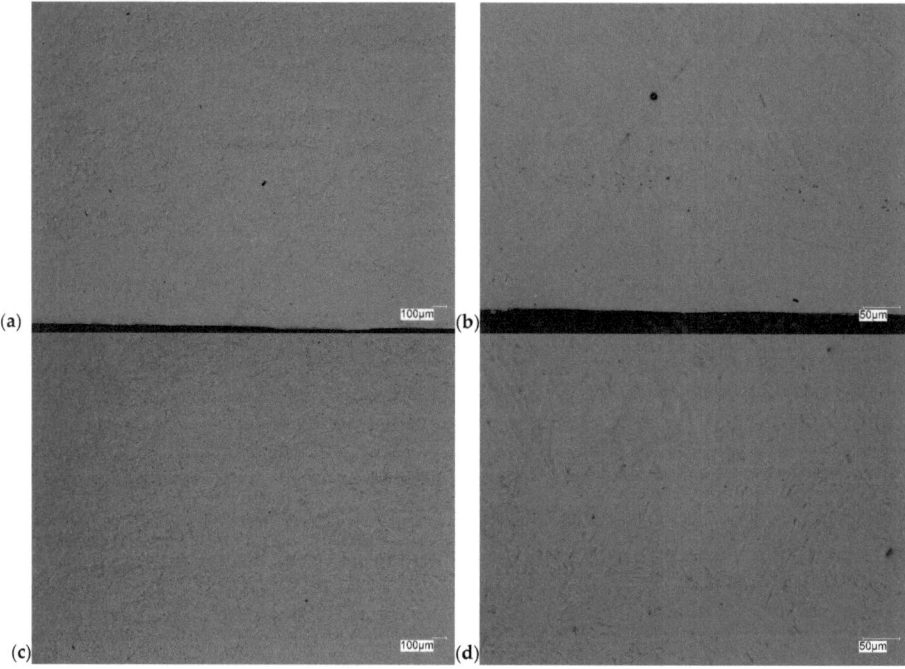

Figure 8. Microscopic view of the surface of sample YY, unetched state: (**a**) 100× edge, (**b**) 500× edge, (**c**) 100× central part, (**d**) 500× central part.

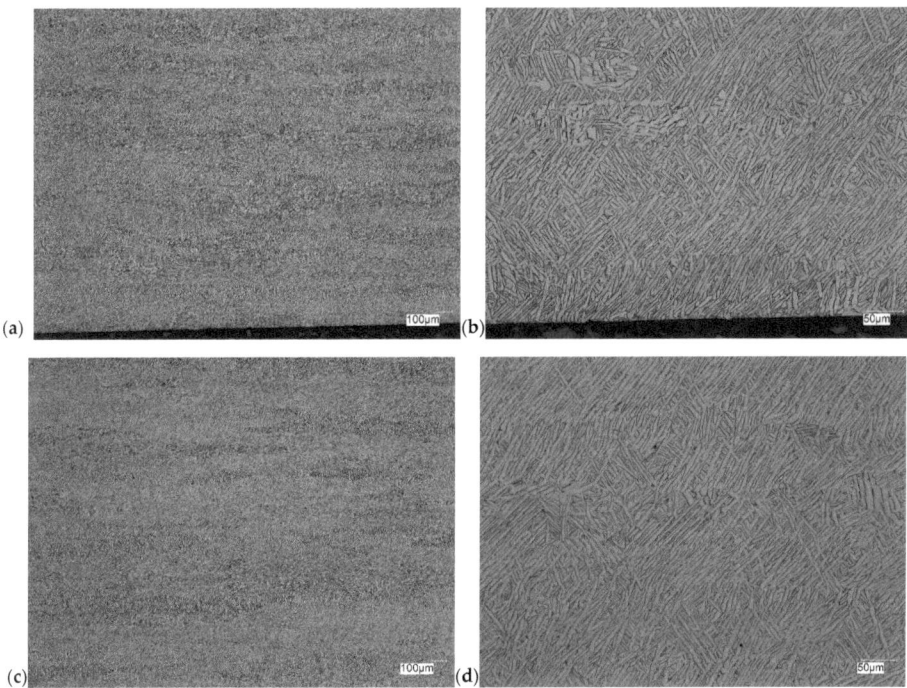

Figure 9. Microscopic view of the surface of sample X, etched state: (**a**) 100× edge, (**b**) 500× edge, (**c**) 100× central part, (**d**) 500× central part.

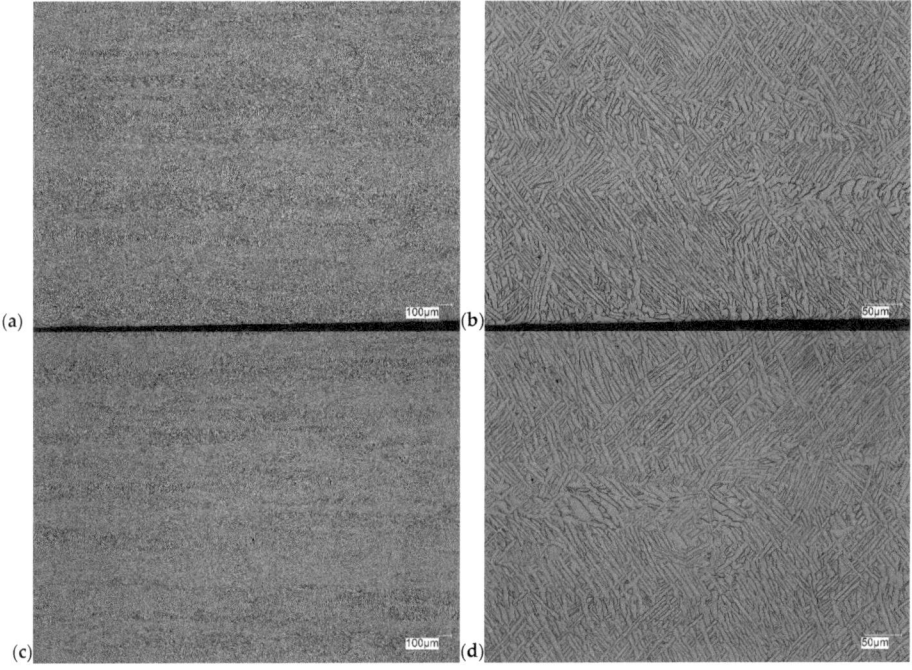

Figure 10. Microscopic view of the surface of sample Y, etched state: (**a**) 100× edge, (**b**) 500× edge, (**c**) 100× central part, (**d**) 500× central part.

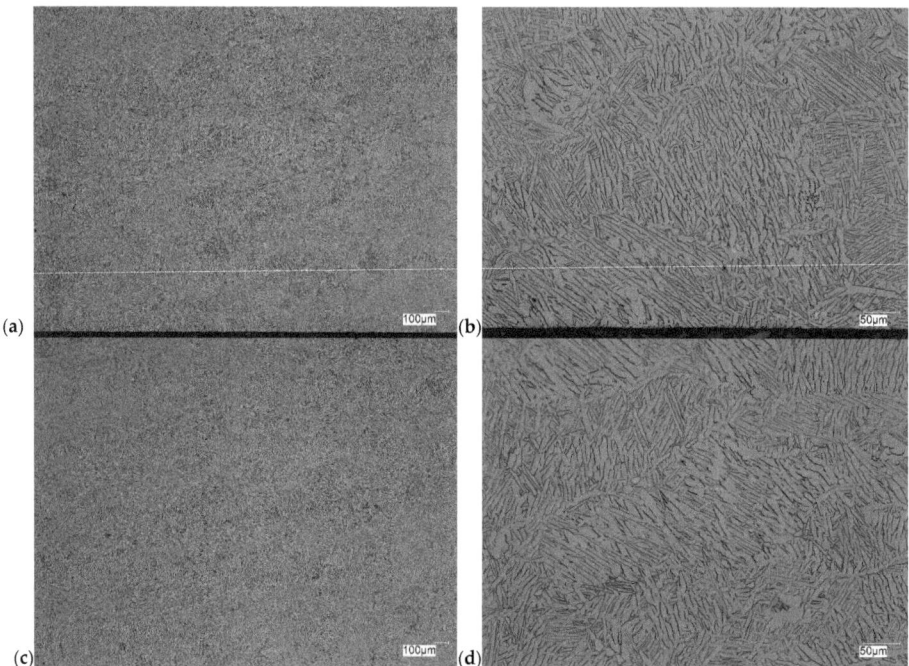

Figure 11. Microscopic view of the surface of sample XX, etched state: (**a**) 100× edge, (**b**) 500× edge, (**c**) 100× central part, (**d**) 500× central part.

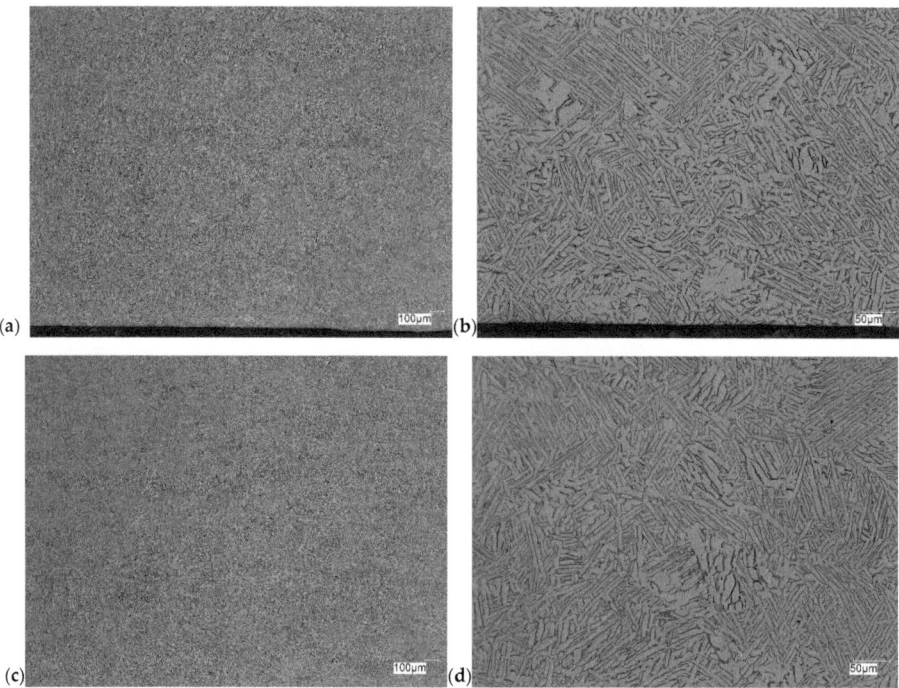

Figure 12. Microscopic view of the surface of sample YY, etched state: (**a**) 100× edge, (**b**) 500× edge, (**c**) 100× central part, (**d**) 500× central part.

Prior to the HIP process, the samples exhibited the presence of monstrosity, most notably in the central region. These pores ranged in size from approximately 15 to 100 μm. They are the effects of the manufacturing process and the high reactivity of titanium. Their number decreases significantly towards the edge of the samples, which should probably be related to faster heat dissipation in these zones. Sample Y displays porosities with markedly more irregular shapes, particularly in the central region of the sample. The morphology and characteristics of these porosities suggest that they result from incomplete welding or gas entrapment at the grain boundaries.

As can be observed after the HIP process, basically all pores were closed. This is due to the long-term effect of temperature and pressure in the volume of the device chamber. This is thanks to the use of Pascal's law, i.e., an even distribution of the pressure force on each fragment. Such parameters facilitate the occurrence of diffusion, which in consequence leads to the closure of porosity. What is also interesting is that the high reactivity of the tested alloy, combined with the significant energy supplied in the HIP process, allows for the initial disclosure of the microstructure, which will be confirmed in the photos of the etched samples presented in Figures 9–12.

The Ti-6Al-4V alloy is a two-phase alloy with a structure ($\alpha + \beta$), used, among others, for orthopedic implants and bone fixation elements. As can be seen in the state after printing, samples X and Y, the material is characterized by a typical bi-modal microstructure (duplex), consisting of α grains in a lamellar matrix ($\alpha + \beta$). An important feature is the clear ordering of the separation in accordance with the directions of layer deposition. Very uniform grain sizes are also visible in both the central and edge parts. After the HIP process, with the increase in temperature and pressure (processing parameters), the average grain size and the volume fraction of the phase mixture ($\alpha + \beta$) increase. The randomness of grain orientation in space also increases, which causes a change in the form of an observed decrease in strength properties.

As can be observed on the surface of the sample, not only are there no porosities in the material visible, but also grain boundaries are visible (even though the sample was not etched) with characteristic visible arrangements in the printing direction.

Strength tests were carried out for both categories of samples (after and without the HIP process). The strength of the samples was assessed by performing a static tensile test at a speed of 5 mm/min using the MTS Criterion Model 43 testing machine (MTS Systems, Eden Prairie, MN, USA) and the Dantec Q400 digital image correlation system (Dantec Dynamics, Ulm, Germany) (Figure 13). A 30 kN force sensor was used for the tests.

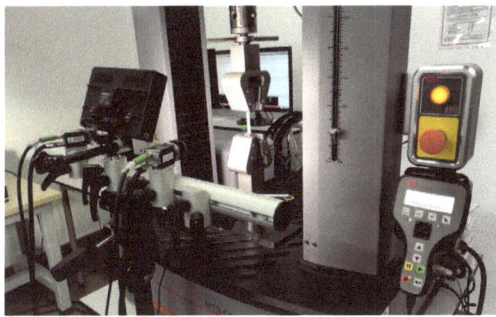

Figure 13. Strength testing machine and digital image correlation system.

The digital image correlation system is an optical measurement method for determining the distribution of object displacements and determining the 3D deformation field. This is a fully non-contact method in which the contour measurement, displacement and deformation of the material are determined during the measurements (using a testing machine). To guarantee the accuracy of the measurements, the system was calibrated using a dedicated calibration plate with characteristic points reflecting the axes of the coordinate system. The digital image correlation system requires appropriate preparation of samples for testing. The surface is covered with matte white paint. The matte tint was intended to prevent light reflections that could distort the recorded image during testing (Figure 14). After the white layer dried, additional black paint in the form of powder was applied to the samples, creating dots on their surface.

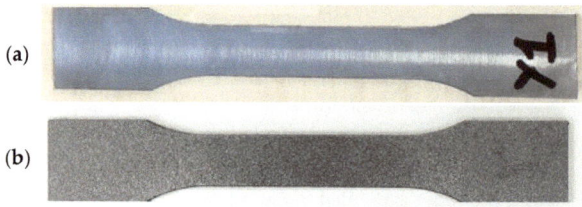

Figure 14. Example of a test sample: (a) before painting, (b) after painting.

The sample was then mounted in special holders of the testing machine, ensuring the stable and axial positioning of the sample during the static tensile test. The sample was then deformed in the direction of the longitudinal axis under quasi-static conditions at a speed of 5 mm/min. The results were recorded by a digital image correlation system at a frequency of 10 Hz. The test was carried out until the sample's failure.

3. Results and Discussion

The values obtained during the tests, such as the maximum force and stress at break, were read from a testing machine, while the strain value in two mutually perpendicular directions were calculated based on Young's modulus obtained in the digital image correlation system. Poisson's ratio was calculated based on the determined strain values. The obtained results were compared between groups of appropriately marked samples, and attention was paid to the division of samples subjected to the hot isostatic pressing process and samples that were not subjected to this process.

3.1. Results for Samples Not Subjected to the HIP Process

The test results obtained for samples not subjected to the HIP process are presented in Table 3. Figure 15 shows the stress–strain curves for the tested samples.

Analyzing the influence of printing direction on the average values of maximum ultimate tensile stress, a higher result was observed when the print was placed horizontally, of 939.7 ± 17.89 MPa compared to 927.5 ± 4.71 MPa vertically. The average maximum force at break was higher for horizontal samples compared to vertical ones, and reached 18.79 ± 0.36 kN and 18.55 ± 0.09 kN, respectively. The difference in the obtained values is small and amounts to only 0.24 MPa. The obtained difference is smaller than the error bar value for samples X, which was 0.36 MPa. The average values of Young's moduli were 115.4 ± 2.7 GPa for samples Y and 115.2 ± 1.6 GPa for samples X, respectively. The difference in the obtained values is very small and amounted to only 0.2 MPa. The Young's modulus determined in the work did not show such changes depending on the direction of printing that were observed by other scientists. They obtained values of 118 GPa and 109 GPa for two printing directions [22]. The obtained strain values at break were the same and amounted to 0.28 mm/mm. The only difference observed was the value of the determined Poisson's number for samples X = 0.34 ± 0.05 and for samples Y = 0.37 ± 0.09.

Table 3. Test results for samples without the HIP process.

Sample Number	Peak Load [kN]	Ultimate Tensile Stress [MPa]	Young's Modulus [GPa]	Strain at Break [mm/mm]	Poisson's Ratio
X1	18.94	946.93	-	0.027	-
X2	19	950.36	116.6	0.027	0.3
X3	18.9	945.26	114.3	0.025	0.4
X4	18.87	943.47	114	0.029	0.4
X5	18.97	948.64	113.9	0.031	0.3
X6	18.07	903.51	117.3	0.027	0.3
Average X	18.79 ± 0.36	939.70 ± 17.89	115.2 ± 1.6	0.028 ± 0.02	0.34 ± 0.05
Y1	18.52	925.75	113.8	0.024	0.3
Y2	18.58	928.81	113.2	0.027	0.4
Y3	18.63	931.34	114.3	0.030	0.5
Y4	18.56	928.11	113.8	0.032	0.3
Y5	18.64	932.13	117.6	0.027	0.3
Y6	18.383	919.16	119.9	0.03	0.4
Average Y	18.55 ± 0.09	927.55 ± 4.71	115.4 ± 2.7	0.028 ± 0.03	0.37 ± 0.09

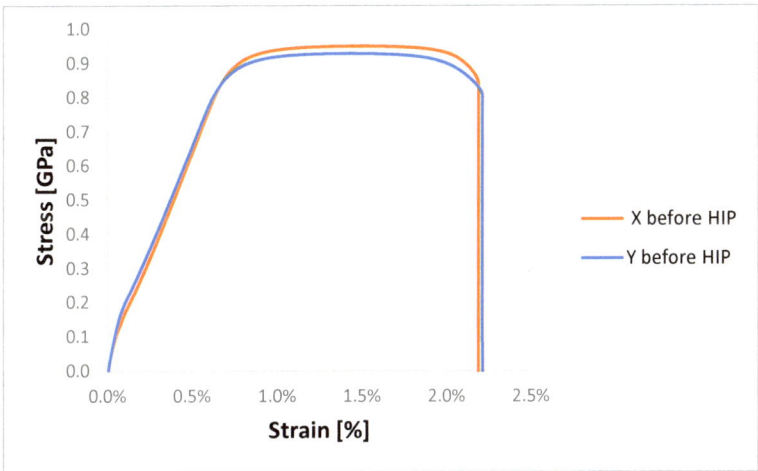

Figure 15. Typical stress–strain curves for samples before HIP.

Analyzing the obtained values, no significant differences were observed between the determined parameters for horizontally printed samples compared to vertically printed samples. It can, therefore, be concluded that the direction of the bend does not have a significant impact on the change in the strength parameters of the prepared samples.

3.2. Results for Samples After the HIP Process

The test results obtained for samples subjected to the HIP process are presented in Tables 4 and 5. Figure 16 shows the stress–strain curves for the tested samples.

Table 4. Test results for samples after the HIP process.

Sample Number	Peak Load [kN]	Ultimate Tensile Stress [MPa]	Young's Modulus [GPa]	Strain at Break [mm/mm]	Poisson's Ratio
X1.1	10.43	501.04	113.58	0.07	0.35
X1.2	15.78	765.04	110.5	0.1	0.31
X1.3	13.91	669.74	120.03	0.05	0.30
X1.4	12.79	622.12	113.17	0.04	0.25
Average (X1.1–X1.4)	13.23 ± 2.24	639.48 ± 109.77	114.32 ± 4.04	0.07 ± 0.03	0.32 ± 0.03
X2.1	15.93	776.87	119.35	0.06	0.33
X2.2	17.57	831.51	117.24	0.06	0.41
X2.3	17.36	831.15	117.53	0.06	0.32
X2.4	17.42	837.29	122.14	0.06	0.28
Average (X2.1–X2.4)	17.07 ± 0.77	819.21 ± 28.36	119.07 ± 2.25	0.06 ± 0	0.34 ± 0.05
X3.1	18.187	887.94	113.88	0.07	0.37
X3.2	19.156	916.12	117.01	0.08	0.42
Average (X3.1–X3.2)	18.67 ± 0.69	902.03 ± 19.93	115.45 ± 2.21	0.08 ± 0.01	0.4 ± 0.04
Average X	16.32 ± 2.80	786.91 ± 134.22	116.28 ± 2.48	0.07 ± 0.01	0.35 ± 0.04
Y1.1	14.48	695.98	126	0.06	0.29
Y1.2	13.39	651.91	102.9	0.06	0.36
Y1.3	15.19	727.97	113.67	0.06	0.42
Average (Y1.1–Y1.3)	14.35 ± 0.91	691.95 ± 38.19	114.19 ± 11.56	0.06	0.36 ± 0.07
Y2.1	18.02	866.15	115.97	0.07	0.32
Y2.2	18.2	868.53	121.99	0.08	0.34
Y2.3	17.94	860.62	111.12	0.07	0.33
Y2.4	18.98	898.72	112.52	0.1	0.34
Average (Y2.1–Y2.4)	18.28 ± 0.47	873.50 ± 17.13	115.4 ± 4.84	0.08 ± 0.01	0.33 ± 0.01
Y3.1	18.53	897.71	131.34	0.07	0.43
Y3.2	18.66	896.56	105.84	0.07	0.29
Y3.3	18.37	890.67	108.5	0.07	0.28
Average (Y3.1–Y3.3)	18.52 ± 0.14	894.98 ± 3.78	115.23 ± 14.02	0.07	0.33 ± 0.08
Average Y	17.05 ± 2.34	820.14 ± 111.54	114.94 ± 0.66	0.07 ± 0.01	0.34 ± 0.01

Table 5. Average test results for samples after the HIP process.

Samples After the HIP Process	$R_{p0.2}$ [MPa]	R_p [MPa]
(X1.1–X1.4)	123.68 ± 0.91	119.33 ± 3.1
(X2.1–X2.4)	122.76 ± 1.57	120.65 ± 5.44
(X3.1–X3.2)	123.53 ± 1.8	122.31 ± 0.2
Average X	123.32 ± 0.49	120.76 ± 1.49
(Y1.1–Y1.3)	123.71 ± 0.49	122.82 ± 1.49
(Y2.1–Y2.4)	-	-
(Y3.1–Y3.3)	122.79 ± 0.35	121.45 ± 1.32
Average Y	123.25 ± 0.65	122.14 ± 0.9

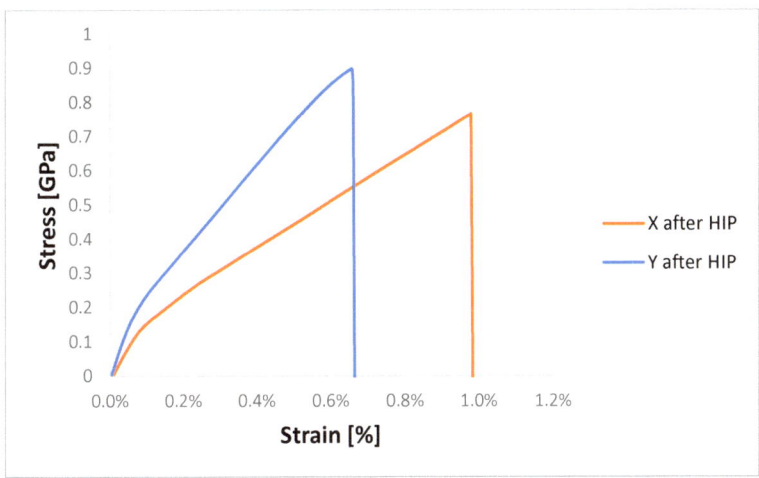

Figure 16. Typical stress–strain curves for samples after HIP.

Analyzing the obtained test results for samples printed horizontally after the HIP process, the highest value of peak load was observed for samples subjected to a pressure of 1500 bar (X3.1–X3.2), which amounted to 18.67 ± 0.69 kN. The smallest one was for samples subjected to a pressure of 500 bar (X1.1–X1.4), which was 13.23 ± 2.24 kN. The highest ultimate tensile stress value of 902.03 ± 19.93 MPa was recorded for samples from the group (X3.1–X3.2), the lowest was 639.48 ± 109.77 MPa and for samples (X1.1–X1.4) subjected to a pressure of 500 bar. In the case of Young's modulus, the highest value was obtained for samples subjected to a pressure of 1000 bar (X2.1–X2.4)—119.07 ± 2.25 GPa, and the lowest for samples (X1.1–X1.4)—114.32 ± 4, 04 GPa. Analyzing the obtained test results for horizontally printed samples after the HIP process, the highest value of peak load was observed for samples subjected to a pressure of 1500 bar (Y3.1–Y3.3), which amounted to 18.52 ± 0.14 kN. The smallest one was for samples subjected to a pressure of 500 bar (Y1.1–Y1.3), which was 14.35 ± 0.91 kN. The highest value of ultimate tensile stress, 894.98 ± 3.78 MPa, was recorded for samples from the group (Y3.1–Y3.3) and the lowest was 691.95 ± 38.19 MPa for samples (Y1.1–Y1.3), subjected to a pressure of 500 bar. The obtained values are smaller than those presented in publication [9]. The authors obtained results in the range between 1226 MPa and 955 MPa. The authors of the work used one different sample printing method (SLM). The method of sample preparation and, more precisely, their structure, is important in the final effect of using HIP, which has already been demonstrated [31]. In the case of Young's modulus, the highest value was obtained for samples subjected to a pressure of 1000 bar (Y2.1–Y2.4)—115.4 ± 4.84 GPa, and the lowest was for samples (Y1.1–Y1.3)—114.19 ± 11, 56 GPa. The obtained values are much higher than in the case of titanium lattice structures subjected to the HIP process [25].

Analyzing the obtained test results after the HIP process, vertically printed samples were characterized by a higher average value of the plasticity pot Rp, which amounted to 122.14 ± 0.9 MPa, than horizontally printed samples, for which this value was 120.76 ± 1.49 MPa.

Analyzing the maximum peak between the group of samples not subjected to the HIP process (first group) (Table 3) and the group of samples subjected to the HIP process (second group) (Table 4), its decrease was noticed for samples after the HIP process. For samples from the first group printed horizontally, X, the average maximum peak load was 18.79 kN, and when samples were printed vertically, Y, its average value was 18.55 kN.

For samples from the second group, the average value of the maximum peak load for samples X was 16.32 kN. The highest value was obtained for samples subjected to the HIP process at a pressure of 1500 bar, 18.67 kN, while the lowest value was 13.23 kN for samples subjected to the HIP process at a pressure of 500 bar. For Y samples from the second group, the average maximum peak load is 17.05 kN. The highest value was observed in the samples heat-treated at a pressure of 1500 bar, measuring 18.52 kN, whereas the lowest value, 14.35 kN, was recorded in the samples subjected to the HIP process at a pressure of 500 bar.

Based on the above data, it can be seen that the hot isostatic pressing process resulted in an increase in the Young's modulus value in several cases, which suggests an increase in the stiffness of the material. The highest increase in Young's modulus of 3.8% was recorded for samples printed in the Y direction and subjected to the HIP process at a pressure of 500 bar. The greatest decrease was observed for samples printed in the Y direction subjected to the HIP process at a pressure of 1500 bar.

The isostatic pressing process worked best at a pressure of 1000 bar for samples printed in the X direction, and at a pressure of 500 bar for samples printed in the Y direction.

In the case of samples printed in the X direction, the highest Young's modulus values of samples after the HIP process were recorded when the process pressure was equal to 1000 bar. The remaining samples from this group were insignificantly affected by the HIP process in the context of Young's modulus.

The obtained test results showed that the hot isostatic pressing process has a significant impact on the deformation value. In each case tested, the deformation decreased by approximately 75%. The largest change of 78.6% occurred in samples printed in the X direction and heat treated at a pressure of 1000 bar, and in samples printed in the Y direction and subjected to a pressure of 500 bar.

Analyzing the results obtained for the conventional yield strength $R_{p0.2}$ (Table 4), it can be seen that the values do not differ significantly for the individual groups of tested samples. Both the variable parameters of the HIP process and the printing direction did not cause significant changes in the determined values. In the case of the yield strength, there is a visible relationship between the direction of printing samples and the parameters of the HIP process. The lowest value of the yield strength was recorded for samples printed in the X direction and subjected to hot isostatic pressing at a pressure of 500 bar. The highest value of the Rp parameter was assigned to the sample printed in the Y direction, heat treated at a pressure of 500 bar.

The yield strength could not be recorded for samples Y2.1–Y3.3. In the case of samples printed in the X direction, a relationship is visible: the higher the pressure of the HIP process, the higher the yield strength.

In order to better illustrate the differences that occurred between the determined parameters for samples before and after the HIP process, Figures 17–19 present the calculated percentage differences between the values for samples before and after the HIP process.

The results show that the isostatic pressing process reduces the value of the maximum force at break. It was noticed that the lower the pressure of the pressing process, the greater the decrease in the maximum peak load value.

The largest percentage differences were observed for samples printed horizontally and subjected to a pressure of 500 bar, which was 29.6%, while the smallest differences were observed for samples printed horizontally and subjected to a pressure of 1500 bar, which was only 0.03%.

In the case of percentage differences for ultimate stress, the situation is similar. The largest percentage differences were observed for samples printed horizontally and subjected to a pressure of 500 bar, which was 31.9%, while for samples printed vertically and subjected to a pressure of 1500 bar, no percentage differences were observed. For a pressure of 1000 bar, samples printed horizontally showed a difference in the ultimate tensile stress value of 12.8%, while for samples printed vertically, the difference was 5.8%.

In the case of Young's modulus, the largest percentage differences were observed for samples printed horizontally and subjected to a pressure of 1000 bar, which were 3.36%. In the case of vertically printed samples, percentage differences practically did not occur.

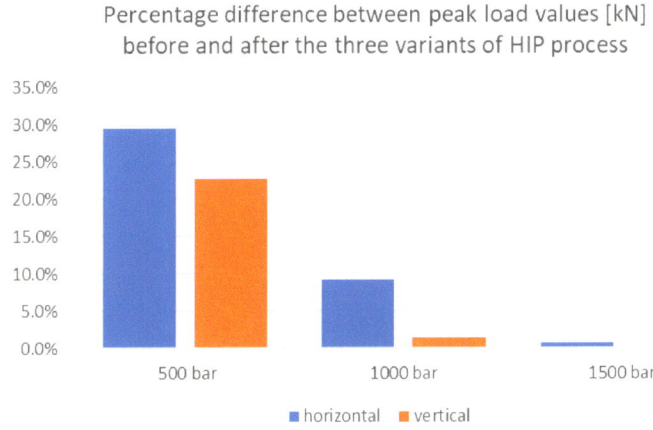

Figure 17. Summary of percentage differences for peak load values before and after the HIP process for three applied pressure values.

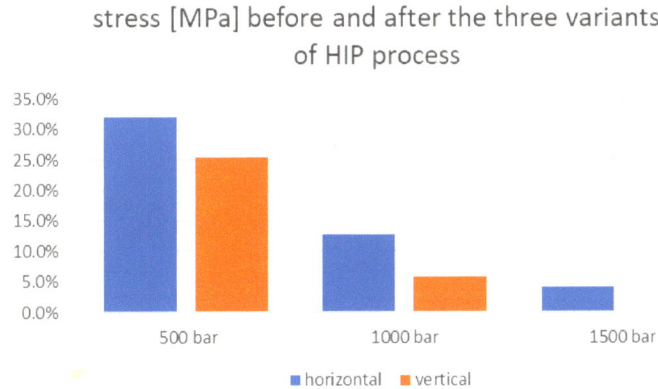

Figure 18. Summary of percentage differences for ultimate tensile stress values before and after the HIP process for three applied pressure values.

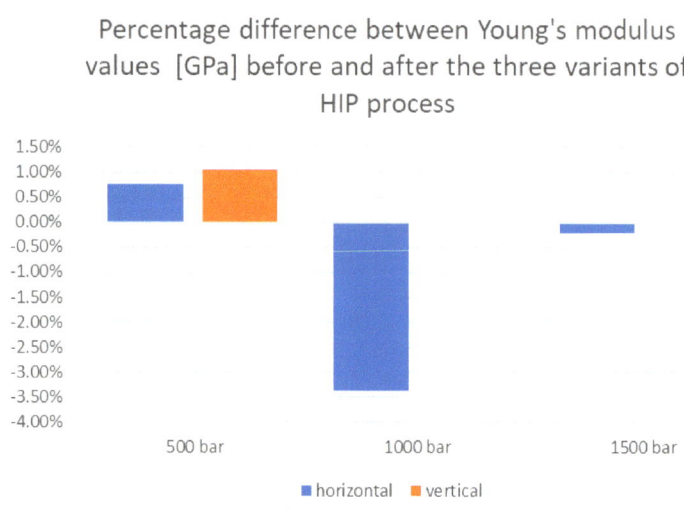

Figure 19. Summary of percentage differences for Young's modulus values before and after the HIP process for three applied pressure values.

4. Conclusions

The paper evaluates the influence of manufacturing and finishing parameters on a biocompatible material, specifically the Ti-6Al-4V alloy. While these studies are preliminary, they have already provided valuable insights into the potential changes in material properties. This is particularly important, as 3D printing technology enables the fabrication of customized prostheses and even damaged tissue fragments. From the perspective of the potential applications of such elements, it is essential to determine the positive or negative impacts of various finishing techniques.

In this study, the influence of hot isostatic pressing (HIP) temperature and pressure on the mechanical properties of the Ti-6Al-4V titanium alloy, produced using the laser engineering net shaping (LENS) method, was assessed. Additionally, the study explored whether the orientation of printing in the LENS process affects the obtained strength properties. Based on the experimental results, the following conclusions can be drawn: the direction of forming the Ti-6Al-4V alloy using the LENS method does not significantly affect the tensile strength properties of the material. The recorded variations in strength were within the margin of error. A comparison of results from samples before and after the HIP process indicated that the process influenced the material's properties. Specifically, higher applied pressures led to higher peak load values. The highest values of Young's modulus were achieved at a pressure of 1000 bar, while the highest ultimate tensile strength (UTS) values were recorded at 1500 bar. The HIP process thus enables the control of both the strength and plasticity of the material. However, it was also observed that variations in HIP parameters significantly affect the final mechanical properties of the samples.

Depending on the applied pressure, some material properties improved while others deteriorated, an outcome that was not anticipated. Typically, the HIP process enhances the strength properties of the materials forming method [6,16,19,20], but this pattern is not always observed. Future research should investigate similar samples, incorporating variations in temperature and a broader range of pressure values during the HIP process. These studies may help establish optimal HIP parameters to improve the strength properties of samples fabricated using the LENS method.

Author Contributions: Conceptualization, B.G.-Z., K.J., and G.S.; formal analysis W.W., A.S., and G.S.; methodology B.G.-Z., K.J., and G.S.; investigations: B.G.-Z., K.J., and A.P.; resources W.W., A.S., and G.S.; supervision W.W., A.S., and G.S.; visualization B.G.-Z., K.J., E.K., A.P., and W.K.; writing—original draft B.G.-Z. and K.J.; writing—review and editing E.K., G.S., and W.K.; funding acquisition: B.G.-Z. All authors have read and agreed to the published version of the manuscript.

Funding: The article processing charge was financed under the European Funds for Silesia 2021–2027 Program co-financed by the Just Transition Fund project entitled "Development of the Silesian biomedical engineering potential in the face of the challenges of the digital and green economy (BioMeDiG)". Project number: FESL.10.25-IZ.01-07G5/23.

Institutional Review Board Statement: Not applicable.

Informed Consent Statement: Not applicable.

Data Availability Statement: Data is contained within the article. Dataset available on request from the authors.

Conflicts of Interest: The authors declare no conflicts of interest.

References

1. Tranquillo, J.; Goldberg, J.; Allen, R. *Chapter 7—Prototyping, Biomedical Engineering Design Biomedical Engineering*; Academic Press: Cambridge, MA, USA, 2023; pp. 197–234. [CrossRef]
2. Gupta, S.; Bit, A. 16-Rapid prototyping for polymeric gels. In *Polymeric Gels Characterization, Properties and Biomedical Applications A volume in Woodhead Publishing Series in Biomaterials*; Woodhead Publishing: Sawston, UK, 2018; pp. 397–439. [CrossRef]
3. Chang, K.-H. *Introduction to e-Design, Product Performance Evaluation with CAD/CAE*; Elsevier: Amsterdam, The Netherlands, 2013. [CrossRef]
4. Atkinson, H.V.; Davies, S. Fundamental aspects of hot isostatic pressing: An overview. *Metall. Mater Trans. A.* **2000**, *31*, 2981–3000. [CrossRef]
5. Vrancken, B.; Thijs, L.; Kruth, J.P.; van Humbeeck, J. Heat treatment of Ti6Al4V produced by Selective Laser Melting: Microstructure and mechanical properties. *J. Alloys Compd.* **2012**, *541*, 177–185. [CrossRef]
6. Plessis, A.; Macdonald, E. Hot isostatic pressing in metal additive manufacturing: X-ray tomography reveals details of pore closure. *Addit. Manuf.* **2020**, *34*, 101191. [CrossRef]
7. Leuders, S.; Tröster, T.; Riemer, A.; Richard, H.A.; Niendorf, T. On the mechanical performance of structures manufactured by selective laser melting: Damage initiation and propagation, Presented at the Additive Manufacturing with Powder Metallurgy Conference. In Proceedings of the AMPM2014 Conference Presentations, Orlando, FL, USA, 18–20 May 2014.
8. Peter, W.; Nandwana, P.; Kirka, M.; Dehoff, R.; Sames, W.; Erdmann, D.; Eklund, A.; Howard, R. Understanding the Role of Hot Isostatic Pressing Parameters on the Microstructural Evolution of Ti-6Al-4V and Inconel 718 Fabricated by Electron Beam Melting. In *Report No. ORNL/TM-2015/77; CRADA/NFE-14-04943*; Oak Ridge National Lab. (ORNL): Oak Ridge, TN, USA, 2015. [CrossRef]
9. Wennerberg, A.; Albrektsson, T. Effects of titanium surface topography on bone integration: A systematic review. *Clin. Oral Implants Res.* **2009**, *20*, 172–184. [CrossRef]
10. Klokkevold, P.; Naishimura, R.; Adachi, M.; Caputo, A. Osteointegration enhanced by chemical etching of the titanium surface. A torque removal study in the Rabbit. *Clin. Oral Implants Res.* **1997**, *8*, 442–447. [CrossRef]
11. Lv, Z.; Li, H.; Che, L.; Chen, S.; Zhang, P.; He, J.; Wu, Z.; Niu, S.; Li, X. Effects of HIP Process Parameters on Microstructure and Mechanical Properties of Ti-6Al-4V Fabricated by SLM. *Metals* **2023**, *13*, 991. [CrossRef]
12. Ramsden, J.; Allen, D.M.; Sephenson, D.; Alcock, J.; Peggs, G.N.; Fuller, G.D.; Goch, G. The design and manufacture of biomedical surfaces. *CIRP Ann.* **2007**, *56*, 687–711. [CrossRef]
13. Götz, H.E.; Müller, M.; Emmel, A.; Holzwarth, U.; Erben, R.G.; Stangl, R. Effect of surface finish on the osseointegration of laser-treated titanium alloy implants. *Biomaterials* **2004**, *25*, 4057–4064. [CrossRef] [PubMed]
14. Roy, M.; Balla, V.K.; Bandyopadhyay, A.; Bose, S. Compositionally graded hydroxyapatite/tricalcium phosphate coating on Ti by laser and induction plasma. *Acta Biomater.* **2011**, *7*, 866–873. [CrossRef]
15. Frazier, W.E. Metal Additive Manufacturing: A Review. *J. Mater. Eng. Perform.* **2014**, *23*, 1917–1928. [CrossRef]
16. Camacho, D.D.; Clayton, P.; O'Brien, W.; Ferron, R.; Juenger, M.; Salamone, S.; Seepersad, C. Applications of additive manufacturing in the construction industry—A prospective review, ISARC. In Proceedings of the International Symposium on Automation and Robotics in Construction, Taipei, Taiwan, 28 June–1 July 2017; Vilnius Gediminas Technical University, Department of Construction Economics & Property: Vilnius, Lithuania, 2017; Volume 34.

17. Dutta, B.; Froes, F.H. The Additive Manufacturing (AM) of titanium alloys. *Met. Powder Rep.* **2017**, *72*, 96–106. [CrossRef]
18. Zhou, Y.; Zeng, W.; Sun, Y.; Zhang, Y. A numerical insight into stress mode shapes of rectangular thin-plate structures. *Mech Based Des Struc.* **2023**, *51*, 2655–2680. [CrossRef]
19. Lario, J.; Vicente, Á.; Amigó, V. Evolution of the Microstructure and Mechanical Properties of a Ti35Nb2Sn Alloy Post-Processed by Hot Isostatic Pressing for Biomedical Applications. *Metals* **2021**, *11*, 1027. [CrossRef]
20. Hijazi, K.M.; Dixon, S.J.; Armstrong, J.E.; Rizkalla, A.S. Titanium Alloy Implants with Lattice Structures for Mandibular Reconstruction. *Materials* **2024**, *17*, 140. [CrossRef] [PubMed]
21. Elfghi, M.A.; Gunay, M. Mechanical Properties of Powder Metallurgy (Ti-6Al-4V) with Hot Isostatic Pressing. *Eng. Technol. Appl. Sci. Res.* **2020**, *10*, 5637–5642. [CrossRef]
22. Chao, C.; Xiangyun, G.; Qing, T.; Raj, K.; Jie, L.; Qingsong, W.; Yusheng, S. Hot isostatic pressing of a near α-Ti alloy: Temperature optimization, microstructural evolution and mechanical performance evaluation. *Mater. Sci. Eng. A.* **2021**, *802*, 14026. [CrossRef]
23. Petrovskiy, P.; Sova, A.; Doubenskaia, M.; Smurov, I. Influence of hot isostatic pressing on structure and properties of titanium cold-spray deposits. *Int. J. Adv. Manuf. Technol.* **2019**, *102*, 819–827. [CrossRef]
24. Guo, R.P.; Xu, L.; Lei, J.F.; Yang, R. Effects of Porosity and Re-HIP on Properties of Ti-6Al-4V Alloy from Atomized Powder. In *Applied Mechanics and Materials*; Trans Tech Publications, Ltd.: Bäch, Switzerland, 2014; Volume 552, pp. 274–277. [CrossRef]
25. Adelmann, B.; Hellmann, R. Mechanical Properties of LPBF-Built Titanium Lattice Structures—A Comparative Study of As-Built and Hot Isostatic Pressed Structures for Medical Implants. *Metals* **2022**, *12*, 2072. [CrossRef]
26. Iwaya, Y.; Machigashira, M.; Kanbara, K.; Miyamoto, M.; Noguchi, K.; Izumi, Y.; Ban, S. Surface properties and biocompatibility of acid-etched titanium. *Dent. Mater. J.* **2008**, *27*, 415–421. [CrossRef] [PubMed]
27. Zhang, K.; Mei, J.; Wain, N.; Wu, X. Effect of hot-isostatic-pressing parameters on the microstructure and properties of powder Ti-6Al-4V hot-isostatically-pressed samples. *Met. Mater. Trans. A.* **2010**, *41*, 1033–1045. [CrossRef]
28. Xu, L.; Guo, R.; Bai, C.; Lei, J.; Yang, R. Effect of hot isostatic pressing conditions and cooling rate on microstructure and properties of Ti–6Al–4V alloy from atomized powder. *J. Mater. Sci. Technol.* **2014**, *30*, 1289–1295. [CrossRef]
29. Cai, C.; Song, B.; Qiu, C.; Li, L.; Xue, P.; Wei, Q.; Zhou, J.; Nan, H.; Chen, H.; Shi, Y. Hot isostatic pressing of in-situ TiB/Ti-6Al-4V composites with novel reinforcement architecture, enhanced hardness and elevated tribological properties. *J. Alloys Compd.* **2017**, *710*, 364–374. [CrossRef]
30. Szafrańska, A.; Antolak-Dudka, A.; Baranowski, P.; Bogusz, P.; Zasada, D.; Małachowski, J.; Czujko, T. Identification of Mechanical Properties for Titanium Alloy Ti-6Al-4V Produced Using LENS Technology. *Materials* **2019**, *12*, 886. [CrossRef] [PubMed]
31. Berger, M.; Jacobs, T.; Boyan, B.; Schwartz, Z. Hot isostatic pressure treatment of 3D printed Ti6Al4V alters surface modifications and cellular response. *J. Biomed. Mater. Res. Part B Appl. Biomater.* **2019**, *108*, 1262–1273. [CrossRef] [PubMed]
32. Antolak-Dudka, A.; Czujko, T.; Durejko, T.; Stępniowski, W.J.; Ziętala, M.; Łukasiewicz, J. Comparison of the Microstructural, Mechanical and Corrosion Resistance Properties of Ti6Al4V Samples Manufactured by LENS and Subjected to Various Heat Treatments. *Materials* **2024**, *17*, 1166. [CrossRef] [PubMed]

Disclaimer/Publisher's Note: The statements, opinions and data contained in all publications are solely those of the individual author(s) and contributor(s) and not of MDPI and/or the editor(s). MDPI and/or the editor(s) disclaim responsibility for any injury to people or property resulting from any ideas, methods, instructions or products referred to in the content.

Review

Advanced Bending and Forming Technologies for Bimetallic Composite Pipes

Hui Li, Yingxia Zhu *, Wei Chen, Chen Yuan and Lei Wang

School of Mechanical Engineering, Jiangsu University, Zhenjiang 212013, China;
2212203054@stmail.ujs.edu.cn (H.L.); chen_wei@ujs.edu.cn (W.C.); 2212303071@stmail.ujs.edu.cn (C.Y.);
2212303045@stmail.ujs.edu.cn (L.W.)
* Correspondence: xia166109@163.com

Abstract: Bimetallic composite pipes, as critical components, effectively integrate the superior properties of diverse materials to meet the growing demand for lightweight, high-strength, and corrosion-resistant solutions. These pipes find extensive applications in petrochemical, power generation, marine engineering, refrigeration equipment, and automotive manufacturing industries. This paper comprehensively reviews advanced bending and forming technologies, with a focus on challenges such as wrinkling, excessive wall thinning, springback, cross-sectional distortion, and interlayer separation. The review combines theoretical analysis, experimental findings, and numerical simulations to provide insights into defect prevention strategies and process optimization. It also evaluates emerging technologies such as artificial neural networks and intelligent control systems, which demonstrate significant potential in enhancing bending accuracy, reducing defects, and improving manufacturing efficiency. Additionally, this work outlines future research directions, emphasizing innovations required to meet the stringent performance standards of bimetallic composite pipe components in high-end applications.

Keywords: bimetallic composite pipe; bending and forming; forming defects; state of the art

Academic Editors: Konrad Laber, Janusz Tomczak, Beata Leszczyńska-Madej, Grażyna Mrówka-Nowotnik and Magdalena Barbara Jabłońska

Received: 27 November 2024
Revised: 20 December 2024
Accepted: 27 December 2024
Published: 30 December 2024

Citation: Li, H.; Zhu, Y.; Chen, W.; Yuan, C.; Wang, L. Advanced Bending and Forming Technologies for Bimetallic Composite Pipes. *Materials* **2025**, *18*, 111. https://doi.org/10.3390/ma18010111

Copyright: © 2024 by the authors. Licensee MDPI, Basel, Switzerland. This article is an open access article distributed under the terms and conditions of the Creative Commons Attribution (CC BY) license (https://creativecommons.org/licenses/by/4.0/).

1. Introduction

Bimetallic composite tubes, as lightweight structural components, effectively address specific challenges faced by conventional pipes. Unlike traditional pipes, which often suffer from material waste and rework due to surface and subsurface flaws (e.g., cracks, trapped air bubbles, and pits) [1,2], bimetallic composite tubes offer enhanced manufacturing precision through advanced bonding and forming techniques. Their dual-material construction minimizes issues like non-uniform thickness and circularity by optimizing the compatibility and properties of inner and outer layers. Moreover, these tubes reduce eccentricity by leveraging tightly controlled forming processes, ensuring structural integrity and reliability in demanding applications such as the petrochemical industry [3], energy power generation [4], ocean engineering [5], refrigeration equipment [6], and the automotive industry [7]. Similar advancements have been made in other bimetallic components, such as bimetallic bolts, which utilize dual-material properties to improve mechanical performance and resistance to harsh environments [8]. While their production demands meticulous control, the ability to integrate multiple material properties in a single component represents a significant improvement over conventional single-material solutions, providing superior strength, corrosion resistance, and adaptability to complex geometries.

The manufacturing processes for bimetallic composite pipes can be broadly categorized into two main types: mechanical bonding and metallurgical bonding. Mechanical bonding processes include drawing, rolling, extrusion, expansion, and spinning, which rely on external forces to join the inner and outer tubes, forming a compact structure. Metallurgical bonding, on the other hand, utilizes the physical and chemical interactions of materials at high temperatures, with key techniques including centrifugal casting, solid-liquid casting, and explosive welding. Mechanical bonding is characterized by high forming efficiency and adaptability, while metallurgical bonding achieves superior interfacial bonding strength, making it suitable for producing composite pipes with high-performance requirements.

To accommodate varying specifications, shapes, materials, and molding tolerances, a range of bending methods for bimetallic composite tubes has been developed. These methods can be categorized based on several criteria. First, considering the forming temperature, the bending process is classified into cold bending and hot bending [9,10]. Second, based on loading conditions, the methods include pure bending, compression bending, push-pull bending, CNC bending, laser bending, and free bending [11–16]. Third, from a material perspective, bimetallic composite tubes encompass combinations such as copper/titanium, aluminum/steel, and aluminum/magnesium [17–19]. Fourth, regarding the shape of the tube, these methods apply to tubes with round, rectangular, and other shaped cross-sections [20,21]. Finally, in terms of bending difficulty, tubes with a large relative radius are easier to bend, whereas those with a small relative bending radius present more challenges [22,23].

During the bending process, the inner and outer layers of composite pipes will be subjected to complex tensile and compressive stress distributions, which may lead to a series of defects or instability phenomena such as wrinkling, excessive wall thinning, cross-section distortion, rebound, and interlayer separation. How to accurately predict and effectively control these phenomena is a long-standing challenge in the field of forming and manufacturing and a focus of researchers' attention [24–26].

At present, for the multiple defects, forming parameters, and process design in the bending and forming of bimetallic composite tubes, researchers have conducted in-depth studies using a variety of methods, which have greatly promoted the technological progress in this field. With the improvement of forming quality requirements for lightweight, high-strength, and corrosion-resistant composite pipe fittings in high-end industries such as petrochemical and offshore engineering, the design of bimetallic composite pipes, control of multiple instabilities, and enhancement of process limits are facing new challenges. To cope with these challenges, this work reviews the common basic problems in bending and forming of bimetallic composite tubes, such as wrinkling instability, wall thickness thinning, interlayer separation, cross-section deformation and rebound, etc. It also evaluates the advantages and disadvantages of the current bending and forming technologies and discusses the future development direction and problems to be solved.

2. Bimetallic Composite Pipe Bending Theory Research Progress

The bimetallic composite pipe features a base-liner structure, which consists of two main components: the inner tube and the outer tube. To meet the diverse property requirements for specific application environments, different materials are typically selected for the inner and outer layers. Common materials include aluminum alloys, magnesium alloys, alloy steels, cast iron, and copper alloys, as detailed in Table 1.

Table 1. Materials and application fields of bimetallic composite pipes in some studies.

Composite Method	Inner Tube Material	Outer Tube Material	Application Field
Mechanical composite	Aluminum Alloys: Al7475, AA5052, Al6063, AA6061, 5A03 Alloy Steel: STKM13C, 316L, SS304	3A21 Aluminum Alloy [26], AZ31 Magnesium Alloy [19], H65 Brass [27], T2 Copper [24], AA1070 Aluminum Alloy [28], 20# Steel [6,25,29]	Chemical pipelines [29], High-temperature cooling walls [30], Solar energy collection [4], Marine development [5], Air conditioning ducts [6], Automotive industry [7]
Metallurgical composite	Aluminum Alloy: Al6061 [18]	1020 Carbon Steel	
	Copper Alloy: T2, C10100	Q235 Carbon Steel [30], Al1070 Aluminum Alloy [31], 6061 Aluminum Alloy [32]	

Investigations [24–26] have shown that the bimetallic composite tube bending process can lead to forming defects such as wrinkling, excessive wall thickness thinning, cross-section distortion, and springback. However, the presence of a double-layer structure results in distinct characteristics that differ from those of single-tube configurations. Consequently, numerous studies have been conducted to examine the bending formation of bimetallic composite tubes, focusing on aspects such as wrinkling, excessive wall thinning, cross-sectional distortion, and springback [9–23]. Understanding the occurrence mechanism, accurately predicting, and effectively controlling these defects are the core issues of the bending theory research of bimetallic composite tubes.

2.1. Wrinkling

2.1.1. Wrinkling Prediction Study

In the early stage of the study of buckling theory, Hill [33] introduced a sufficient condition for the uniqueness of elastic-plastic materials, which served as a foundational tool for analyzing wrinkling behavior and laid the groundwork for preliminary bifurcation theory. Building on this, Hutchison [34] extended Hill's theory by applying it to the wrinkling analysis of thin shells, thereby providing critical theoretical support for understanding the buckling behavior of thin shell structures. In the context of the bimetallic composite tube bending process, the interaction and wrinkling behavior of the inner and outer walls can be effectively described using plastic bifurcation theory. This theory, which incorporates the plastic deformation properties of materials, offers a more accurate representation of the buckling and wrinkling phenomena occurring in the inner liner pipe under high bending strains [35]. Yuan et al. [36] further advanced this approach by employing the J2 plastic deformation theory (Equation (1)) to model the inelastic behavior of materials and predict wrinkling in bimetallic composite pipes. Their findings revealed that the bending flexural strain in the inner pipe was significantly lower than that observed in single-layer pipes.

$$\sigma_{ij} = \frac{E_s}{(1+v_s)} \left\{ \frac{v_s}{(1-2v_s)} \delta_{ij}\delta_{kl} + \frac{1}{2}\left[\delta_{ik}\delta_{jl} + \delta_{il}\delta_{jk}\right] \right\} \varepsilon_{ij} \qquad (1)$$

where σ_{ij} is the stress tensor component, E_s is the modulus of elasticity of the material, v_s is the Poisson's ratio, and ε_{kl} is the strain tensor component. δ_{ij} is the identity matrix filter symbol used in tensor operations, which equals 1 when the subscripts are identical and 0 otherwise. Different combinations of subscripts (e.g., $\delta_{kl}\delta_{ik}\delta_{jl}\delta_{il}\delta_{jk}$) are used to represent various stress components in volumetric deformation and shear deformation, respectively.

Unlike bifurcation theory, the energy principle (Equation (2)) has been widely applied due to its simplicity and computational efficiency. For forming processes with relatively simple boundary conditions, instability prediction models can be quickly derived based on energy criteria [37,38]. For example, Zhang et al. [10] investigated the deformation behavior of ultra-thin-walled tubes under the hydraulic push-pull bending process. Through finite element simulations and experimental validation, they demonstrated the effectiveness of the energy principle in wrinkling prediction. Specifically, by analyzing the variation in the system's total energy during deformation, the critical conditions for wrinkling were determined. When the internal pressure was insufficient, the inner wall of the tube exhibited significant fluctuations in wall thickness, resulting in pronounced wrinkling. By applying appropriate internal pressure, wrinkling could be effectively suppressed, leading to a more uniform wall thickness distribution during the forming process.

$$U_{min} = W_{ex} \qquad (2)$$

where U_{min} is the minimum energy required for wrinkling, W_{ex} is the plastic deformation energy in the deformation zone, and wrinkling occurs when Wex reaches a critical point.

Naderi [39] developed an energy-based 3D finite element model for the prediction of bending wrinkling of thin-walled tubes based on the principle of energy, considering geometrical nonlinearities, material defects, inelastic material behavior, and interfacial contact, which was based on a carbon steel/stainless steel composite tube. It was found that geometrical and thickness defects lead to an increase in the probability of the occurrence of wrinkles in the inner tube.

2.1.2. Influencing Factors and Control of Wrinkling

In the process of bending composite pipes, effective strategies for controlling wrinkles encompass the selection of appropriate materials, optimization of the bending process, implementation of mandrel filling techniques, and enhancement of the accuracy in initial preparation (Table 2).

Table 2. Strategies for controlling wrinkling during bending of composite tubes.

Wrinkle Control Strategies	Descriptive
Selection of inner tube material [40,41]	Adoption of inner tube materials with better bending processability and optimization of the ratio of inner and outer tube wall thicknesses
Optimization of process parameters [42,43]	Adjustment of process parameters such as bending radius, bending speed, and temperature to regulate the material stress state
Filled with mandrels [44,45]	Bending process with various filled mandrels (polymers), granular media (sand), or liquid media (water, etc.)
Improvement of initial preparation accuracy [46]	Use of high-precision machining equipment to control the initial preparation process of composite tubes to reduce initial defects and improve molding accuracy

Currently, although the wrinkling phenomenon that occurs during bending of bimetallic composite tubes is well understood, its effective avoidance remains a major challenge to improve the bending limit and to achieve accurate forming, especially in composite tube applications with small relative bending radius. The literature [22] points out that the increase in the diameter difference between the inner and outer materials of bimetallic composite tubes, as well as the change in the contact stress distribution between the inner and outer layers, and the expansion of the critical wrinkling region, increase the risk of wrinkling significantly. During the bending process of composite pipes, the inner and outer materials exhibit distinct stress and strain responses, which can lead to challenges in maintaining structural integrity. A critical issue arises from the delamination or separation at the contact interface between the inner and outer materials, which significantly contributes to the formation of wrinkles. Therefore, enhancing the control and stability of the contact interface between these materials is essential to mitigate the risk of wrinkling and ensure the mechanical performance of composite pipes.

2.2. Excessive Wall Thickness Reduction and Cross-Section Distortion

During the bending process of composite pipes, excessive wall thinning often occurs, primarily due to the uneven distribution of stress between the inner and outer wall materials. As shown in Figure 1, the inner wall is subjected to compressive stress, which can result in localized yielding and plastic deformation of the material. Simultaneously, the outer wall experiences tensile stress, leading to localized deformation and stretching. This disparity in stress distribution between the inner and outer walls is the fundamental cause of the wall thinning phenomenon observed during bending molding. This non-uniformity in stress distribution not only leads to wall thickness thinning but may also cause distortion of the cross-section. The prediction of wall thickness thinning and cross-section distortion is influenced by material properties, geometry, and process parameters, so researchers have continued to explore the laws that influence them to achieve more precise control. These research results provide an important theoretical basis for improving the molding process of composite pipes, enhancing product quality, and ensuring structural reliability.

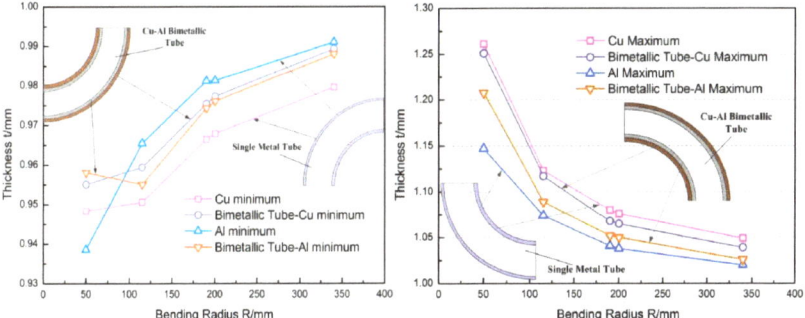

Figure 1. Wall thickness variation of the tube under different bending radii [47].

2.2.1. Theoretical Research on Wall Thickness Thinning and Cross-Section Distortion

Tang et al. [48] pointed out that during the bending process, the outer wall of the tube undergoes tensile thinning, while the inner wall experiences compressive thickening (as shown in Figure 2). To accurately predict the variation in stress distribution during bending, mesh convergence analysis plays a critical role in finite element simulations. By progressively reducing element size, mesh convergence analysis evaluates the relationship between estimation errors and mesh size, thereby yielding precise stress distribution results [49]. Studies have shown that when the element size reaches a certain threshold, the error in the calculated stress distribution significantly decreases, and prediction accuracy improves markedly. The mesh convergence curve effectively validates the reliability of the model and the accuracy of the computational results. Additionally, the necking instability of the material is closely related to wall thickness variation during the bending process. During tension, localized stress concentration leads to necking, which manifests as localized thickening or thinning during bending. By comparing experimental calibration with model predictions, the relationship between stress distribution and wall thickness variation can be further validated, and model parameters can be optimized to enhance prediction accuracy. Moreover, Guo et al. [47] proposed a wall thickness variation prediction model based on plasticity theory (Equation (3)). Incorporating Hencky's stress–strain relationship, the model reveals the coupled effect of material properties on wall thickness variation and stress distribution. This model enables prediction of the overall wall thickness variation of composite pipes and elucidates the coupling relationship between the thickness changes of the inner and outer layers, as well as the shift in the strain-neutral layer during free bending.

$$0 = \frac{4}{3}\int_{\rho_a}^{\rho_N} \text{sgn}(\delta_1) \cdot Y_1 \left[\left(\frac{\Delta t_1}{t_{10}}\right)^+\right] d\rho/\rho + \frac{4}{3}\int_{\rho_N}^{\rho_b} \text{sgn}(\delta_2) \cdot Y_2 \left[\left(\frac{\Delta t_2}{t_{20}}\right)^+\right] d\rho/\rho \quad (3)$$

where ρ is the radial position, ρ_a is the inner radius, ρ_b is the outer radius, ρ_N is the radius of the neutral layer, $\text{sgn}(\delta_i)$ is the sign function, Y_1 and Y_2 are the flow stress functions of copper and aluminum, respectively, Δt_1 and Δt_2 are the thickness variations of copper and aluminum, respectively, and t_{10} and t_{20} are the initial thicknesses of copper and aluminum, respectively.

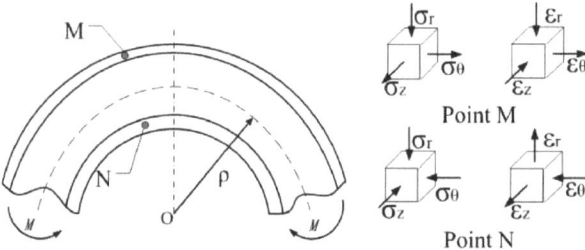

Figure 2. Pipe bending process: external wall structure diagram [47].

When analyzing the cross-sectional flattening of single-layer tubes, traditional methods usually treat them as elliptical [50], focusing only on the outermost layer at the bending site and ignoring the cross-sectional shape between the outermost layer and the geometric center. However, in the case of bimetallic tubes, the deformation trend of each layer is often inconsistent, and the coupling between the layers leads to a deformation behavior that cannot follow the original mechanical properties. Fu [51] innovatively proposed a physics-driven B-spline curve fitting method to characterize the full cross-sectional deformation of Cu-Al bimetallic tubes during the rotary draw bending process. In this process, the characteristic curve of cross-sectional deformation can be described using the B-spline fitting method. The equation for the characteristic curve $h(u)$ on the outer side of the cross-section is as follows:

$$h(u) = \sum_{j=0}^{4} d_j N_{j,3}(u), u \in [0,1] \quad (4)$$

Here, d_j represents the control points, $N_{j,3}(u)$ is the third-order B-spline basis function, and the distribution of the control points is closely related to the variation in the outer wall thickness. The equation for the characteristic curve on the inner side of the cross-section is expressed as follows:

$$h_i(\theta) = r_0 + t_0 + \Delta t \cos(\theta) + f(\alpha, \beta) \quad (5)$$

Here, Δt represents the wall thickness variation, θ denotes the positional angle, and $f(\alpha, \beta)$ is the fitting function coupled with curvature and contact pressure.

Figure 3 clearly illustrates the fitting of the outer and inner characteristic curves, particularly highlighting the distribution characteristics of wall thickness variation. The curves accurately describe the geometric features of cross-sectional deformation during the bending process by capturing the coupling relationship between control points and positional angles. Through the application of physical constraint conditions and the principle of energy minimization, the B-spline fitting method achieves precise modeling of the full cross-sectional deformation.

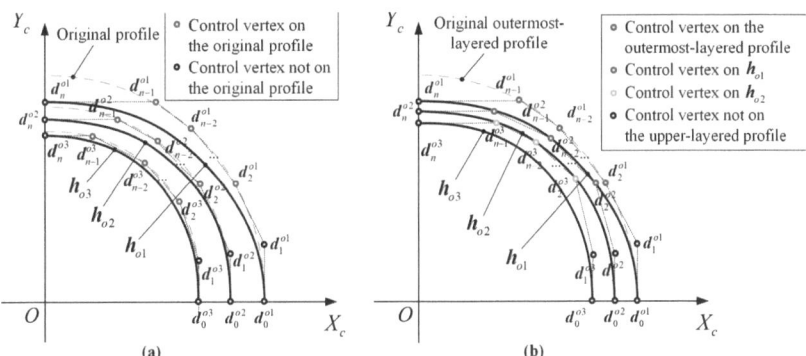

Figure 3. Setting of control vertices according to the convex hull property [51]. (a) Control vertices on original profile; (b) Control vertices on upper-layered profile.

2.2.2. Influence and Control of Wall Thickness Reduction and Cross-Section Distortion

In the composite pipe bending process, effective control of wall thickness reduction and cross-section distortion strategies include the selection of material parameters, the use of mandrel filling, process parameters, and bending method optimization. The relevant studies are shown in Table 3.

Table 3. Strategies for controlling wall thickness reduction and cross-section distortion in the composite pipe bending process.

Wrinkle Control Strategies	Descriptive
Control of material parameters [52,53]	Optimize the ratio of inner and outer tube material and wall thickness to balance the deformation and reduce cross-section distortion
Filled with mandrels [54–56]	Filling of bending areas with PVC cores, rigid cores, etc., prevents instability of the cross-section shape
Optimization of process parameters [57,58]	Adjustment of process parameters such as bending angle, lubrication effect, boost speed, bending radius, etc.
Improved bending [59,60]	The use of multi-stage bending, free bending, and other advanced technology to reduce local stress concentration and control wall thickness changes

These research results provide a solid theoretical basis for improving wall thickness thinning and cross-section distortion of bimetallic composite tubes in the bending process. Although neither can be completely avoided in the bending process, these instabilities can be effectively controlled by the above optimized conditions, keeping them within acceptable tolerances. This ensures the quality and reliability of the composite tube.

2.3. Springback

In the bending process of bimetallic composite pipes, springback refers to the shape deviation caused by the release of residual stress after the removal of external forces. Springback theories are typically based on the following assumptions: the material is considered an ideal elastoplastic body, the elastic stage follows Hooke's law, and the plastic stage conforms to specific hardening laws. However, in practical applications, the strain-hardening effect of materials significantly influences springback behavior. Moreover, traditional theories assume a fixed neutral layer position. However, for bimetallic composite pipes, the neutral layer undergoes significant displacement due to the differing elastic moduli and yield strengths of the inner and outer materials, leading to reduced prediction accuracy. Additionally, conventional springback theories often neglect the effects of strain rate and temperature. In actual production, particularly during high-speed bending or when dealing with temperature-sensitive materials (such as aluminum and magnesium alloys), variations in strain rate and temperature significantly impact springback behavior. These deviations in assumptions limit the applicability of theoretical models under complex conditions. Therefore, in engineering applications, experimental calibration and finite element simulation are essential for more accurate springback prediction and control, ensuring that the forming quality meets practical requirements.

2.3.1. Research on Springback Theory

In the study of single pipe springback theory, Al-Qureshi [61] derived the elastic-plastic deformation equation during bending, highlighting that the location of the elastic-plastic transition within the cross-section significantly influences the required bending force, springback degree, and residual stress after unloading. This theory assumes a material model without strain hardening, simplifying the analysis but limiting its ability to reflect real-world strain-hardening behavior during bending. To address this, some studies [62] incorporated a hardening law in the form of a power function into the analytical model, enabling more accurate predictions of springback by accounting for material strain hardening. For composite pipes, traditional single-pipe springback theories are insufficient due to differing material properties across layers. Zhang [11] introduced the concept of a composite elastic modulus Ec and neutral layer shift $S\varepsilon$, as shown in Equation (6), to enhance the springback model. This approach assumes that the neutral layer shift is a critical factor and simplifies its behavior for calculation, achieving better agreement between theoretical predictions and experimental results (Figure 4). However, such assumptions, including linearized hardening and uniform material properties, limit the applicability of these models under non-ideal boundary conditions or non-uniform material distributions commonly encountered in industrial applications, emphasizing the need for further refinements through experimental calibration and advanced simulations.

$$D_\varepsilon = \frac{D_k + D_{k+1} + \cdots + D_{k+m}}{m+1}, k \in N^*, k \leq n, m \in N^* \tag{6}$$

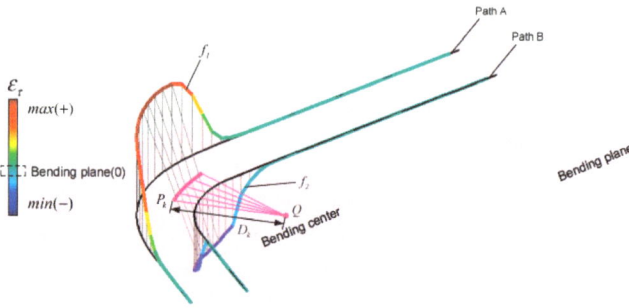

Figure 4. Process diagram of neutral layer shifting extraction [14].

$$S_\varepsilon = R_0 - D_\varepsilon \tag{7}$$

where D_ε is the neutral layer shift, D_k is the distance from the neutral layer position to the bending center, S_ε is the relative displacement of the neutral layer, and R_0 is the initial bending radius.

2.3.2. Influence and Control of Rebound

Table 4 delineates various significant factors that impact pipe springback, thereby affecting the behavior and specific patterns of springback in pipes. A comprehensive understanding of these factors enables more precise predictions and regulation of springback during the bending process, ultimately facilitating the optimization of processing parameters.

Table 4. Influence factors of springback during bending of composite tubes.

Factor	Law of Influence
Material Properties [63–65]	The higher the modulus of elasticity and yield strength, the more pronounced the rebound is
Bending angle [66,67]	The larger the bending angle, the more significant the rebound effect is
Frictional conditions [68,69]	Higher coefficients of friction between the tool and the material result in greater rebound

Investigations in the literature [70,71] have investigated the impact of employing a composite tube filled with a plastic mandrel on the resilience angle during rotational bending. The findings indicate that the presence of a plastic mandrel within the composite tube establishes an approximately linear correlation between the resilience angle and the bending angle. Furthermore, the application of the overbending compensation method has been demonstrated to effectively regulate the bending angle, thereby enhancing both the precision and efficiency of the bending process.

The above research results have a key role in realizing the precision bending of composite pipes. However, the current effective control of springback mainly relies on empirical and experimental methods. There are many factors affecting the rebound and the fluctuation in the rebound amount is large, which together lead to the challenging task of accurately predicting the rebound angle.

2.4. Interface Strengthening

Interfacial bond strength determines the service performance of bimetallic composite pipes. Therefore, interlayer separation is an important molding defect in bimetallic composite pipes. Defects such as wrinkling of the inner tube, thinning of the outer tube, and cross-section distortion will aggravate the interlayer separation defects to a certain extent. At present, overcoming the interlayer separation defect still relies on improving the interfacial bonding strength during the initial preparation of bimetallic composite pipes. The initial preparation methods of bimetallic composite pipes mainly include two types: mechanical bonding and metallurgical bonding.

2.4.1. Control of Interlayer Separation in the Mechanical Composite Process

By optimizing spinning parameters [25,39,72], such as the press-in volume, feed rate, and roller angle, the interfacial bonding strength in spinning processes can be significantly enhanced. Residual contact pressure serves as a critical factor in preventing interlayer separation (Figure 5). Maximizing the residual contact pressure during spinning enables synergistic deformation of the inner and outer materials during bending, mitigating stress concentration and slip at the interface. Furthermore, studies on 20/316L bimetallic composite pipes have demonstrated that when the indentation depth is 0.14 mm, the orientation angle is 2.5°, and the feed rate is 0.3 mm/rev, residual contact stress is effectively improved, ensuring interfacial bonding quality. As illustrated in Figure 6, these optimized parameters facilitate a uniform and higher residual contact stress distribution at the interface, significantly enhancing the stability and performance of composite pipes during subsequent bending processes.

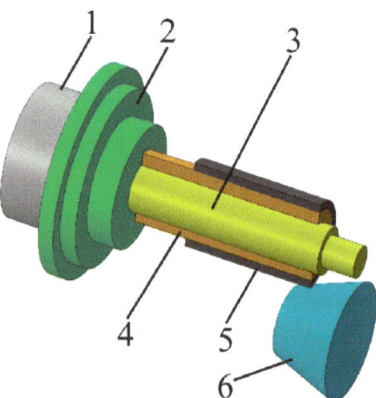

Figure 5. Spinning process (1, spindle; 2, flange; 3, mandrel; 4, inner tube; 5, outer tube; 6, spinning wheel).

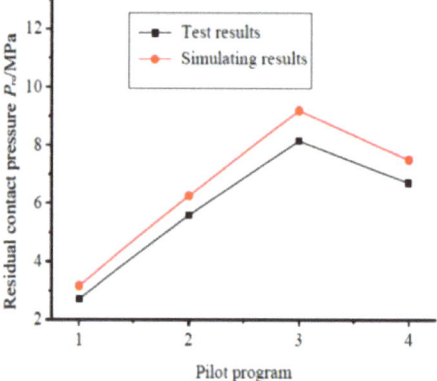

Figure 6. Comparison of residual contact pressure [29].

The drawing method combines the inner and outer materials through mold deformation by applying axial force to the composite pipe (Figure 7). In the study, it is found that the cone size directly affects the bonding strength of the composite pipe interface. A larger cone diameter can effectively increase the interface contact area, thereby improving the bond strength and reducing the separation phenomenon caused by stress concentration. By optimizing the cone size in the drawing process, the bonding quality of the inner and outer layers can be significantly improved, and the risk of interlayer separation can be reduced (Figure 8) [73–75].

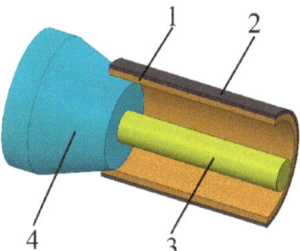

Figure 7. Drawing process (1, inner tube; 2, outer tube; 3, mandrel; 4, drawing die).

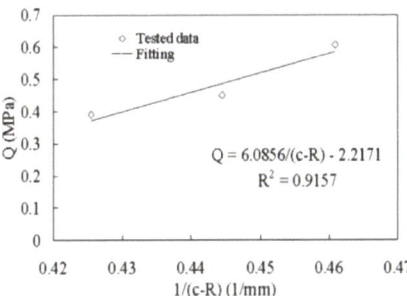

Figure 8. Effect of cone size on interfacial bonding strength [73].

2.4.2. Control of Layer Separation in the Metallurgical Composite Processes

Centrifugal casting utilizes high-speed rotational centrifugal forces to uniformly distribute molten metal along the mold wall, forming a composite structure (Figure 9). Controlling the pouring temperature and mold rotational speed is critical to improving the interfacial bonding quality [76–78]. As shown in Figure 10, research on 40Cr/Q345B bimetallic ring blanks in the literature [77] indicates that when the outer metal pouring temperature is 1570 °C and the mold rotational speed is 800 r/min, the axial temperature difference at the interface can be effectively reduced (approximately 75.6 °C), promoting more uniform metallurgical bonding. Furthermore, when the inner metal pouring temperature is 1600 °C and the pouring interval is 221 s, a stable metallurgical bonding layer is formed at the interface, significantly reducing interlayer separation and shrinkage defects. By optimizing these parameters, the diffusion of elements between the inner and outer metals can be effectively enhanced, forming a uniform metallurgical bonding layer at the interface and preventing defects and microcracks caused by excessively fast or slow cooling rates. In addition, reasonable pouring intervals and mold rotational speeds help minimize interface separation caused by thermal stress differences, further enhancing the interfacial metallurgical bonding strength (Table 5).

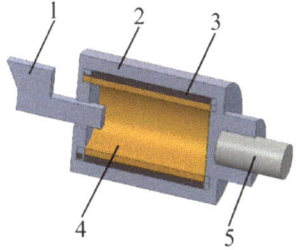

Figure 9. Centrifugal casting process (1, pouring gate; 2, mold; 3, outer tube; 4, inner tube; 5, pindle).

Figure 10. Interfacial microstructure of 40Cr/Q345B bimetallic ring blank [77].

Table 5. Strategies for controlling interlayer separation by different processes.

Crafts	Category	Methods
Spin molding	mechanical compound	Optimization of spinning parameters such as press-in volume, feed rate, etc. to increase residual contact
Drawing		Increased cone diameter for improved interfacial bonding
Centrifugal Casting	Metallurgical composite	Enhanced metallurgical bonding by controlling pouring temperature and mold speed Adjustment of casting intervals and mold speed to improve interfacial element diffusion
Solid-liquid casting		Optimize casting temperature to improve interfacial metallurgical bonding Introduction of Sn-Bi intermediate layer to improve wettability and interfacial bonding
Explosion Welding		Enhanced interfacial bonding through high-pressure conditions created by explosive welding Optimize collision speed and angle, fine crystal enhancement to improve interface integration

The solid-liquid casting method forms a bimetallic structure by bonding molten metal with a solid substrate material (Figure 11). Studies [79,80] have shown that introducing an Sn-Bi interlayer can effectively improve interfacial wettability and enhance bonding strength. As observed in Figure 12 [79], the use of an Sn-Bi alloy as an interlayer improves the contact condition between the inner and outer metals, reduces interfacial porosity and inclusions, and prevents interlayer separation caused by interfacial defects.

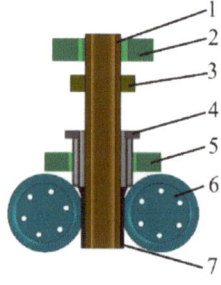

Figure 11. Solid rolling composite process. (1, inner tube; 2 and 5, heating device; 3, guiding device; 4, casting device; 6, cooling rolls; 7, outer tube).

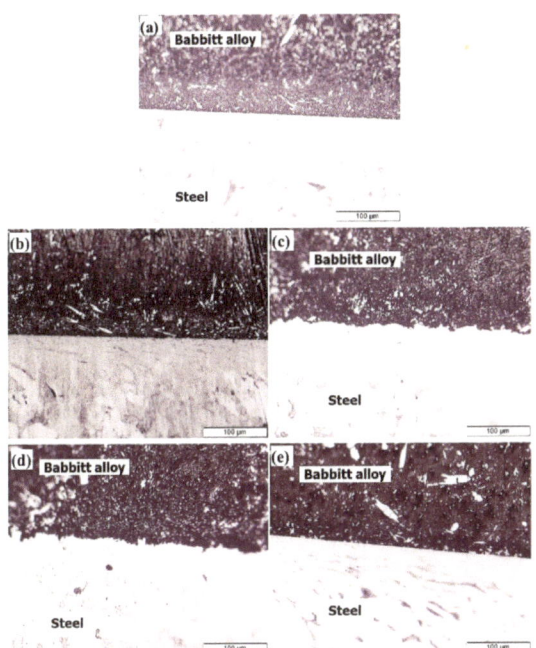

Figure 12. Edge interfacial thickness of Babbitt–steel composite with Sn and Sn–Bi interlayers, (**a**) bimetal with Sn interlayer, (**b**) bimetal with Sn + 1% Bi interlayer, (**c**) bimetal with Sn + 2% Bi interlayer, (**d**) bimetal with Sn + 3% Bi interlayer, and (**e**) bimetal with Sn + 4% Bi interlayer [79].

Explosive welding tightly bonds two metals through shock waves and pressure generated by high-energy explosions. Utilizing this technique, effective bonding between the inner and outer materials is achieved under extreme pressure and temperature conditions (Figure 13). As shown in Figure 14, a study [80] on the microstructure of Fe/Al at different detonation points revealed that insufficient detonation energy fails to achieve effective bonding between the materials, while at the terminal position, Fe and Al are effectively bonded. The high pressure and grain refinement induced by explosive welding not only enhance interfacial bonding strength but also effectively suppress the initiation and propagation of interfacial cracks, reducing the occurrence of interlayer separation. Furthermore, optimizing the detonation angle and velocity further improves the metallurgical bonding quality at the welding interface, significantly enhancing the mechanical properties of the composite pipe [30–32,81].

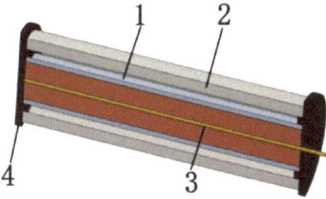

Figure 13. Explosive molding. (1, inner tube; 2, outer tube; 3, detonating cord; 4, piston).

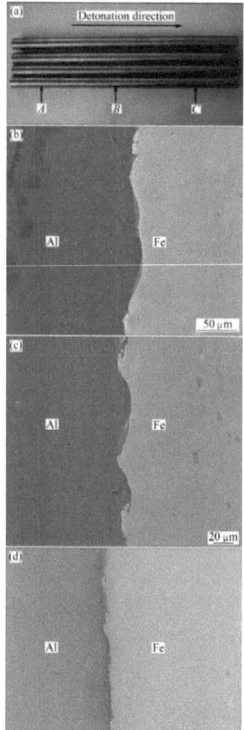

Figure 14. Secondary electron images of interface with good bonding properties: (**a**) sampling positions, (**b**) interface at position A, (**c**) interface at position B, and (**d**) interface at position C [81].

2.5. Process Optimization Design

Traditional tube bending techniques primarily rely on empirical formulas and limited experimental data, making it difficult to handle complex material and process variations, particularly when addressing nonlinear springback behavior. For example, traditional methods often require extensive trials and adjustments to meet the bending demands of different materials and geometries, which is both time-consuming and costly. Although finite element methods (FEMs) have achieved considerable success in predicting tube bending and forming, their computational complexity limits efficiency, especially in large-scale industrial applications. To overcome the limitations of traditional methods, artificial neural networks (ANNs) and database technologies have been gradually introduced into tube bending processes [82]. ANNs possess the capability to handle complex nonlinear relationships [61], enabling high-precision springback prediction and process optimization, by learning from large datasets of historical data. However, these datasets also present certain limitations: first, the data is primarily derived from simulation results, which, although experimentally validated, may lack sufficient representativeness under extreme conditions; second, the data generation focuses predominantly on specific materials (e.g., aluminum alloys), limiting generalization to other materials; finally, the variation ranges of certain parameters in the datasets are relatively narrow, which may constrain the model's prediction accuracy, particularly when faced with novel geometries or process conditions (Table 6).

Wu et al. developed a predictive model for the bending deformation of welded thin-walled aluminum alloy square tubes using a backpropagation (BP) neural network [83]. By training the model with 270 sets of finite element simulation data, they successfully demonstrated its ability to predict welding-induced deformation under complex nonlinear conditions with high accuracy. This method not only enhances prediction precision but also minimizes the reliance on physical experiments, thereby significantly reducing production costs. Building on this foundation, Wang et al. highlighted the advantages of artificial neural networks in addressing challenges posed by complex geometries and variable process conditions [84]. They introduced a high-precision rebound prediction model based on a graph neural network (GNN), which achieves an impressive prediction accuracy of 99.993%. The GNN model excels in capturing geometric and process variations in real time during the bending process. Moreover, its real-time feedback capability enables automatic adjustment of process parameters during production, effectively lowering the defect rate and further optimizing manufacturing efficiency.

Database technology played a key role in this process. By constructing and maintaining a database of the pipe bending process, researchers are able to efficiently manage experimental and production data and provide powerful data support for model training. Wu et. al. generated a large amount of training data through finite element simulation and successfully created a dynamically updatable database to optimize the training of the neural network model [83]. This database-driven model training ensures sustainable optimization of the model to maintain highly accurate predictions under changing process conditions.

With the continuous advancement of machine learning and big data technologies, tube bending processes are anticipated to become increasingly intelligent and automated. In particular, the integration of ANN with databases is expected to extend beyond the optimization of production processes to encompass the early stages of design. This development will enable engineers to make real-time predictions and adjustments during product prototyping, thereby enhancing the efficiency and accuracy of product development. Furthermore, tube bending technology is likely to evolve into a more intelligent and interconnected system. Future production processes will seamlessly incorporate real-time monitoring, data analysis, and model optimization, leading to significant improvements in both production efficiency and product quality. By continuously learning from new data, ANN models will be capable of autonomously adjusting production parameters, ensuring that the tube bending process becomes increasingly precise and reliable over time.

Table 6. Advantages of the combination of ANN and database.

Advantage	Descriptive
Improved prediction accuracy [85]	Accurate prediction of bending deflection and springback under complex geometric and material conditions
Real-time feedback and process optimization [86]	Provide real-time monitoring and feedback to dynamically adjust process parameters
Reduced costs [87]	Reduces the need for physical experimentation and lowers production and R&D costs

3. Bimetallic Composite Pipe Bending Technology

At present, the most widely used bending technologies for bimetallic composite pipes include free bending and CNC bending forming. Compared with monometallic tubes, bimetallic tubes face more complex stress–strain interactions during the forming process, primarily due to the distinct material properties of the inner and outer layers. These differences lead to significant challenges, such as uneven stress distributions, increased springback, and interfacial separation, particularly in high-strength or multi-layered materials. For example, the inner layer may undergo severe compressive deformation, resulting in wrinkling, while the outer layer experiences tensile deformation that increases the risk of cracking. These challenges necessitate strict control of processing parameters [14,66], including neutral layer shift, mold clearance, friction coefficient, and bending angle. In high-strength materials, additional difficulties arise from their higher elastic modulus and yield strength, which exacerbate springback and make precise forming more complex. For multi-layer materials, weak interfacial bonding or differences in thermal and mechanical properties between layers can lead to delamination or interfacial slip during bending.

The free bending technique is particularly suitable for composite tube space forming [47] by flexibly adjusting the bending radius and angle [88]. This technique mainly consists of a bending die (Bending Die), a pressing device (Pressing), a guiding device (Guider), and bearings (Bearing). Different bending radii are achieved by replacing or adjusting the radius of the bending die, while the bending angle is controlled by adjusting the position and magnitude of the acting force P_U (as shown in Figure 15). Free bending controls the bending forming process by gradually applying the bending force, which reduces springback and cross-sectional deformation and achieves a more uniform wall thickness distribution without the need for complex molds [89–91]. This technique is particularly suitable for the bending forming of bimetallic composite pipes with larger bending radii and smaller diameters, which can significantly improve the forming accuracy and material utilization (Table 7).

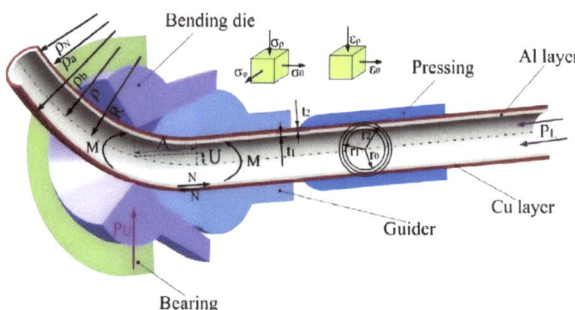

Figure 15. Free bending technology pipeline model [47].

Table 7. Advantages of free bending.

Evaluation Criteria	Technical Advantages
Accuracy of bending mechanics models	Experimental and theoretical model validation shows that the mechanical model for free bending achieves a prediction deviation of ≤5%, demonstrating high precision and suitability for parameter optimization in complex forming tasks [47,90]
Complex curve forming capability and flexibility	The dynamic adaptability of five-axis and six-axis free bending equipment is high, enabling precise forming of complex three-dimensional paths and multi-radius bends, with errors controlled within ≤0.5 mm [89,91]
Springback prediction and compensation effectiveness	Based on U-R experimental data, springback prediction and compensation techniques for free bending significantly improve forming accuracy. The compensation curve reduces the deviation between the springback angle and the target angle to ≤1° [90]

The CNC bending technology for bimetallic composite pipes integrates traditional rotary bending methods with advanced CNC technology, as illustrated in Figure 16. The mold structure used in this process comprises several key components, including the Pressure Die, Wiper Die, Clamping Die, and Rotary Bending Die. The bending force is applied through the Rotary Bending Die to achieve the desired pipe formation. To better understand the forming mechanism and optimize the bending process, extensive research has been conducted using Cu-Ti and Cu-Al bimetallic tubes as experimental models. These studies have demonstrated that the precise formation of bimetallic composite tubes, particularly those with small bending radius and complex three-dimensional geometries, can be achieved by rigorously controlling critical processing parameters. Such parameters include the neutral layer shift, die clearance, friction coefficient, and bending angle [11,44]. As a result, this technology is emerging as a pivotal innovation in the manufacturing of bimetallic composite pipes, offering robust technical support for the large-scale industrial production of intricate pipe fittings [92]. (Table 8).

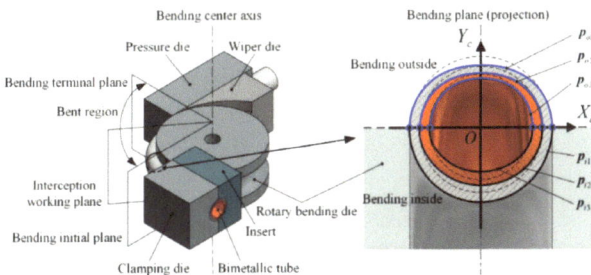

Figure 16. A model combining rotary stretching and bending with numerical control technology [51].

Table 8. Advantages of CNC bending.

Evaluation Criteria	Technical Advantages
Cross-section deformation control capability	The cross-sectional characterization model based on B-spline curve fitting significantly reduces the error compared to traditional elliptic models, with the average cross-sectional deformation error controlled within 1.82% [51]
Bending axis accuracy	The bending axis deviation is controlled within 0.5 mm [51]
Real-time prediction and multitask learning	By integrating multi-source input multitask learning (MTL) with digital twin (DT) technology, real-time prediction of springback and defect classification during the bending process is achieved, enhancing both prediction efficiency and accuracy [92]

4. Trends in Composite Pipe Bending and Forming

In order to meet the urgent manufacturing needs of high-performance, lightweight bending pipe fittings for industries such as the aerospace and automotive industries, bending theory and technology are showing the following development trends:

(1) Complex geometry and high precision requirements

With the increasing demand for composite fittings, the geometry of pipes also tends to be complex. In recent years, more and more application scenarios require composite fittings to meet the requirements of small bending radii, thin walls, large diameters, and complex three-dimensional shapes. For example, in the automotive and aerospace fields, more complex fluid passages are required. These applications require composite tubes to not only have high mechanical properties but also to enable high-precision bending and forming [93,94]. Therefore, one of the development trends of the composite pipe bending is to improve forming accuracy, especially in the bending process of complex shapes, to maintain uniform wall thickness, stable cross-sectional shapes, and consistent bending radii.

(2) Intelligent process control and automation

Modern manufacturing increasingly relies on intelligent decision-making and automation control, particularly in processes such as composite pipe bending and forming. By integrating advanced technologies, including artificial intelligence (e.g., neural networks), big data analysis, and database systems, it is possible to achieve real-time monitoring and feedback optimization throughout the manufacturing process. For instance, prediction models based on ANN can analyze historical data and dynamically adjust process parameters to maintain high-quality forming, even under complex and variable conditions [95]. This application of intelligent technology not only enhances the efficiency of the bending process but also significantly reduces the scrap rate and lowers production costs [96]. Looking ahead, the deep integration of artificial intelligence with composite pipe bending processes is expected to become the dominant trend in the evolution of composite pipe forming technology.

(3) The development of non-traditional bending technology

In order to cope with the limitations of traditional bending technology, new bending methods are constantly being developed. For example, laser-assisted hot bending, free bending, and CNC bending technology have been gradually applied to the forming process of composite pipe parts [97–99]. Laser-assisted hot bending can precisely control the bending angle through localized heating and avoid deformation and defects caused by mechanical contact. Free bending, on the other hand, is able to achieve more flexible tube forming without the use of complex molds, which is particularly suitable for small-batch, high-complexity production needs. With the continuous optimization and improvement of these technologies, they will further improve the flexibility and reliability of the composite pipe forming process.

5. Challenges in Composite Pipe Bending and Forming

In response to the above pipe bending trends, the challenges that need to be addressed are summarized below:

(1) Challenges of complex geometry and high precision requirements

With the wide application of bimetallic composite tubes in aerospace, automotive, and other high-precision manufacturing areas, the complex geometry and precision requirements of tube bending have become more stringent. In the bending and forming of small bending radii, thin walls, large diameters, and complex three-dimensional shapes, the differences in stress distribution between the inner and outer materials make the tubes highly susceptible to defects, especially wrinkling of the inner wall, thinning of the outer wall, and cross-sectional distortion [100]. Complex geometric bending increases the difficulty of predicting and controlling these defects, and current theoretical models and process tools are significantly inadequate in dealing with these problems. In addition, as the demand for complex geometries increases, accurately controlling strain-neutral layer offsets during the forming process, as well as coping with inhomogeneous deformations of each layer of material during bending, remain the main challenges for improving forming accuracy.

(2) Challenges of intelligent process control and automation

Although intelligent technologies and automation control are gradually being applied to the bending and forming processes of composite pipes, the realization of complete intelligence still faces many difficulties. Although current artificial intelligence (such as neural networks) and big data analysis have been used to a certain extent for process optimization, these methods require a large amount of data accumulation and algorithm training. They also face the problems of high cost and long cycle times in actual production. Especially when dealing with complex geometries or different material combinations, existing intelligent systems struggle to cope with changing production demands and lack sufficient adaptive capabilities [101]. In addition, there are technical bottlenecks in real-time monitoring and dynamic adjustment of process parameters by automated equipment, which makes it difficult to quickly respond to sudden process changes [102]. Further development of intelligent algorithms and control systems is needed in the future to ensure the efficient and stable operation of process automation.

(3) Development and challenges of non-traditional bending technologies

New bending technologies, such as laser-assisted hot bending, free bending, and CNC bending around the bend, offer new possibilities for the bending and forming of composite pipes. However, practical applications of these technologies still face many problems. Laser-assisted hot bending technology can accurately control the bending angle through non-contact heating, avoiding the deformation defects caused by traditional mechanical contact. However, it has a high sensitivity to material properties and process parameters, especially in bimetallic composite pipe materials with complex interlayer stresses and interfacial bonding. This sensitivity can easily lead to problems such as delamination or cracking [103]. Free bending technology is suitable for small-lot, high-complexity production, but there are still major limitations in high-precision control. Although the CNC bending winding technique is superior in mass production and high efficiency, the precise control of complex multilayer materials and cross-sectional variations still needs to be further optimized. Therefore, how to make breakthroughs in improving the flexibility and adaptability of these new technologies has become a key challenge for the wide application of non-traditional bending technologies.

6. Conclusions

This review highlights the critical challenges in bimetallic composite pipe bending, focusing on the limitations of existing methods in addressing defects such as wrinkling, wall thinning, and springback under complex geometries. Compared to previous studies, advancements in computational modeling, including hybrid finite element method-artificial neural network (FEM-ANN) approaches, have significantly improved defect prediction and process optimization accuracy. However, these advancements remain limited by issues such as the quality of input data, the adaptability of algorithms, and challenges in practical implementation.

To overcome these limitations, future research should investigate alternative materials, including lightweight alloys with improved formability, and integrate advanced computational tools like multitask learning (MTL) and digital twin (DT) frameworks for real-time monitoring and process optimization. Additionally, the development of adaptive artificial intelligence (AI)-driven systems capable of dynamic learning under varying manufacturing conditions holds great promise for transforming composite pipe bending techniques. Such innovations could achieve higher precision, efficiency, and product quality, supporting diverse industrial applications.

Author Contributions: Data curation, investigation, writing—review and editing, H.L.; conceptualization, methodology, supervision, Y.Z.; investigation, writing—review and editing, W.C.; software, visualization, C.Y.; software, visualization, L.W. All authors have read and agreed to the published version of the manuscript.

Funding: The authors gratefully acknowledge the financial support from the National Natural Science Foundation of China (51601070, 51875263).

Conflicts of Interest: The authors declare no conflicts of interest.

References

1. Zhang, Z.Y.; Chen, C.; Li, H.; Tang, P. Research on mixing law of liquid fertilizer injected into irrigation pipe. *Horticulturae* **2022**, *8*, 200. [CrossRef]
2. Gao, K.; Li, G.Y.; Cao, Y.P.; Li, C.Q.; Chen, D.; Wu, G.; Du, Q.S.; Wang, F.; Alexander, F.; Che, F.; et al. Permafrost thawing caused by the China-Russia Crude oil pipeline based on multi-type data and its impacts on geomorphological reshaping and water erosion. *CATENA* **2024**, *242*, 108134. [CrossRef]
3. Zhang, S.X.; Ma, Q.Z.; Xu, C.F.; Li, L.F.; Wang, M.F.; Zhang, Z.; Wang, S.; Li, L. Root cause analysis of liner collapse and crack of bi-metal composite pipe used for gas transmission. *Eng. Fail. Anal.* **2022**, *132*, 105942. [CrossRef]
4. Pérez-Álvarez, R.; Montoya, A.; López-Puente, J.; Santana, D. Solar power tower plants with Bimetallic receiver tubes: A thermomechanical study of two-and three-layer composite tubes configurations. *Energy* **2023**, *283*, 129170. [CrossRef]
5. Shifler, D.A. Marine and Offshore Piping Systems. In *Laque's Handbook of Marine Corrosion*; Wiley: Hoboken, NJ, USA, 2022; pp. 667–689.
6. Xu, W.B.; Li, Y. Mechanical analysis of 20/316L bimetal composite pipe formed by spinning. *SN Appl. Sci.* **2022**, *4*, 55. [CrossRef]
7. Swarnkar, R.; Karmakar, S.; Pal, S.K. An investigation of bimetallic tube fabrication through a novel friction stir extrusion based technology for automotive applications. *Mater. Today Commun.* **2023**, *35*, 106363. [CrossRef]
8. Zhou, Z.M.; Ding, Y. A study on the fatigue performance and corrosion resistance of 304/45 bimetallic composite bolts. *Materials* **2023**, *16*, 4454. [CrossRef]
9. Barenyi, I.; Slany, M.; Kouril, K.; Zouhar, J.; Kolomy, S.; Sedlak, J.; Majerik, J. Processing of Bimetallic Inconel 625-16Mo3 Steel Tube via Supercritical Bend: Study of the Mechanical Properties and Structure. *Materials* **2023**, *16*, 6796. [CrossRef] [PubMed]
10. Sas-Boca, I.-M.; Iluțiu-Varvara, D.-A.; Tintelecan, M.; Aciu, C.; Frunz, D.I.; Popa, F. Studies on hot-rolling bonding of the al-cu bimetallic composite. *Materials* **2022**, *15*, 8807. [CrossRef]
11. Zhang, Z.C.; Wu, J.J.; Xu, X.L.; Yang, Z.K.; Wu, W.; Liu, L. Mechanical modeling of tube bending considering elastoplastic evolution of tube cross-section. *Materials* **2022**, *15*, 3997. [CrossRef] [PubMed]
12. Yan, Y.; Wang, H.B.; Li, Q. The inverse parameter identification of Hill 48 yield criterion and its verification in press bending and roll forming process simulations. *J. Manuf. Process.* **2015**, *20*, 46–53. [CrossRef]
13. Zhang, X.; Zhao, C.C.; Du, B.; Chen, D.; Li, Y.; Han, Z.J. Research on hydraulic push-pull bending process of ultra-thin-walled tubes. *Metals* **2021**, *11*, 1932. [CrossRef]
14. Zhang, S.Y.; Fu, M.Y.; Wang, Z.L.; Lin, Y.C.; He, C. Spring-back prediction of the bi-layered metallic tube under CNC bending considering neutral layer shifting extraction. *Appl. Sci.* **2020**, *10*, 4978. [CrossRef]
15. Safari, M.; Alves de Sousa, R.; Joudaki, J. Recent advances in the laser forming process: A review. *Metals* **2020**, *10*, 1472. [CrossRef]
16. Wang, W.; Abd El-Aty, A.; Bai, X.; Sun, J.; Lee, M.-G.; Wei, W.B.; Chen, H.; Guo, X.Z.; Tao, J. Theoretical analysis, finite element modelling, and experimental investigation of manufacturing convoluted spiral tubes through free bending forming technology. *Int. J. Adv. Manuf. Technol.* **2021**, *117*, 279–293. [CrossRef]
17. Zhu, Y.X.; Chen, W.; Tu, W.B.; Guo, Y.; Chen, L. Three-dimensional finite element modeling of rotary-draw bending of copper-titanium composite tube. *Int. J. Adv. Manuf. Technol.* **2020**, *106*, 2377–2389. [CrossRef]
18. Bembalge, O.B.; Singh, B.; Panigrahi, S.K. Magnetic pulse welding of AA6061 and AISI 1020 steel tubes: Numerical and experimental investigation. *J. Manuf. Process.* **2023**, *101*, 128–140. [CrossRef]
19. Zhan, L.Q.; Wang, G.; Yang, J.L.; Kong, D.H.; Zhang, W.C.; Wang, G.F. Study on gas bulging forming and contradictive cooling bonding of AZ31/Al7475 bimetal composite tube. *J. Mater. Eng. Perform.* **2020**, *29*, 4652–4658. [CrossRef]
20. Zhao, J.X.; Zhang, H.L.; Zhang, W.; Zhao, H.; Li, Y.; Qin, X.; Hu, H.J.; Ou, Z.W. Influence of extrusion method on formation of magnesium-aluminum bimetallic composite tube. *J. Mater. Eng. Perform.* **2023**, *32*, 7134–7148. [CrossRef]
21. Tajyar, A.; Masoumi, A. Experimental analysis of bonding strength in shape rolling of Al–Cu bimetallic circular pipes into square tubes. *Proc. Inst. Mech. Eng. Part C J. Mech. Eng. Sci.* **2017**, *231*, 4087–4098. [CrossRef]
22. Li, H.; Yang, H.; Yan, J.; Zhan, M. Numerical study on deformation behaviors of thin-walled tube NC bending with large diameter and small bending radius. *Comput. Mater. Sci.* **2009**, *45*, 921–934. [CrossRef]
23. Liu, H.; Liu, Y. Cross section deformation of heterogeneous rectangular welded tube in rotary draw bending considering different yield criteria. J Manuf Process 61: 303–310. *J. Manuf. Process.* **2021**, *61*, 303–310. [CrossRef]
24. Jin, K.; Yuan, Q.W.; Tao, J.; Domblesky, J.; Guo, X.Z. Analysis of the forming characteristics for Cu/Al bimetal tubes produced by the spinning process. *Int. J. Adv. Manuf. Technol.* **2019**, *101*, 147–155. [CrossRef]

25. Guo, X.Z.; Yu, Y.H.; Tao, J.; Wang, H.; El-Aty, A.A.; Wang, C.; Luo, X.Y.; Kim, N. Maximum residual contact stress in spinning process of SS304/20 bimetallic pipe. *Int. J. Adv. Manuf. Technol.* **2020**, *106*, 2971–2982. [CrossRef]
26. Xu, W.C.; Zhang, Z.P.; Huang, K.; Shan, D.B. Effect of heat treatment and initial thickness ratio on spin bonding of 3A21/5A03 composite tube. *J. Mater. Process. Technol.* **2017**, *247*, 143–157. [CrossRef]
27. Wang, C.; Zhang, B.K.; Yao, D.F.; Tian, Z.Z.; Zhao, C.J. Study of copper/aluminum bimetallic tube rotary ring spinning composite forming characteristics. *Appl. Sci.* **2023**, *13*, 4727. [CrossRef]
28. Kajikawa, S.; Kawaguchi, H.; Kuboki, T.; Akasaka, I.; Terashita, Y.; Akiyama, M. Tube drawing process with diameter expansion for effectively reducing thickness. *Metals* **2020**, *10*, 1642. [CrossRef]
29. Xu, W.B.; Yang, Y.Y.; Dai, C.W.; Xie, J.G. Optimization of spinning parameters of 20/316L bimetal composite tube based on orthogonal test. *Sci. Eng. Compos. Mater.* **2020**, *27*, 272–279. [CrossRef]
30. Ding, L.; Xu, J.F.; Ma, H.H.; Rui, T.N.; Zhang, B.Y.; Shen, Z.W.; Wang, B.; Tian, J.; Xu, Q.M.; Zhao, Y.; et al. A novel dynamic self-constrained explosive welding method for manufacturing copper-steel bimetallic tube. *J. Mater. Res. Technol.* **2023**, *24*, 7229–7241. [CrossRef]
31. Yu, Y.; Xu, P.B.; Li, K.; Zhao, J.; Dai, Y.T.; Ma, H.H.; Yang, M. Study on modalities and microdefects of Al-Cu bimetallic tube by underwater explosive cladding. *Arch. Civ. Mech. Eng.* **2019**, *19*, 1390–1398. [CrossRef]
32. Jiang, L.; Luo, N.; Liang, H.L.; Zhao, Y. Microstructure and texture distribution in the bonding interface of Cu/Al composite tube fabricated by explosive welding. *Int. J. Adv. Manuf. Technol.* **2022**, *123*, 3021–3031. [CrossRef]
33. Hill, R. A general theory of uniqueness and stability in elastic-plastic solids. *J. Mech. Phys. Solids* **1958**, *6*, 236–249. [CrossRef]
34. Hutchinson, J.W. Plastic buckling. *Adv. Appl. Mech.* **1974**, *14*, 67–144.
35. Pan, C.; Cheng, C.; Abd El-Aty, A.; Wang, J.H.; Tao, J.; Liu, C.M.; Guo, X.Z.; Hu, S.H. Predicting the wrinkling in AA5052 seamless tubes manufactured by free bending forming technology. *J. Manuf. Process.* **2023**, *101*, 1065–1079. [CrossRef]
36. Yuan, L.; Kyriakides, S. Plastic bifurcation buckling of lined pipe under bending. *Eur. J. Mech. -A/Solids* **2014**, *47*, 288–297. [CrossRef]
37. Wang, X.; Cao, J. Wrinkling limit in tube bending. *J. Eng. Mater. Technol.* **2001**, *123*, 430–435. [CrossRef]
38. He, Y.; Jing, Y.; Mei, Z.; Heng, L.; Yongle, K. 3D numerical study on wrinkling characteristics in NC bending of aluminum alloy thin-walled tubes with large diameters under multi-die constraints. *Comput. Mater. Sci.* **2009**, *45*, 1052–1067. [CrossRef]
39. Naderi, G.; Torshizi, S.E.M.; Dibajian, S.H. Experimental-numerical study of wrinkling in rotary-draw bending of Tight Fit Pipes. *Thin-Walled Struct.* **2023**, *183*, 110428. [CrossRef]
40. Qian, L.Y.; Cui, Z.G.; Sun, C.Y.; Geng, S.; Sun, Z.H. Investigation of deformation compatibility and power consumption during KOBO extrusion of bimetallic composite tube. *Int. J. Adv. Manuf. Technol.* **2022**, *118*, 3477–3486. [CrossRef]
41. Zhao, H.; Yuan, T.; Sun, Z.W.; Hong, X.; Hu, H.J.; Ou, Z.W. Numerical simulation and experimental study of AL/MG bimetallic extrusion shear forming. *J. Mater. Eng. Perform.* **2024**, *33*, 8057–8065. [CrossRef]
42. Liang, Y.F.; Chen, Z.F.; Wang, W.; Wang, C.Y. The effect of friction coefficient on wrinkles of lined pipe under bending. *Int. J. Solids. Struct.* **2024**, *288*, 112615. [CrossRef]
43. Zhao, T.F.; Hu, Z.H. Numerical analysis of detaching and wrinkling of mechanically lined pipe during its spooling-on stage to the reel. *Theor. Appl. Mech. Lett.* **2015**, *5*, 205–209. [CrossRef]
44. Sun, H.; Li, H.; Gong, F.K.; Liu, Y.L.; Li, G.J.; Fu, M.W. Filler parameters affected wrinkling behavior of aluminum alloy double-layered gap tube in rotary draw bending process. *Int. J. Adv. Manuf. Technol.* **2022**, 1–16. [CrossRef]
45. Tian, Y.; Hu, H.J.; Zhang, D.F. A novel severe plastic deformation method for manufacturing Al/Mg bimetallic tube. *Int. J. Adv. Manuf. Technol.* **2021**, *116*, 2569–2575. [CrossRef]
46. Zheng, J.B.; Shu, X.D.; Li, Z.X.; Xiang, W.; Xu, Y.C. The deformation behavior and microstructure evolution of cladding tube during spinning composite process of MG/AL tube. *J. Mater. Eng. Perform.* **2024**, 1–14. [CrossRef]
47. Guo, X.Z.; Wei, W.B.; Xu, Y.; Abd El-Aty, A.; Liu, H.; Wang, H.; Luo, X.Y.; Tao, J. Wall thickness distribution of Cu–Al bimetallic tube based on free bending process. *Int. J. Mech. Sci.* **2019**, *150*, 12–19. [CrossRef]
48. Tang, N.C. Plastic-deformation analysis in tube bending. *Int. J. Press. Vessel. Pip.* **2000**, *77*, 751–759. [CrossRef]
49. Bettaieb, A.B.; Tuninetti, V.; Duchene, L. Numerical simulation of T-bend of multilayer coated metal sheet using solid-shell element. In Proceedings of the 14th International Conference on Metal Forming, Metal Forming 2012, Krakow, Poland, 16–19 September 2012; Wiley-VCH Verlag: Krakow, Poland, 2012; pp. 1311–1314.
50. Cheng, X.; Wang, H.; Abd El-Aty, A.; Tao, J.; Wei, W.B.; Qin, Y.; Guo, X.Z. Cross-section deformation behaviors of a thin-walled rectangular tube of continuous varying radii in the free bending technology. *Thin-Walled Struct.* **2020**, *150*, 106670. [CrossRef]
51. Fu, M.Y.; Wang, Z.L.; Zhang, S.Y.; Liu, X.J.; Lin, Y.C.; Wang, L. Full-cross-section deformation characterization of Cu/Al bimetallic tubes under Rotary-Draw-Bending based on physics-driven B-spline curves fitting. *Mater. Des.* **2022**, *215*, 110493. [CrossRef]
52. Karthikeyan, M.; Jenarthanan, M.P.; Giridharan, R.; Anirudh, S.J. Effects of Wall Thinning Behaviour During Pipe Bending Process—An Experimental Study. In *Innovative Design, Analysis and Development Practices in Aerospace and Automotive Engineering (I-DAD 2018) Volume 1*; Springer: Singapore, 2019; pp. 315–324.

53. Cheng, X.; Zhao, Y.X.; Abd El-Aty, A.; Guo, X.Z.; You, S.M. Deformation behavior of convolute thin-walled AA6061-T6 rectangular tubes manufactured by the free bending forming technology. *Int. J. Adv. Manuf. Technol.* **2021**, 1–16. [CrossRef]
54. Zhu, Y.X.; Wan, M.M.; Wang, C.; Tu, W.B.; Xu, F. Influence of mandrel-cores filling on size effect of cross-section distortion of bimetallic thin-walled composite bending tube. *Chin. J. Aeronaut.* **2023**, *36*, 421–435. [CrossRef]
55. Shi, S.C.; Yang, L.F.; Guo, C. Deformation behaviors of tube rotary-draw bending filled with steel balls. *Adv. Mat. Res.* **2011**, *328*, 1403–1407. [CrossRef]
56. Elyasi, M.; Rami, F.T. Comparison of resistance rotary draw bending of CP-Ti tube with different mandrels. *J. Mech. Sci. Technol.* **2024**, *38*, 1835–1841. [CrossRef]
57. Li, H.P.; Liu, Y.L.; Zhu, Y.X.; Yang, H. Global sensitivity analysis and coupling effects of forming parameters on wall thinning and cross-sectional distortion of rotary draw bending of thin-walled rectangular tube with small bending radius. *Int. J. Adv. Manuf. Technol.* **2014**, *74*, 581–589. [CrossRef]
58. Xie, W.L.; Jiang, W.H.; Wu, Y.F.; Song, H.W.; Deng, S.Y.; Lăzărescu, L.C.; Zhang, S.H.; Banabic, D. Process parameter optimization for thin-walled tube push-bending using response surface methodology. *Int. J. Adv. Manuf. Technol.* **2022**, *118*, 3833–3847. [CrossRef]
59. Periasamy, G.; Arun, C.S. Investigating Ovality and Wall-Thinning Behaviors in Free-Form Bending of Thin Pipe Bends with Varying Guider Length and Roller Die Diameter. *J. Braz. Soc. Mech. Sci. Eng.* **2023**, *11*, 605.
60. Li, J.; Wang, Z.L.; Zhang, S.Y.; Lin, Y.C.; Wang, L.; Sun, C.; Tan, J.R. A novelty mandrel supported thin-wall tube bending cross-section quality analysis: A diameter-adjustable multi-point contact mandrel. *Int. J. Adv. Manuf. Technol.* **2023**, *124*, 4615–4637. [CrossRef]
61. Al-Qureshi, H.A. Elastic-plastic analysis of tube bending. *Int. J. Mach. Tools Manuf.* **1999**, *39*, 87–104. [CrossRef]
62. El Megharbel, A.; El Nasser, G.; El Domiaty, A. Bending of tube and section made of strain-hardening materials. *J. Mater. Process. Technol.* **2008**, *203*, 372–380. [CrossRef]
63. Wang, W.; Hu, S.H.; Abd El-Aty, A.; Wu, C.; Yang, Q.C.; Chen, H.; Shen, Y.Z.; Tao, J. Springback analysis of different A-values of Cu and Al tubes in free bending forming technology: Experimentation and finite element modeling. *Int. J. Adv. Manuf. Technol.* **2021**, *113*, 705–719. [CrossRef]
64. Sun, C.; Wang, Z.L.; Zhang, S.Y.; Wang, L.; Tan, J.R. Physical logic enhanced network for small-sample bi-layer metallic tubes bending springback prediction. In Proceedings of the CAAI International Conference on Artificial Intelligence, Beijing, China, 27–28 August 2022; Springer: Cham, Switzerland, 2022; pp. 124–135.
65. Zhang, H.; Hu, Y. Research on the axial elongation and springback law of thick-walled tubes in cold bending forming. *Int. J. Adv. Manuf. Technol.* **2022**, *120*, 669–689. [CrossRef]
66. Zhang, S.Y.; Fu, M.Y.; Wang, Z.L.; Fang, D.Y.; Lin, W.M.; Zhou, H.F. Springback prediction model and its compensation method for the variable curvature metal tube bending forming. *Int. J. Adv. Manuf. Technol.* **2021**, *112*, 3151–3165. [CrossRef]
67. Wang, Z.L.; Lin, Y.C.; Qiu, L.M.; Zhang, S.Y.; Fang, D.Y.; He, C.; Wang, L. Spatial variable curvature metallic tube bending springback numerical approximation prediction and compensation method considering cross-section distortion defect. *Int. J. Adv. Manuf. Technol.* **2022**, *118*, 1811–1827. [CrossRef]
68. Ben-Elechi, S.; Bahloul, R.; Chatti, S. Investigation on the effect of friction and material behavior models on the springback simulation precision: Application to automotive part B-Pillar and material TRIP800 steel. *J. Braz. Soc. Mech. Sci. Eng.* **2022**, *44*, 380. [CrossRef]
69. Farhadi, A.; Nayebi, A. Springback analysis of thick-walled tubes under combined bending-torsion loading with consideration of nonlinear kinematic hardening. *Prod. Eng.* **2020**, *14*, 135–145. [CrossRef]
70. Zhu, Y.X.; Liu, Y.L.; Li, H.P.; Yang, H. Springback prediction for rotary-draw bending of rectangular H96 tube based on isotropic, mixed and Yoshida–Uemori two-surface hardening models. *Mater. Des.* **2013**, *47*, 200–209. [CrossRef]
71. Liu, C.M.; Liu, Y.L.; Yang, H. Influence of different mandrels on cross-sectional deformation of the double-ridge rectangular tube in rotary draw bending process. *Int. J. Adv. Manuf. Technol.* **2017**, *91*, 1243–1254. [CrossRef]
72. Jiang, S.Y.; Zhang, Y.Q.; Zhao, Y.N.; Zhu, X.M.; Sun, D.; Wang, M. Investigation of interface compatibility during ball spinning of composite tube of copper and aluminum. *Int. J. Adv. Manuf. Technol.* **2017**, *88*, 683–690. [CrossRef]
73. Zheng, M.; Zhao, T.Y.; Gao, H.; Teng, H.; Hu, J. Effect of cone size on the bonding strength of bimetallic composite pipes produced by drawing approach. *Arch. Metall. Mater.* **2018**, *63*, 451–456. [CrossRef]
74. Lee, S.K.; Jeong, M.S.; Kim, B.M.; Lee, S.K.; Lee, S.B. Die shape design of tube drawing process using FE analysis and optimization method. *Int. J. Adv. Manuf. Technol.* **2013**, *66*, 381–392. [CrossRef]
75. Kim, S.M.; Lee, S.K.; Lee, C.J.; Kim, B.-J.; Jeong, M.-S. Process design of multi-pass shape drawing considering the drawing stress. *Trans. Mater. Process.* **2012**, *21*, 265–270. [CrossRef]
76. Wang, Y.J.; Qin, F.C.; Qi, H.P.; Qi, H.Q.; Meng, Z.B. Interfacial bonding behavior and mechanical properties of a bimetallic ring blank subjected to centrifugal casting process. *J. Mater. Eng. Perform.* **2022**, *31*, 3249–3261. [CrossRef]

77. He, Y.C.; Qin, F.C.; Deng, X.J.; Wang, Y.B. Study on interface bonding behavior and quality control of 40CR/Q345B bimetallic ring blank in vertical centrifugal casting. *J. Mater. Eng. Perform.* **2024**, *33*, 10310–10323. [CrossRef]
78. Li, G.G.; Jiang, W.M.; Guan, F.; Zhu, J.W.; Yu, Y.; Fan, Z.T. Microstructure evolution, mechanical properties and fracture behavior of Al-xSi/AZ91D bimetallic composites prepared by a compound casting. *J. Magnes. Alloys* **2024**, *12*, 1944–1964. [CrossRef]
79. Fathy, N. Interfacial microstructure and shear strength improvements of babbitt–steel bimetal composites using sn–bi interlayer via liquid–solid casting. *Sustainability* **2023**, *15*, 804. [CrossRef]
80. Gui, H.L.; Hu, X.T.; Liu, H.; Zhang, C.; Li, Q.; Hu, J.H.; Chen, J.X.; Gou, Y.J.; Shuang, Y.H.; Zhang, P.Y. Research on the solid–liquid composite casting process of incoloy825/p110 steel composite pipe. *Materials* **2024**, *17*, 1976. [CrossRef]
81. Sun, X.J.; Tao, J.; Guo, X.Z. Bonding properties of interface in Fe/Al clad tube prepared by explosive welding. *Trans. Nonferrous Met. Soc. China* **2011**, *21*, 2175–2180. [CrossRef]
82. Tuninetti, V.; Forcael, D.; Valenzuela, M.; Martínez, A.; Ávila, A.; Medina, C.; Pincheira, G.; Salas, A.; Oñate, A.; Duchêne, L. Assessing feed-forward backpropagation artificial neural networks for strain-rate-sensitive mechanical modeling. *Materials* **2024**, *17*, 317. [CrossRef] [PubMed]
83. Wu, C.B.; Wang, C.; Kim, J.W. Bending deformation prediction in a welded square thin-walled aluminum alloy tube structure using an artificial neural network. *Int. J. Adv. Manuf. Technol.* **2021**, *117*, 2791–2805. [CrossRef]
84. Wang, Z.L.; Wang, C.C.; Zhang, S.Y.; Qiu, L.M.; Lin, Y.C.; Tan, J.R.; Sun, C. Towards high-accuracy axial springback: Mesh-based simulation of metal tube bending via geometry/process-integrated graph neural networks. *Expert Syst. Appl.* **2024**, *255*, 124577. [CrossRef]
85. Liu, S.L.; Wu, T.Y.; Liu, J.H.; Wang, X.; Jin, P.; Huang, H.; Liu, W. A universal, rapid and accurate measurement for bend tubes based on multi-view vision. *IEEE Access* **2019**, *7*, 78758–78771. [CrossRef]
86. Görüş, V.; Bahşı, M.M.; Çevik, M. Machine learning for the prediction of problems in steel tube bending process. *Eng. Appl. Artif. Intel.* **2024**, *133*, 108584. [CrossRef]
87. Cao, H.Q.; Yu, G.C.; Liu, T.; Fu, P.C.; Huang, G.Y.; Zhao, J. Research on the curvature prediction method of profile roll bending based on machine learning. *Metals* **2023**, *13*, 143. [CrossRef]
88. Abd El-Aty, A.; Guo, X.Z.; Lee, M.G.; Tao, J.; Hou, Y.; Hu, S.H.; Li, T.; Wu, C.; Yang, Q.C. A review on flexibility of free bending forming technology for manufacturing thin-walled complex-shaped metallic tubes. *Int. J. Lightweight Mater. Manuf.* **2023**, *6*, 165–188. [CrossRef]
89. Hong, M.; Zhang, W.W. Springback prediction of free bending based on experimental method. *Appl. Sci.* **2023**, *13*, 8288. [CrossRef]
90. Cheng, X.; Guo, X.Z.; Tao, J.; Xu, Y.; Abd El-Aty, A.; Liu, H. Investigation of the effect of relative thickness (t0/d0) on the formability of the AA6061 tubes during free bending process. *Int. J. Mech. Sci.* **2019**, *160*, 103–113. [CrossRef]
91. Hu, S.H.; Cheng, C.; Abd El-Aty, A.; Zheng, S.; Wu, C.; Luo, H.R.; Guo, X.Z.; Tao, J. Forming characteristics of thin-walled tubes manufactured by free bending process-based nontangential rotation bending die. *Thin-Walled Struct.* **2024**, *194*, 111313. [CrossRef]
92. Sun, C.; Wang, Z.L.; Zhang, S.Y.; Zhou, T.T.; Li, J.; Tan, J.R. Digital-twin-enhanced metal tube bending forming real-time prediction method based on multi-source-input MTL. *Struct. Multidiscip. Optim.* **2022**, *65*, 296. [CrossRef]
93. Wang, Z.L.; Xiang, Y.Z.; Zhang, S.Y.; Liu, X.J.; Ma, J.; Tan, J.R.; Wang, L. Physics-informed springback prediction of 3D aircraft tubes with six-axis free-bending manufacturing. *Aerosp. Sci. Technol.* **2024**, *147*, 109022. [CrossRef]
94. Yang, H.; Li, H.; Zhang, Z.Y.; Zhan, M.; Liu, J.; Li, G.J. Advances and trends on tube bending forming technologies. *Chin. J. Aeronaut.* **2012**, *25*, 1–12. [CrossRef]
95. Zhang, Z.K.; Wu, J.J.; Liang, B.; Wang, M.Z.; Yang, J.Z.; Muzamil, M. A new strategy for acquiring the forming parameters of a complex spatial tube product in free bending technology. *J. Mater. Process. Technol.* **2020**, *282*, 116662. [CrossRef]
96. Zhou, X.Y.; Ma, J.; Zhou, W.B.; Welo, T. Forming-based geometric correction methods for thin-walled metallic components: A selective review. *Int. J. Adv. Manuf. Technol.* **2023**, *128*, 17–39. [CrossRef]
97. Campbell, F.C. *Manufacturing Technology for Aerospace Structural Materials*, Elsevier Science: Oxford, UK, 2011.
98. Ancellotti, S.; Fontanari, V.; Slaghenaufi, S.; Cortelletti, E.; Benedetti, M. Forming rectangular tubes into complicated 3D shapes by combining three-roll push bending, twisting and rotary draw bending: The role of the fabrication loading history on the mechanical response. *Int. J. Mater. Form.* **2019**, *12*, 907–926. [CrossRef]
99. İşler, B.; Uzun, G. Defects in Rotary Draw Bending and Their Effects on Formability. *Gazi Üniversitesi Fen Bilim. Derg. Part C Tasarım Ve Teknol.* **2024**, *12*, 714–723. [CrossRef]
100. Rajhi, W.; Ayadi, B.; Khaliq, A.; Al-Ghamdi, A.; Ramadan, M.; Al-shammrei, S.; Boulila, A.; Aichouni, M. Constitutive behavior and fracture of intermetallic compound layer in bimetallic composite materials: Modeling and application to bimetal forming process. *Mater. Des.* **2021**, *212*, 110294. [CrossRef]
101. Li, G.Y.; Jiang, W.M.; Guan, F.; Zhang, Z.; Wang, J.L.; Yu, Y.; Fan, Z.T. Preparation, interfacial regulation and strengthening of Mg/Al bimetal fabricated by compound casting: A review. *J. Magnes. Alloys* **2023**, *11*, 3059–3098. [CrossRef]

102. Nikhare, C.P. Effect of Metal-Composite Layer Thickness on Springback After U-Bending. In Proceedings of the ASME 2020 International Mechanical Engineering Congress and Exposition, Portland, OR, USA, 16 March 2020; American Society of Mechanical Engineers: New York, NY, USA, 2020.
103. Hosseini, S.M.; Roostaei, M.; Mosavi Mashhadi, M.; Jabbari, H.; Faraji, G. Fabrication of Al/Mg Bimetallic Thin-Walled Ultrafine-Grained Tube by Severe Plastic Deformation. *J. Mater. Eng. Perform.* **2022**, *31*, 4098–4107. [CrossRef]

Disclaimer/Publisher's Note: The statements, opinions and data contained in all publications are solely those of the individual author(s) and contributor(s) and not of MDPI and/or the editor(s). MDPI and/or the editor(s) disclaim responsibility for any injury to people or property resulting from any ideas, methods, instructions or products referred to in the content.

Article

Flow Behavior Analysis of the Cold Rolling Deformation of an M50 Bearing Ring Based on the Multiscale Finite Element Model

Wenting Wei [1,2,3,4], Zheng Liu [1,2], Qinglong Liu [5], Guanghua Zhou [1,2], Guocheng Liu [1,2,*], Yanxiong Liu [1,2,*] and Lin Hua [1,2,3]

Academic Editor: Alexander Yu Churyumov

Received: 10 December 2024
Revised: 23 December 2024
Accepted: 25 December 2024
Published: 27 December 2024

Citation: Wei, W.; Liu, Z.; Liu, Q.; Zhou, G.; Liu, G.; Liu, Y.; Hua, L. Flow Behavior Analysis of the Cold Rolling Deformation of an M50 Bearing Ring Based on the Multiscale Finite Element Model. *Materials* 2025, 18, 77. https://doi.org/10.3390/ma18010077

Copyright: © 2024 by the authors. Licensee MDPI, Basel, Switzerland. This article is an open access article distributed under the terms and conditions of the Creative Commons Attribution (CC BY) license (https://creativecommons.org/licenses/by/4.0/).

1. Hubei Key Laboratory of Advanced Technology for Automotive Components, Wuhan University of Technology, Wuhan 430070, China; wei_wt@whut.edu.cn (W.W.); liu_zh@whut.edu.cn (Z.L.); zhoughjixi@163.com (G.Z.); hualin@whut.edu.cn (L.H.)
2. Hubei Collaborative Innovation Center for Automotive Components Technology, Wuhan University of Technology, Wuhan 430070, China
3. Hubei Longzhong Laboratory, Xiangyang 441000, China
4. The State Key Laboratory of Refractories and Metallurgy, Wuhan University of Science and Technology, Wuhan 430081, China
5. China Railway Construction Heavy Industry Corporation Liminted, Changsha 410000, China; liuql_whut@126.com
* Correspondence: liugch@whut.edu.cn (G.L.); liuyx@whut.edu.cn (Y.L.)

Abstract: Through the ferrite single-phase parameters of M50 bearing steel obtained based on nanoindentation experiments and the representative volume element (RVE) model established based on the real microstructure of M50, this paper established a multiscale finite element model for the cold ring rolling of M50 and verified its accuracy. The macroscale and mesoscale flow behaviors of the ring during the cold rolling deformation process were examined and explained. The macroscopic flow behavior demonstrated that the stress distribution was uniform following rolling. The equivalent plastic strain (PEEQ) grew stepwise over time, with the raceway showing the highest PEEQ. The mesoscopic simulation revealed that the stress was concentrated in the cementite, and the maximum occurred at the junction of the ferrite and cementite. The largest PEEQ was found in the ferrite matrix positioned between the two adjacent cementites. The cementite flew with the deformation of the ferrite. The radial displacement of the cementite decreased from the edge of the raceway to both ends and decreased from the inner to the outer surface. Its axial displacement was basically the same on the inner surface and decreased from the inner to the outer surface. Its circumferential displacement decreased from the inner and outer surfaces to the intermediate thickness region.

Keywords: M50 bearing ring; cold rolling; multiscale model; flow behavior

1. Introduction

M50 bearing steel is a widely utilized high-temperature bearing steel in the aerospace industry [1,2]. M50 is classified as a high-carbon bearing steel, and the presence of chromium (Cr), molybdenum (Mo), and vanadium (V) within its composition facilitates the formation of various types of cementite, which are diffusely distributed throughout the matrix [3]. This characteristic is crucial for ensuring its exceptional red hardness and fatigue life. Ring rolling technology is an advanced and efficient continuous partial plastic forming technology, which is mainly used for manufacturing seamless rings. Compared with the traditional hot-rolled rings, the products obtained by cold rolling have good surface quality, high dimensional accuracy, high mechanical strength, and longer fatigue life. With the wide application of the cold ring rolling process on 100Cr6 bearing steel, in recent years, there are also scholars who study using the cold ring rolling process to process M50 bearing steel [4–7]. Through experimentation and numerical simulation, Baoshou Sun [4] examined the impact of mandrel roll feeding rate during cold ring rolling of GCr15 bearing steel. Additionally, there was good agreement between the two approaches' outcomes. S. Deng et al. [8] used a combination of cold-ring-rolled finite element simulation and testing to study GCr15 bearing steel. They examined the evolution of the weave after different deformation rates and the microstructural changes in cold ring rolling. Using experimental techniques, Feng Wang et al. [6] examined the microstructure evolution and tempering transformation kinetics of M50 bearing steel during cold ring rolling and came to insightful findings.

In the context of high-carbon bearing steels, it has become increasingly acknowledged that cold deformation can induce damage or even microcracking within the material [9–11]. If the microcracks resulting from cold deformation are not entirely mitigated through heat treatment and machining processes, they may lead to significant issues during service, potentially serving as a critical factor in ring failure [12–15]. Current research on cold ring rolling technology primarily focuses on macroflow behavior, stress–strain distribution, and formability, often neglecting the meso- and microlevels. Therefore, this study of the second phase and the matrix in the mesolevel change in cold rolling technology has great importance for the bearing rolling process and the prevention of bearing cracks.

Multiscale approaches are widely used in the research of metallic and composite materials as an important way to bridge macroscale and mesoscale processes. Piotr Maciol et al. [16] explored the potential for parallelizing multiscale models within the agile multiscale methodology framework by integrating finite element method (FEM)-based continuous macroscopic models with concurrently computed mesoscale models. Yutai Su et al. [17] introduced a novel rapid method for generating representative volume elements (RVEs) of randomly distributed metal matrix nanocomposites characterized by quasi-continuous volume distributions, utilizing a combination of the Gaussian filtering algorithm and cutting quantile functions, and they subsequently validated the accuracy of their model. Lingwei Kong et al. [18] developed a new constitutive model that accounts for the mesoscopic damage and plasticity mechanisms of quasi-brittle materials through the coupling of micro-planes and micromechanics. Zhang et al. [19] proposed a multiscale MCCPFEM framework to simulate microscale thermal interface grooving (TIG) welding and mesoscale deformation anisotropy, effectively capturing and analyzing the interplay between deformation anisotropy and TIG-induced tissue evolution during the isothermal compression of the IMI834 alloy with lamellar clusters. For M50 bearing steel, multiscale modeling can solve the problems of the accurate characterization of its microstructure, controlling the morphology and distribution of cementite during the annealing process as well as reducing the number of physical experiments and saving costs. Consequently, multiscale analysis proves to be particularly valuable for investigating the connections between macroscopic and mesoscale phenomena in the cold rolling process of M50 bearing rings. However, there are still very few reports that specifically cover this subject.

The RVE is defined as a mesoscopic volume element within a macroscopic body of material. Through the study of this element, the linkage between the macroscopic and mesoscopic deformation mechanisms of the material is established. In metallic materials represented by dual-phase steels, the method of establishing RVE models based on the real microstructure of the material has matured, and the accuracy of the method has been verified [20,21]. However, in metal matrix composites and in metallic materials, such as GCr15 and M50, where granular phases are present, conventional RVE modeling is generally based on microscopic characterization statistics and stochastic algorithms [17,22]. Although this method is able to reflect the interaction of multiple meso-structures under macro-deformation, it ignores the microstructural features of the material and does not provide a completely accurate description of the mesoscale changes in the material. Therefore, an RVE model that accurately reflects the meso-structure of the second phase is needed in the multiscale analysis of such materials.

This paper presented a multiscale approach to the investigation of the cold ring rolling process. Firstly, a macro-finite element model of cold ring rolling was developed, grounded in actual working conditions. Subsequently, a sub-model was employed to facilitate the transfer of mechanical responses between the macroscale and mesoscale. Then, utilizing image processing techniques, we established an RVE that accurately reflected the mesoscopic changes. This framework enabled the multiscale numerical simulation of the cold ring rolling process for M50 bearing steel, allowing for an in-depth examination of the macro- and mesoflow behaviors associated with the cold ring rolling process.

2. Materials and Methods

2.1. M50 Bearing Steel

This research employed ring blanks made from commercially accessible M50 bearing steel, with the chemical compositions of the initial material detailed in Table 1. The microstructural characteristics of the bearing steel, examined using a scanning electron microscope (SEM) (JSM-IT800(HL), Nippon Electronics Corporation, Tokyo, Japan) after undergoing spheroidal annealing heat treatment, are depicted in Figure 1. In this figure, the bright white regions represented cementite, whereas the other areas were classified as ferrite.

Table 1. Chemical composition of M50 bearing steel (wt.%).

C	Cr	Mo	V	Ni	Mn	Si	Fe
0.8	4.02	4.2	0.93	0.05	0.29	0.16	Bal.

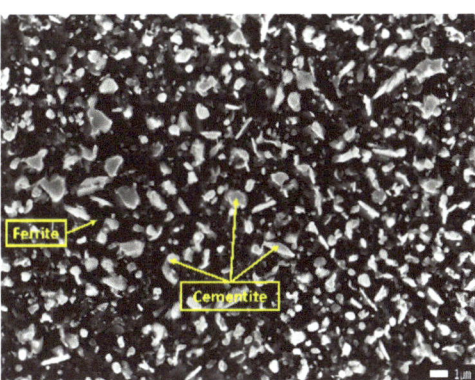

Figure 1. SEM image of M50 bearing steel in spheroidized annealed condition.

2.2. Macro-Finite Element Modelling

In this study, the commercial finite element software ABAQUS (ABAQUS 614) was employed to develop a macrocold rolling finite element model of the M50 bearing ring, as shown in Figure 2. The model components, including the driven roller, mandrel roller, ring blank, and guide roller, were constructed based on parameters derived from actual working conditions, as detailed in Table 2. However, the stiffness of the mold is often much larger than the stiffness of the ring material, so the mold was generally set as a rigid body in the simulation. The driven roller was constrained to rotate solely around the z-axis, while the mandrel roller was permitted to rotate under applied forces and translate in the negative direction of the x-axis. The guide rollers were capable of translation within the XOY plane and rotation around the z-axis. The friction coefficient between the driven roller, mandrel roller, and ring was set to 0.3, whereas the interface between the guide roller and the ring was assumed to be smooth. For the ring blank, a mesh size of 0.5 mm was utilized, employing an eight-node linear hexahedron (C3D8R) mesh type, which incorporates linear interpolation and selective reduced integration. In addition, the Arbitrary Lagrangian–Eulerian (ALE) adaptive mesh technique is applied to the ring components to automatically adjust the coarseness and refinement of the mesh to achieve more accurate results and higher computational efficiency. As for the mold, no meshing is required for it, as it is set as a rigid body.

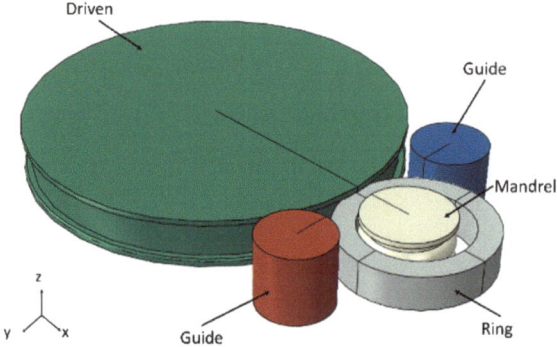

Figure 2. Macrocold rolling finite element model of the M50 bearing ring.

Table 2. M50 bearing ring cold rolling parameters.

Parameter	Value
Blank size/(mm × mm × mm)	Φ41 × 29 × 10
Ring size/(mm × mm × mm)	Φ55 × 45 × 10
Depth of groove ball pressing in/mm	2
Driven roller radius/mm	25
Rotate speed of driven roller/rad·s^{-1}	6.3
Mandrel roller radius/mm	10
Feeding speed of mandrel roller/mm·s^{-1}	2
Guide roller radius/mm	10

2.3. Meso-Finite Element Modelling

This paper presented an RVE model derived from the actual microstructure of the material, which effectively reflected the meso-structure of cementite and provided a more accurate representation of the correlations involved. Figure 3 shows the schematic diagram for constructing the RVE model. The SEM image of M50 bearing steel was systematically processed using digital image processing (DIP) techniques at the pixel level to differentiate the matrix from the second phase and to compile the pixel point information matrix. Each pixel point was treated as an element with a size of 0.005 μm, thereby establishing a connection between elements and nodes. To distinguish the matrix from the second phase, separate sets were created, and the RVE model that corresponded to the actual microstructure was developed and formatted as an input file. Finally, the generated input file was imported into ABAQUS software to create the corresponding model.

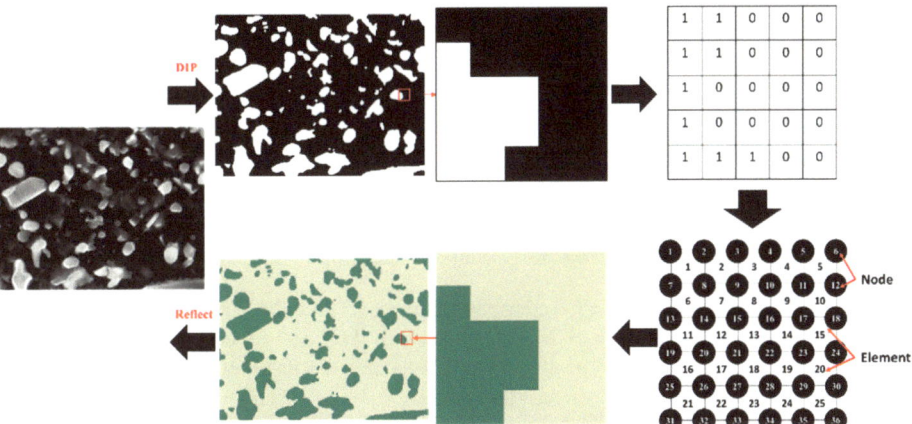

Figure 3. The schematic diagram for constructing the RVE model.

To ensure the accuracy of the RVE model, it is essential to obtain single-phase stress–strain curves for M50 bearing steel, specifically for the ferrite and cementite phases. The mechanical properties of ferrite can be determined through nanoindentation experiments and inverse analyses of the load–displacement curves [23–26]. In contrast, cementite is typically regarded as an elastic or brittle material due to its minimal or negligible deformation during the loading process [27,28]. The mechanical parameters of ferrite were derived using the method established by Oliver and Pharr [29]. A commercial indenter (TI980 (NULL, Hangzhou, China)) with a Berkovich-type diamond (1140 GPa modulus, 0.07 Poisson's ratio) was used in nanoindentation experiments to produce load–displacement (P-H) curves. The indenter had a radius of 50 nm, an inclination angle of 65.3°, and an indentation depth of 150 nm. The stress–strain curves for the ferrite single phase were calculated from the experimental load–depth data. Subsequently, a three-dimensional finite element model of the nanoindentation process was established in accordance with the experimental conditions, as illustrated in Figure 4. In this model, the indenter was represented as a three-dimensional discrete rigid body, constrained to move −150 nm along the z-axis. The base's lower end, which was assigned the single-phase parameters obtained from the aforementioned calculations, was fixed in place. Mesh refinement was conducted in the contact region, utilizing a refined mesh size of 100 nm and employing the C3D8R element type. Finally, a comparison was made between the P-H curves obtained from both experimental and simulation data to ascertain the stress–strain curve of the ferrite phase. Detailed results are presented in Section 3.2.

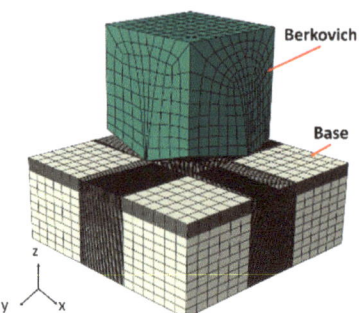

Figure 4. Three-dimensional finite element modeling of nanoindentation.

Following the construction of the RVE model, it is imperative to verify the model's reliability. Ma, S. M. [30,31], Zhao, K. [32], and colleagues assessed the validity of the RVE model by applying a unidirectional tensile boundary condition and analyzing the discrepancies between the resulting stress–strain curve and the macroscopic tensile curve. As illustrated in Figure 5, symmetry boundary conditions were applied to the left node of the RVE model, while displacement variations were introduced on the right side. The stress–strain curves were computed using a first-order homogenization strategy, after which these curves were compared to the macroscopic tensile curves to ascertain the model's reliability. Detailed results are presented in Section 3.2.

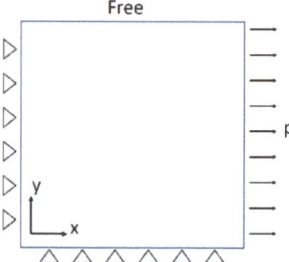

Figure 5. Schematic view of the RVE unidirectional stretching condition.

After verifying the reliability of the RVE model under tensile conditions, a multiscale numerical simulation of the cold rolling process for the ring can be conducted. The multiscale finite element modeling is shown in Figure 6. The sub-model was utilized to extract and refine the macroscopic element individually and to establish links between macroscopic and mesoscale elements. The mesoscale model utilized the element strains from the sub-model as the driving conditions. Specifically, the three-dimensional strain was decomposed into a radial–axial strain and a radial–circumferential strain; the change in the RVE in the radial–axial direction was mainly observed, and the change in the radial–circumferential direction was used to support the illustration. The strains $\varepsilon 11$, $\varepsilon 12$, and $\varepsilon 22$ in the radial–axial and radial–circumferential directions, transmitted by the sub-model, were applied as loads to the RVE model, individually. The selection of the RVE model was predicated on an analysis of the images captured under SEM in the preceding period, thereby facilitating the acquisition of the composition and distribution characteristics of the material's secondary phase. Subsequently, a representative image was selected from the existing array of images to serve as the foundation for developing the RVE model. A 4-node bilinear plane strain quadrilateral element was selected for the mesh type to effectively simulate the plane stress state. Constraints were implemented to connect the nodes at the edges of the RVE to a reference point, facilitating the application of loads. The establishment of periodic boundary conditions (PBCs) allowed the RVE model to maintain consistent deformation states for the upper and lower elements, as well as the left and right elements, thereby enabling the simulation of the material's actual flow behavior. Python software (Python 3.10) was used to carry out batch operations on the input files for all of the previously listed procedures. Furthermore, the locations of the cementite, as indicated in the figure, were selected for subsequent flow analyses.

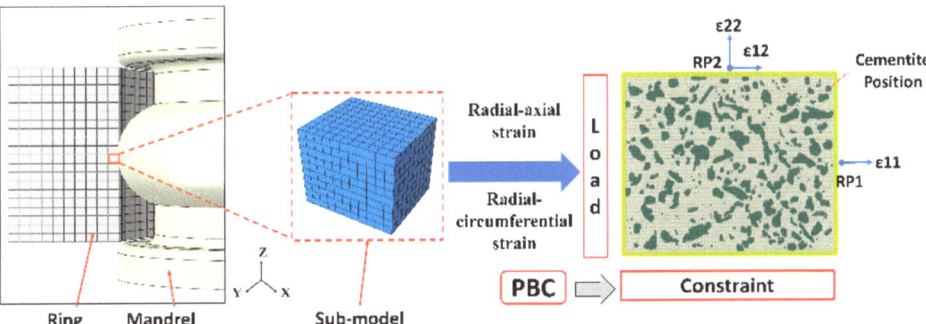

Figure 6. Multiscale finite element model.

3. Results and Discussion

3.1. Macroscopic Flow Behavior Analysis of M50 Cold-Rolled Bearing Rings

Figure 7 presents the schematic view illustrating the macrocold rolling deformation results of the M50 bearing ring. Specifically, Figure 7a depicts the axial view of the forging at the end of the rolling process, and a quarter of the ring was intercepted for display. Figure 7b illustrates the stress distribution at the same time period as well as the partial magnification at the raceway. Table 3 provides a comparison of the simulated and expected dimensions of the forging post-rolling. Analysis of Figure 7a and Table 3 indicates that the roundness of the forging at the end of the rolling process is satisfactory, with minimal discrepancies between the simulation results and the anticipated dimensions. Comparing the expected dimensions to the highest outer direction (OD) error, the difference is only 0.375%. The inner diameter (ID) inaccuracy can be as much as 0.32%. There is just a 0.5% maximum inaccuracy in the ring thickness. Furthermore, as evidenced in Figure 7b, the stress distribution across the ring component at the conclusion of the rolling process was found to be uniform. The minimum stress values were observed at the upper and lower end surfaces of the ring, whereas the maximum stress values were noted sporadically at the outer surface of the ring and the raceway, with negligible evidence of stress concentration.

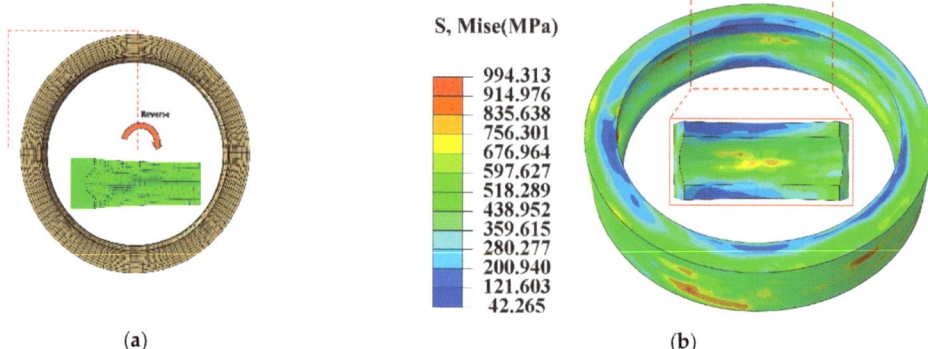

Figure 7. Schematic view of macrocold rolling deformation results of the M50 bearing ring. (**a**) Axial view of the forging at the end of rolling; (**b**) stress distribution of the ring.

Table 3. Finite element simulation dimensions of the forging at the end of rolling and expected dimensions.

Parameters	Simulation Size (mm)	Expected Size (mm)	Difference (%)
OD	54.794~54.915	55	−0.375~−0.015
ID	44.856~44.949	45	−0.032~−0.013
Thickness	4.975~4.996	5	−0.5~−0.008

Figure 8 illustrates the equivalent plastic strain (PEEQ) distribution during the cold rolling of the ring. During the initial phase of rolling, the ring was clearly subject to forces from the driven roll as well as the convex mold of the mandrel roll. This interaction led to the emergence of the PEEQ, primarily on the outer surface of the ring and at the raceway. As the mandrel roll advanced, it established full contact with the inner surface of the ring, resulting in variations in the PEEQ on both the outer surface and the entire inner surface of the ring, which subsequently extended into the intermediate thickness region. By the conclusion of the rolling process, the distribution of the PEEQ within the ring was characterized by a non-uniform pattern, exhibiting a gradual transition from lower to higher strain values and moving from the middle thickness region towards the surface region. The maximum PEEQ was observed at the raceway on the inner surface of the ring, while the minimum PEEQ was found in the middle region of the upper-end face of the ring.

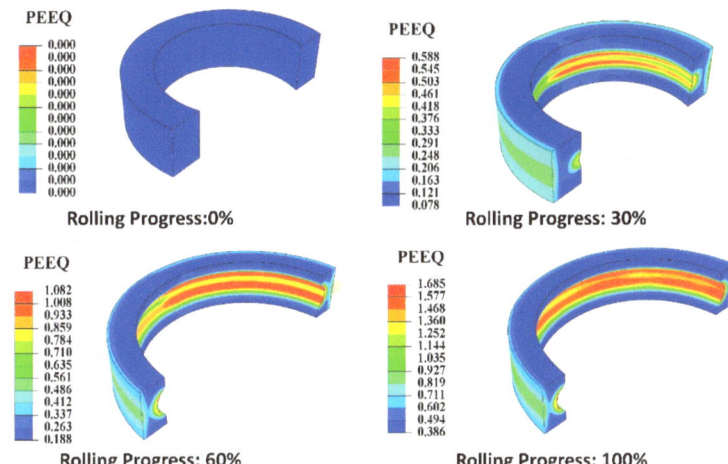

Figure 8. Distribution of the PEEQ during the cold rolling of the ring.

Figure 9 illustrates the trajectory of pickup points utilized for the analysis of the PEEQ at the end of the rolling process. This analysis encompassed the axial selection of locations at the upper-end face, the mid-height region, and the lower-end face of the ring, as well as the radial selection of positions at the outer surface, the mid-thickness region, and the inner surface of the ring. The variation curves of the PEEQ for the aforementioned locations are presented in Figure 10. It was observed that the PEEQ at each position on the ring followed a descending order. The raceway of the inner surface exhibits the highest strain, followed by the outer surface and then the intermediate thickness region. Additionally, compared to the top and lower-end face sections, the PEEQ in the ring's mid-height zone was higher. In the radial direction, the disparity in diameter between the driven roll and the mandrel roll resulted in a varying PEEQ on the inner and outer surfaces of the ring. In the axial direction, the presence of a closed hole pattern, along with the symmetry of the mandrel roller at the top and bottom, led to a PEEQ that was symmetrical about the mid-height. Notably, the PEEQ at the raceways on the inner surface was significantly greater than observed at other locations.

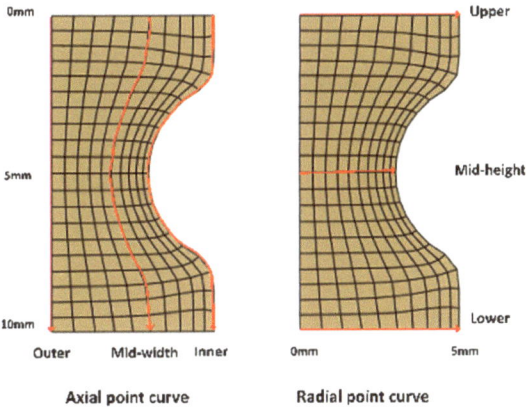

Figure 9. PEEQ cross-section pickup point path diagram at the end of rolling.

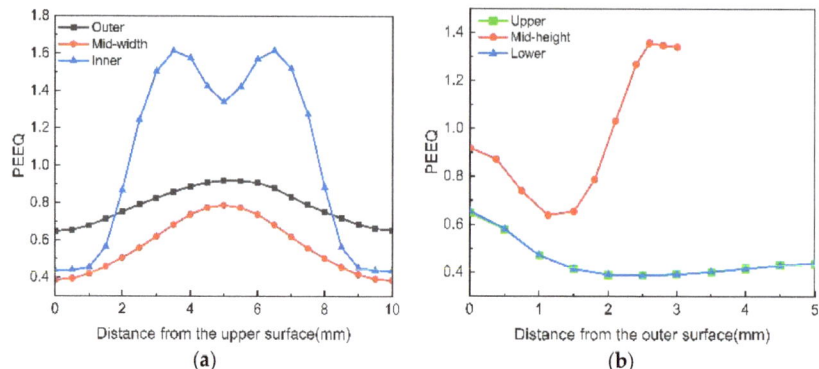

Figure 10. Distribution of the PEEQ after rolling. (**a**) Axial distribution; (**b**) radial distribution.

At the conclusion of the rolling process, the radial deformation of the ring serves as an indicator of the extent of deformation experienced by the ring, which is crucial for directing the rolling operation. Figure 11a illustrates the radial–axial cross-section of the ring component both prior to and following the rolling process. The radial deformation at each position at the end of rolling was determined based on the radial dimensions of the ring, with the calculation formula provided in Equation (1).

$$\Delta = \frac{|B - B_0|}{B_0} \times 100\% \tag{1}$$

where Δ is the radial deformation, B_0 is the ring thickness before rolling, and B is the ring thickness after rolling.

Figure 11b presents the curve depicting the variation in radial deformation of the ring subsequent to the rolling process. It was apparent that the radial deformations of the ring displayed considerable disparities due to the influence of the raceway; nonetheless, these deformations maintained a symmetrical profile along the central height region, a characteristic attributed to the design of the closed hole pattern. The radial deformation exhibited a gradual decline from the raceway towards both lateral extremities of the ring. The peak deformation was noted at the mid-height of the ring, reaching a maximum of 49%, whereas the minimum deformation was observed at the end face of the ring, recorded at 17%.

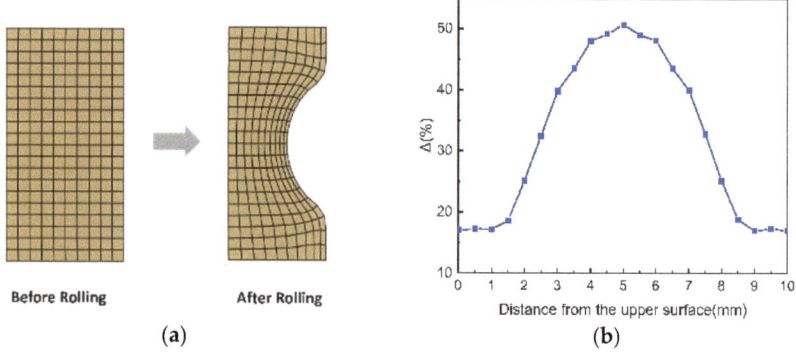

Figure 11. Radial deformation of the ring. (**a**) Cross-section of the ring before and after rolling; (**b**) curve of the radial deformation of the ring.

To accurately depict the progression of the PEEQ at various feature locations on the ring during the rolling process, nine feature elements were identified, as shown in Figure 12a, corresponding to points A through I. Figure 12b illustrates the changes in the PEEQ for each feature element throughout the rolling process. It was apparent that due to the presence of a reserved gap, no PEEQ was observed at the initiation of the rolling process. As the mandrel roller progressed, the PEEQ in each section of the ring exhibited a stepwise increase, primarily resulting from the continuous movement of the plastic deformation zone within the ring. Upon the ring's entry into the radial hole pattern, substantial plastic deformation occurred, as evidenced by the ascending steps of deformation depicted in the figure. In contrast, minimal plastic deformation was noted when the ring exited the hole pattern, which corresponded to the horizontal phase of the effect transformation illustrated in the figure. Additionally, the PEEQ gradient curves revealed that at the conclusion of the rolling process, the PEEQ was maximized at the raceway (corresponding to points F and I in the figure) and minimized at the mid-thickness zone of the upper-end face and the non-raceway zone (corresponding to points B and C in the figure).

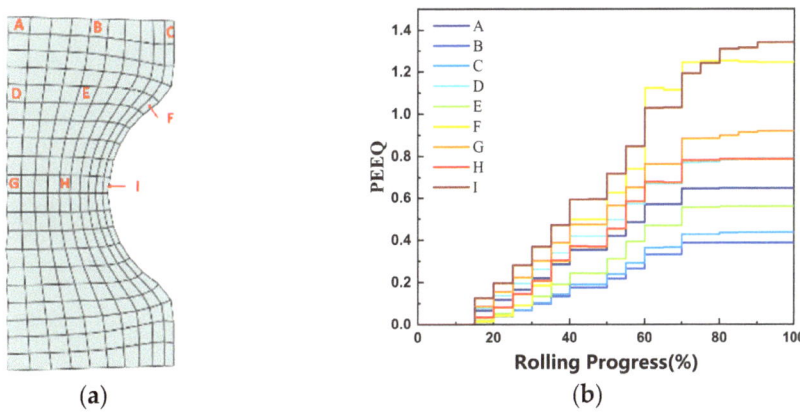

Figure 12. PEEQ curve of the ring. (a) Element selection position; (b) PEEQ gradient change curves.

3.2. Mesoscopic Flow Behavior Analysis of M50 Cold-Rolled Bearing Rings

In this investigation, the stress–strain curve of ferrite in M50 bearing steel was established through the application of the nanoindentation experimental inversion technique. Figure 13a displays the indentation map generated from the nanoindentation experiment. As can be seen in the figure, due to the random nature of the indentation, not all of them are pressed on ferrite (dark gray area in the figure); they are also on cementite (bright gray area in the figure) or at the junction of the two phases. Moreover, each indentation was separated by a distance of at least 8 μm [33] in order to avoid mutual influence between the indentations. Figure 13b depicts the results of the finite element simulation corresponding to the nanoindentation process. Furthermore, Figure 13c provides a comparison between the load–displacement curves obtained from the simulation and those derived from experimental data. The findings revealed that the simulation curve demonstrated a comparable trend to the experimental curves and corresponded with the experimental load at an indentation depth of 150 nm. Consequently, it can be inferred that the stress–strain curves of ferrite in M50 bearing steel, as determined by the power law model, were fundamentally accurate. In order to ascertain whether the curve predictions met the benchmark requirements, we proposed the concept of goodness of fit, which is now widely used to assess the accuracy of simulation results. The specific method and equation are as follows. The goodness of fit of the simulated or experimental curves to the experimental mean curve was first calculated. Then, the results were compared. The specific results are displayed in the figure. The calculated goodness of fit of the simulated curve to the average curve was the largest (0.998), which exceeded the goodness of fit of the other experimental curves, providing substantial evidence for the accuracy of the curves. However, the plastic and elastic work generated by the indenter cannot be well illustrated by simulation due to the different paths of loading and unloading. Therefore, only the loading process was simulated in this paper. For additional information, please consult Equation (3).

$$R_i^2 = 1 - \frac{\sum(y - \hat{y}_i)^2}{\sum(y - \bar{y})^2} \tag{2}$$

where R_i^2 is the goodness of fit of the experimental or simulation curve relative to the experimental mean curve, $i = 1\sim 6$. When $i = 1\sim 5$, it is five sets of experimental curves. When $i = 6$, it is the simulation curve. y is the load value of the experimental mean curve, $\hat{y}_i$ is the load value of the experimental or simulation curve, and $\bar{y}$ is the mean value of the load of the experimental mean curve.

$$\sigma = \begin{cases} E \cdot \varepsilon, \sigma < \sigma_p \\ R \cdot \varepsilon_p^n, \sigma \geq \sigma_p \end{cases} \tag{3}$$

where the modulus of elasticity is 205 GPa, the yield strength is 376.5 MPa, the hardening coefficient R is 1155.75, and the hardening index n is 0.178.

As noted by Bhadeshia [34], the modulus of elasticity for cementite was reported to be 230 GPa, with a Poisson's ratio of 0.3. The stress–strain behavior of cementite can be characterized by a non-linear equivalent stress–strain relationship, which was mathematically represented in an exponential form, as delineated in Equation (4) below.

$$\sigma_p = \sigma_p^0 + (\sigma_p^1 - \sigma_p^0)(1 - \exp^{-\frac{E_\theta}{\sigma_p^1 - \sigma_p^0}\varepsilon_{nl}}) \tag{4}$$

where σ_p^0 can be taken as the yield strength of the steel, which is 444.52 MPa, as seen in Figure 14b. According to the simulations of the cementite by Hu et al. [35], σ_p^1 is 5.2 GPa, ε_{nl} is the inelastic strain, and σ_p is the equivalent stress in the cementite. The ultimate stress–strain curve for ferrite and cementite is depicted in Figure 14a. Figure 14b provides a comparative analysis of the curves obtained from unidirectional stretching conditions applied to the RVE alongside the corresponding experimental curves. As illustrated, there was a notable degree of overlap between the two sets of curves, which served to validate the efficacy of the RVE model utilized in this research.

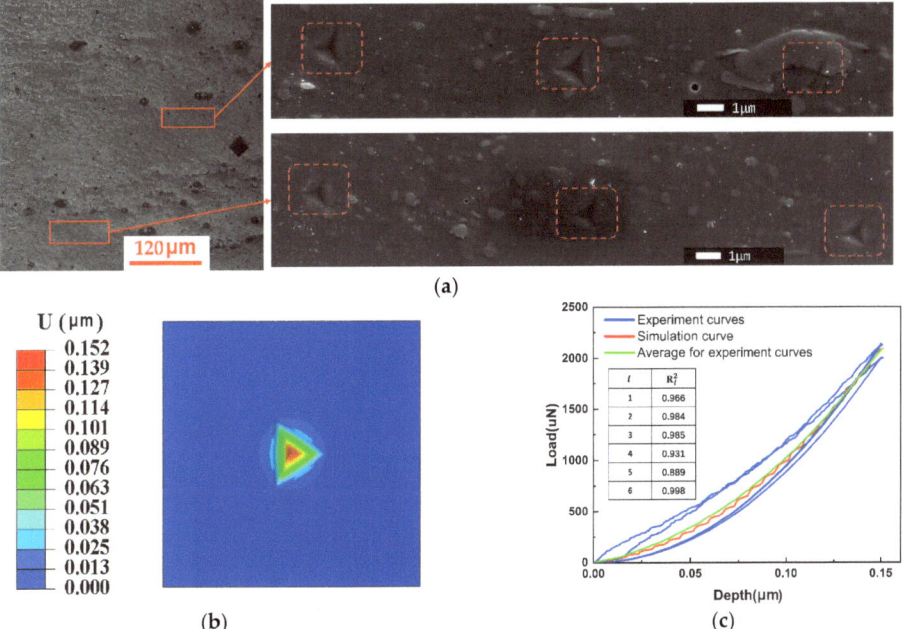

Figure 13. Experimental and simulation results of nanoindentation. (**a**) High magnification of nanoindentation; (**b**) finite element results of nanoindentation; (**c**) load–displacement curves in a nanoindentation test at 150 nm.

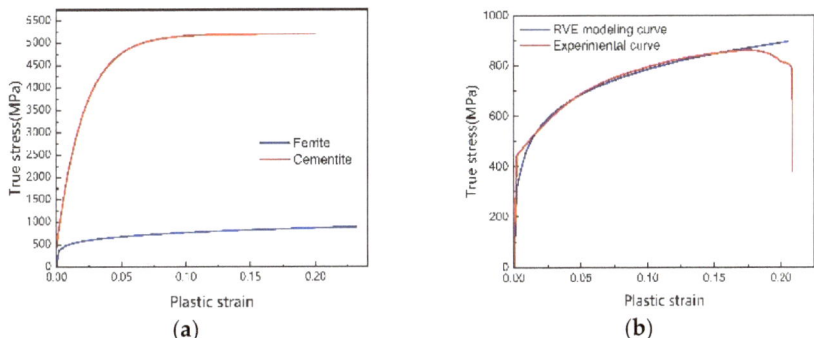

Figure 14. M50 bearing steel parameter curves. (**a**) Single-phase stress–strain curve; (**b**) RVE single-phase tensile validation plot.

Figures 15 and 16 illustrate the mesoscale stress–strain distribution corresponding to various positions on the M50 cold-rolled bearing ring, specifically aligned with the nine locations identified in Figure 12. As depicted in Figure 15, significant disparities in stress distribution between the ferrite and cementite phases were noted. The highest concentration of stress was observed at the interface of these two phases, which may facilitate the development of micropores and microcracks within the material matrix during the rolling deformation process [36]. According to Figure 16, deformation first happened in the ferrite close to the cementite–ferrite interface, and its rate increased as the distance from the cementite decreased. The ferrite farther away from this interaction was then affected by the deformation. Most of the plastic strains were found in the ferrite, specifically in the area with the highest strain, which was mostly found in the ferrite matrix between two neighboring cementite particles. This area was likely to encounter significant challenges related to deformation coordination. Additionally, the degree of RVE deformation at each feature location showed variability, as shown in Figures 15 and 16. The RVE's deformation was greater in the ring's middle thickness region than it was at the upper-end face. The radial height of both areas is the same. This observation aligns with observations made at the macroscopic level. Additionally, the deformation of the RVE in the inner surface region of the ring at the same axial height was more pronounced than that at the outer surface, with the maximum radial deformation of the RVE consistently occurring at the raceway.

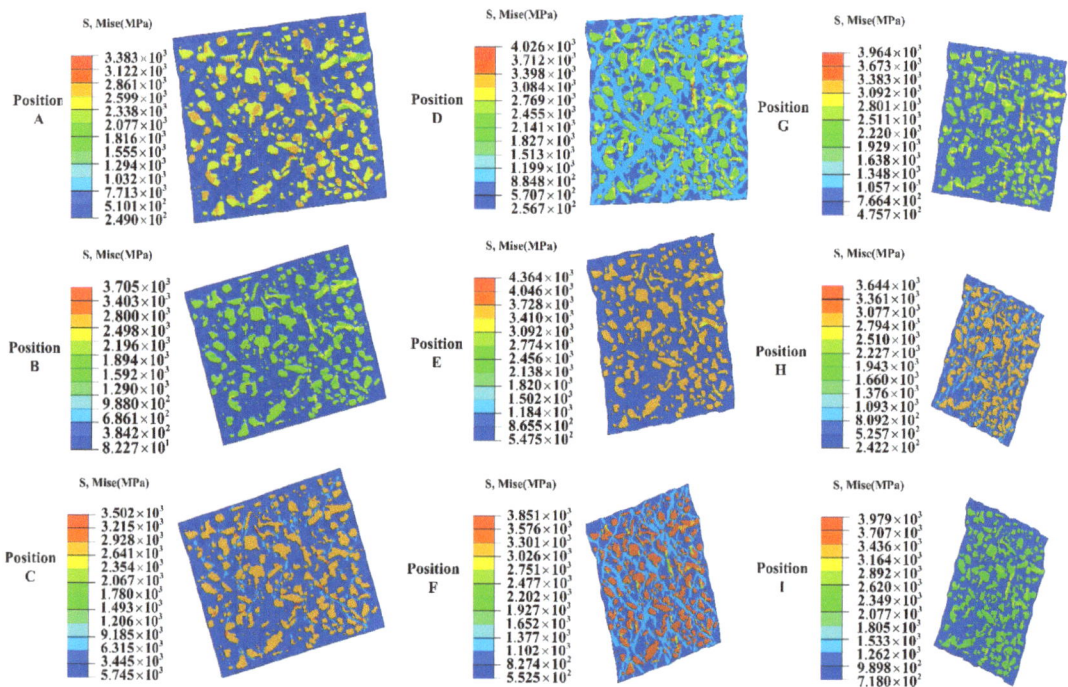

Figure 15. RVE stress distribution at different locations of the M50 bearing ring at the end of rolling.

The average PEEQ variation curves for ferrite in the RVE at various places (A~I) of the M50 bearing ring are displayed in Figure 17. The deformation results at the end of rolling were basically consistent with the macro-PEEQ curves. The PEEQ at each location was practically zero because the ferrite was still in the elastic deformation stage at the beginning of the rolling process. As the rolling process continued, the soft ferrite phase began to gradually experience plastic deformation. However, the average strain of ferrite in the entire RVE varies due to the various strains applied to each RVE. After rolling, the ring's upper surface position had the highest radial average strain of the outer surface element, while the other locations had the highest average PEEQ of the inner surface element. The medium thickness area element showed the lowest average strain in all positions. The reason why the upper surface position appeared different from the others could be that the material flow caused the area to not fit the mandrel roll entirely, which decreased the amount of rolling force delivered to the area and, as a result, the degree of deformation. Axial patterns were observed on the outer surfaces (A, D, G) and middle thickness (B, E, H), with the PEEQ decreasing as the distance from the ring's center increased. On the inner surface of the ring element, the strain was highest at F and the lowest at C. The F spot was situated at the raceway's edge, where the shear force was higher and a greater shear deformation took place. This led to the conclusion that shear was another significant element in the ring's deformation.

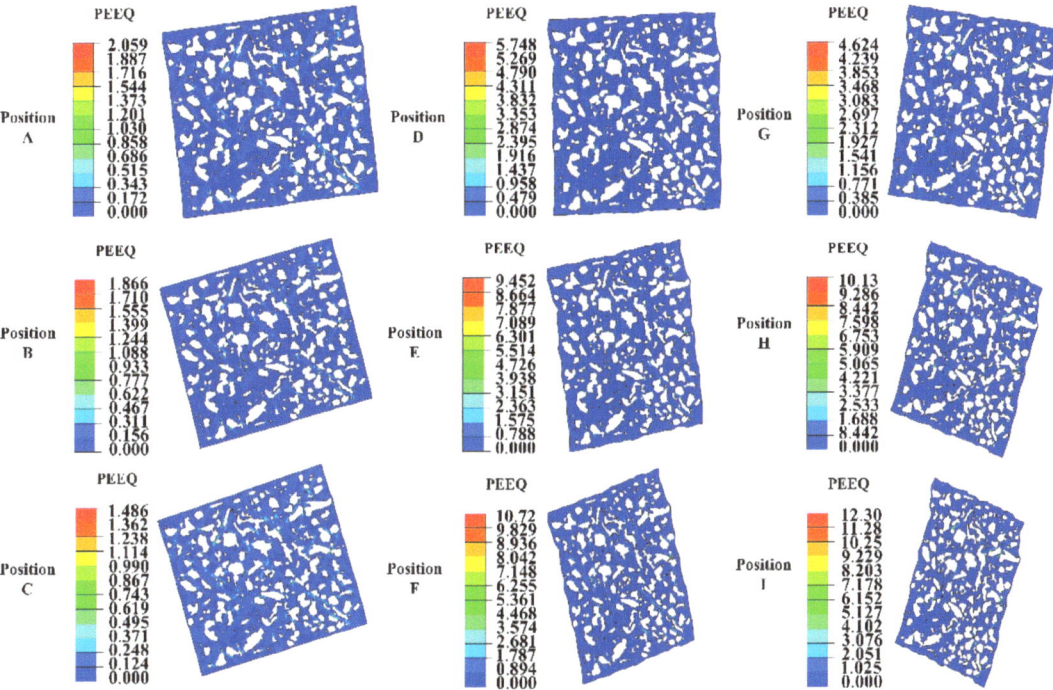

Figure 16. RVE strain distribution at different locations of the M50 bearing ring at the end of rolling.

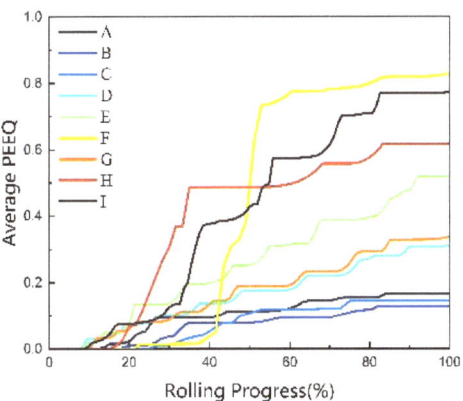

Figure 17. Ferrite mean PEEQ curve.

Based on the findings of the aforementioned analyses, the elements in the medium-height region and those on the inner surface of the ring were chosen for investigation. Figure 18 shows the cementite flow curves and deformation of the RVE, where Figure 18a,b,e show the axial elements of the ring and Figure 18c–e show the radial elements. As shown in the figure, the cementite essentially did not experience displacement changes at the beginning of the rolling process because the ferrite was still in the elastic deformation stage. As rolling continued, the ferrite gradually experienced plastic deformation and flowed. For axial and radial movement, the ferrite was wrapped around the cementite. The RVE displayed a compressed state for the axial element, and the cementite flowed radially in the direction of the RVE compression. Displacement was observed to be maximal at the edge of the raceway and minimal at the upper surface of the ring. The shear force exerted a considerable influence on the RVE in the axial direction, leading to a specific angle of deflection and reduced stretching. The cementite at the upper-end face and the edge of the raceway showed the flow trend along the deflection direction, whereas the cementite in the middle of the raceway was subjected to the opposite shear force and showed the opposite flow trend. All three had nearly identical axial displacements. The flow behavior for axial and radial elements was similar in the radial direction. As the RVE was subjected to shear forces in the same direction in the axial direction, the cementite flowed in the shear direction. The inner surface and the middle width region showed notable cementite displacements, whereas the outer surface showed only slight displacements. This phenomenon can be attributed to the outer surface of the ring being subjected to greater radial forces and lesser shear forces, which resulted in minimal changes in the RVE in the axial direction. However, the RVE deformed more severely in the axial direction at other places of the ring due to significant shear stresses.

Figure 19 shows the distribution and displacement change curves of the RVE in the radial–circumferential direction for each of the above positions. As observed in the image, similar to the radial–axial direction, the flow of cementite basically did not occur in the early stages of ring rolling, and cementite was wrapped by ferrite for flow in the middle and late stages of rolling. The axial elements (Figure 19a,b,e) experienced both circumferential and radial forces. In the radial direction, the displacement of the cementite at the center of the raceway was clearly the largest, but the displacement at other positions was relatively small. The displacement of the cementite in the circumferential direction showed a tendency to increase gradually from the upper-end surface of the ring to the center of the ring raceway. This was because the ring's circumferential flow increased with proximity to the raceway. For the radial elements (Figure 19c–e), the cementite's circumferential displacement showed a tendency to gradually decrease from the inner and outer surfaces to the middle thickness region of the ring, while its radial displacement showed a tendency to gradually increase from the outer surface to the inner surface of the ring.

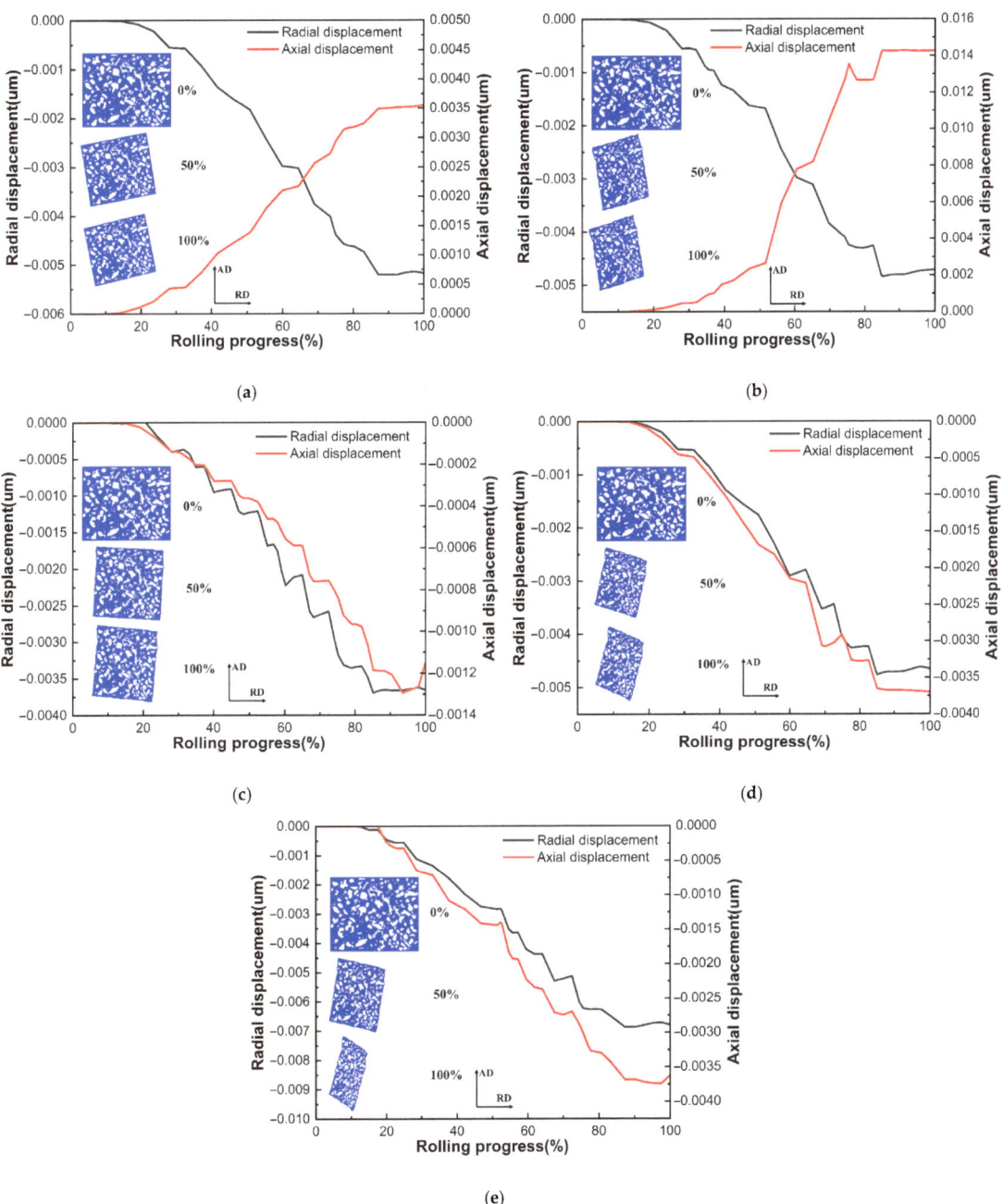

Figure 18. Variation curves of cementite displacement in the radial–axial plane: (**a**) position C; (**b**) position F; (**c**) position G; (**d**) position H; (**e**) position I.

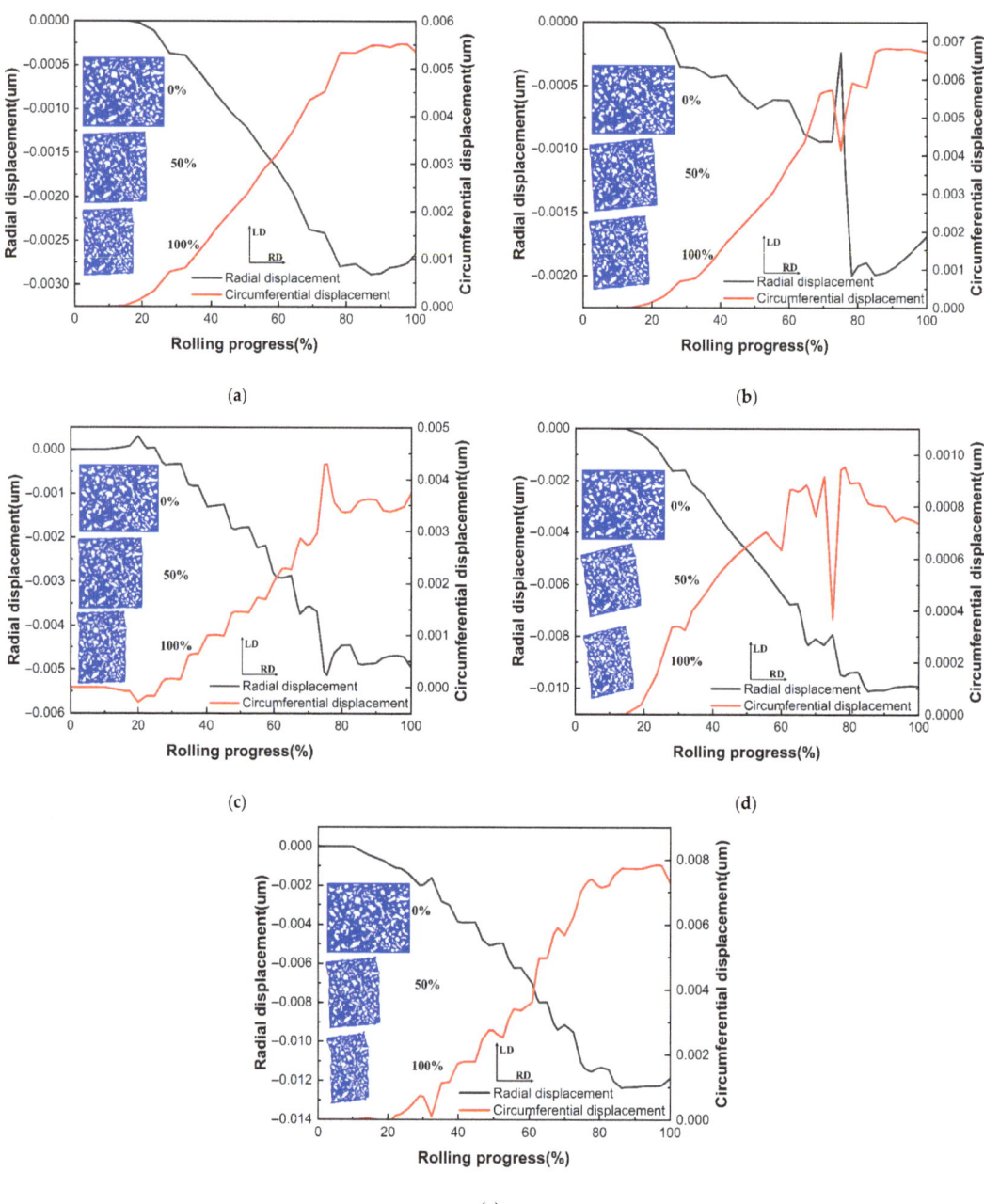

Figure 19. Variation curves of cementite displacement in the radial–circumferential plane: (**a**) position C; (**b**) position F; (**c**) position G; (**d**) position H; (**e**) position I.

3.3. Mechanisms of Macroscale and Mesoscale Flow Behavior Evolution

Figure 20 shows the mechanism of macroscopic and mesoscopic flow behavior during cold rolling. It is clear that the cementite dispersion is irregular with the mandrel roll feed during the ring rolling process. As seen in Figure 20a, the cementite displacement on the inner surface of the ring is substantially larger than on the outer surface of the ring in the radial direction. In the axial direction, the cementite at the ring raceway exhibits a large displacement because of the high shear force. Additionally, when the cementite advances away from this region, its axial displacement reduces. As illustrated in Figure 20b, the displacement of cementite on the inner surface of the ring is larger than that on the outer surface in the radial direction in the radial–circumferential plane. In the circumferential direction, the figure shows more cementite at the ring's inner and outer surfaces than in the middle region due to greater displacement at the surfaces. Some shapes are defined in Figure 20c, where the circle indicates the position of the initial cementite, the red arrow indicates the direction of displacement, and the length of the line segment indicates the relative magnitude of the displacement distance. As shown in Figure 20c, the displacement of the cementite shows a certain pattern. The flow of the cementite is most intense in the radial direction in the raceway edge region, followed by the raceway center region, and finally, other regions. And, at the middle height of the ring, the displacement of the cementite gradually decreases from the inner surface to the outer surface. The displacement of cementite at the ring's middle height gradually decreases from the inner to the outer surface. However, on the inner surface, the displacement remains essentially unchanged along the axial direction. The circumferential variation of the cementite is shown by the fact that the cementite on the inner and outer surfaces of the ring maintains approximately the same displacement, while the cementite in the middle thickness region of the ring is less displaced.

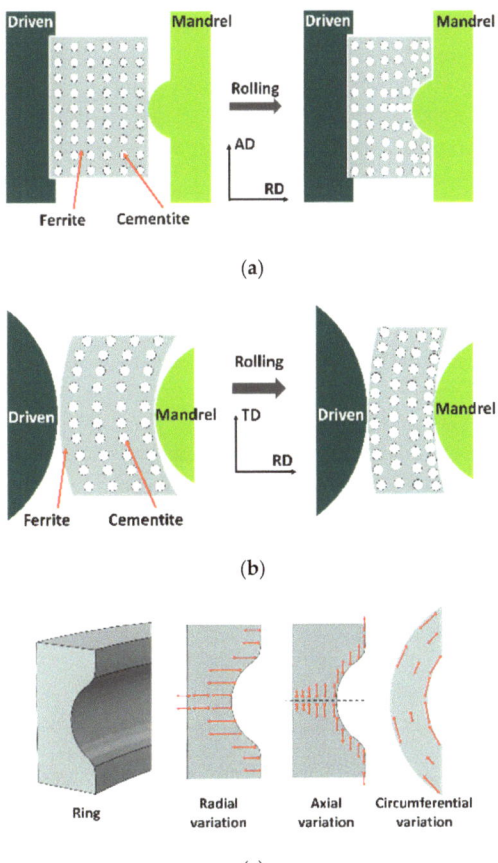

Figure 20. Mechanism of macro- and microflow behavior during the ring rolling process. (**a**) Radial–axial direction; (**b**) radial–circumferential direction; (**c**) schematic of the flow behavior after rolling.

This paper successfully combines macro- and mesoviews to reflect the flow trend of cementite in the cold ring rolling process. The method is of great significance in guiding the bearing-forming process. Through this method, the macroscopic rolling and forming process can be designed so that different areas of the bearing can be pretreated differently at the early stage of rolling or even at the stage of raw material acquisition. Then, the ideal bearing properties can be obtained in the area through the rolling process so as to facilitate the subsequent heat treatment process. Specifically, the initial material can be optimized according to the stress and strain distribution. Through reasonable heat treatment of the initial material, the stress concentration phenomenon in the cold ring rolling process can be reduced, and the damage in the cold ring rolling process can be reduced or avoided. In addition, this method is not only applicable to the cold ring rolling process of M50 bearing steel but also to the forming process of other high-carbon steels.

4. Conclusions

This paper established a multiscale model from the macroscale to the mesoscale. Based on the actual working conditions, a macroscopic cold-rolled three-dimensional finite element model of the M50 bearing ring was established. The sub-model was applied to transfer the response between macroscale and mesoscale. Based on the actual microstructure of the material, the RVE model was developed and verified. Through nanoindentation studies, the material's single-phase characteristics were determined. A mesoscale finite element model was finally developed. Analysis was performed on the cold ring rolling macroscale and mesoscale flow characteristics of M50 bearing steel. This approach significantly lowers the time and computational resources needed for direct full-scale modeling while preserving the capture of the high-precision results in crucial regions using the sub-modeling technique to decouple the macroscopic and mesoscale computations. The conclusions of this paper are as follows:

1. Through the macroscopic cold ring rolling three-dimensional finite element simulation, it was found that the stress distribution was relatively uniform. The PEEQ showed a stepwise growth. In the axial direction, the PEEQ showed a trend of high in the middle and low on both sides, and in the radial direction, it showed a trend of high in the middle of the two ends. The maximum PEEQ appeared in the raceway.
2. The mesoscopic simulation revealed that the stress was concentrated in the cementite, and the maximum stress occurred at the junction of the two phases. The greatest strain was found in the ferrite matrix positioned between the two adjacent cementites, while the strain first appeared in the matrix at the cementite–ferrite junction.
3. The cementite flew with the deformation of the matrix. The radial displacement of the cementite decreased from the edge of the raceway to both ends and decreased from the inner to the outer surface. Its axial displacement was basically the same on the inner surface of the ring and decreased from the inner to the outer surface. Its circumferential displacement decreased from the inner and outer surfaces to the intermediate thickness region.

Author Contributions: Conceptualization, W.W. and Z.L.; methodology, W.W., Z.L. and Q.L.; software, Z.L. and G.Z.; validation, W.W. and Z.L.; formal analysis, W.W., Z.L., Q.L., G.Z., G.L. and Y.L.; investigation, W.W., Z.L. and Q.L.; resources, G.L., Y.L. and L.H.; writing—original draft Preparation, Z.L.; writing—review and editing, W.W., Z.L. and Q.L.; supervision, G.L., Y.L. and L.H.; project administration, L.H.; funding acquisition, G.L. and Y.L. All authors have read and agreed to the published version of the manuscript.

Funding: This research was funded by the Open Foundation of The State Key Laboratory of Refractories and Metallurgy (grant number No. G202403), the National Key R&D Program of China (grant number 2020YFA0714900), the 111 Project (grant number B17034), and the Open Foundation of The State Key Laboratory of Materials Processing and Die & Mould Technology (grant number P2024-026).

Institutional Review Board Statement: Not applicable.

Informed Consent Statement: Not applicable.

Data Availability Statement: The original contributions presented in this study are included in the article. Further inquiries can be directed to the corresponding authors.

Conflicts of Interest: Author Qinglong Liu was employed by the company China Railway Construction Heavy Industry Corporation Liminted. The remaining authors declare that the research was conducted in the absence of any commercial or financial relationships that could be construed as a potential conflict of interest.

References

1. Rejith, R.; Kesavan, D.; Chakravarthy, P.; Murty, S.N. Bearings for aerospace applications. *Tribol. Int.* **2023**, *181*, 108312. [CrossRef]
2. Bhadeshia, H.K.D.H. Steels for bearings. *Prog. Mater. Sci.* **2012**, *57*, 268–435.
3. Zaretsky, E.V. Rolling bearing steels—A technical and historical perspective. *Mater. Sci. Technol.* **2012**, *28*, 58–69. [CrossRef]
4. Sun, B.; Xu, J.; Xing, C. Numerical and experimental investigations on the effect of mandrel feeding speed for high-speed rail bearing inner ring. *Int. J. Adv. Manuf. Technol.* **2019**, *100*, 1993–2006. [CrossRef]
5. Qian, D.; He, Y.; Wang, F.; Chen, Y.; Lu, X. Microstructure and mechanical properties of M50 steel by combining cold rolling with austempering. *Metals* **2020**, *10*, 381. [CrossRef]
6. Wang, F.; Qian, D.; Xie, L.; Dong, Z.; Song, X. Microstructure evolution and tempering transformation kinetics in a secondary hardened M50 steel subjected to cold ring rolling. *ISIJ Int.* **2021**, *61*, 361–371. [CrossRef]
7. Wang, F.; Qian, D.; Hua, L.; Lu, X. The effect of prior cold rolling on the carbide dissolution, precipitation and dry wear behaviors of M50 bearing steel. *Tribol. Int.* **2019**, *132*, 253–264. [CrossRef]
8. Deng, S.; Hua, L.; Shi, D. Effect of cold rolling on plastic deformation and microstructure of bearing ring. *Mater. Sci. Technol.* **2017**, *33*, 984–990. [CrossRef]
9. Li, W.; Sakai, T.; Li, Q.; Lu, L.T.; Wang, P. Reliability evaluation on very high cycle fatigue property of GCr15 bearing steel. *Int. J. Fatigue* **2010**, *32*, 1096–1107. [CrossRef]
10. Wu, H.N.; Xu, D.S.; Wang, H.; Yang, R. Molecular dynamics simulation of tensile deformation and fracture of γ-TiAl with and without surface defects. *J. Mater. Sci. Technol.* **2016**, *32*, 1033–1042. [CrossRef]
11. Gadalińska, E.; Baczmański, A.; Braham, C.; Gonzalez, G.; Sidhom, H.; Wroński, S.; Buslaps, T.; Wierzbanowski, K. Stress localisation in lamellar cementite and ferrite during elastoplastic deformation of pearlitic steel studied using diffraction and modelling. *Int. J. Plast.* **2020**, *127*, 102651. [CrossRef]
12. Lu, X.; Zhou, L.; Lei, C.; Liu, H.; Lan, H. Formation mechanism and microstructure evolution of light etched region during rolling contact fatigue in M50 steel. *J. Mater. Res. Technol.* **2023**, *24*, 8955–8966. [CrossRef]
13. Zhang, C.; Peng, B.; Wang, L.; Ma, X.; Gu, L. Thermal-induced surface damage of M50 steel at rolling-sliding contacts. *Wear* **2019**, *420*, 116–122. [CrossRef]
14. Guan, J.; Wang, L.; Zhang, Z.; Shi, X.; Ma, X. Fatigue crack nucleation and propagation at clustered metallic carbides in M50 bearing steel. *Tribol. Int.* **2018**, *119*, 165–174. [CrossRef]
15. Ganti, S.; Turner, B.; Kirsch, M.; Anthony, D.; McCoy, B.; Trivedi, H.; Sundar, V. Three-dimensional (3D) analysis of white etching bands (WEBs) in AISI M50 bearing steel using automated serial sectioning. *Mater. Charact.* **2018**, *138*, 11–18. [CrossRef]
16. Macioł, P.; Michalik, K. Parallelization of fine-scale computation in Agile Multiscale Modelling Methodology. *AIP Conf. Proc.* **2016**, *1769*, 060009.
17. Su, Y.; Shen, Z.; Long, X.; Chen, C.; Qi, L.; Chao, X. Gaussian filtering method of evaluating the elastic/elasto-plastic properties of sintered nanocomposites with quasi-continuous volume distribution. *Mater. Sci. Eng. A* **2023**, *872*, 145001. [CrossRef]
18. Kong, L.; Xie, H.; Li, C. Coupled microplane and micromechanics model for describing the damage and plasticity evolution of quasi-brittle material. *Int. J. Plast.* **2023**, *162*, 103549. [CrossRef]
19. Zhang, J.; Li, H.; Sun, X.; Zhan, M. A multi-scale MCCPFEM framework: Modeling of thermal interface grooving and deformation anisotropy of titanium alloy with lamellar colony. *Int. J. Plast.* **2020**, *135*, 102804. [CrossRef]
20. Zhou, J.; Gokhale, A.M.; Gurumurthy, A.; Bhat, S.P. Realistic microstructural RVE-based simulations of stress–strain behavior of a dual-phase steel having high martensite volume fraction. *Mater. Sci. Eng. A* **2015**, *630*, 107–115. [CrossRef]
21. Ji, H.; Song, Q.; Gupta, M.K.; Liu, Z. A pseudorandom based crystal plasticity finite element method for grain scale polycrystalline material modeling. *Mech. Mater.* **2020**, *144*, 103347. [CrossRef]

22. Chawla, N.; Sidhu, R.S.; Ganesh, V.V. Three-dimensional visualization and microstructure-based modeling of deformation in particle-reinforced composites. *Acta Mater.* **2006**, *54*, 1541–1548. [CrossRef]
23. Jang, J.I.; Shim, S.H.; Komazaki, S.; Sugimoto, T. Correlation between microstructure and nanohardness in advanced heat-resistant steel. *Key Eng. Mater.* **2006**, *326*, 277–280. [CrossRef]
24. Yan, F.K.; Zhang, B.B.; Wang, H.T.; Tao, N.R.; Lu, K. Nanoindentation characterization of nano-twinned grains in an austenitic stainless steel. *Scr. Mater.* **2016**, *112*, 19–22. [CrossRef]
25. Hu, X.; Yang, L.; Wei, X.; Wang, H.; Fu, G. Molecular dynamics simulation on nanoindentation of M50 bearing steel. *Materials* **2023**, *16*, 2386. [CrossRef] [PubMed]
26. Cheng, G.; Choi, K.S.; Hu, X.; Sun, X. Determining individual phase properties in a multi-phase Q&P steel using multi-scale indentation tests. *Mater. Sci. Eng. A* **2016**, *652*, 384–395.
27. Young, M.L.; Almer, J.D.; Daymond, M.R.; Haeffner, D.R.; Dunand, D.C. Load partitioning between ferrite and cementite during elasto-plastic deformation of an ultrahigh-carbon steel. *Acta Mater.* **2007**, *55*, 1999–2011. [CrossRef]
28. Taupin, V.; Pesci, R.; Berbenni, S.; Berveiller, S.; Ouahab, R.; Bouaziz, O. Lattice strain measurements using synchrotron diffraction to calibrate a micromechanical modeling in a ferrite–cementite steel. *Mater. Sci. Eng. A* **2013**, *561*, 67–77. [CrossRef]
29. Oliver, W.C.; Pharr, G.M. An improved technique for determining hardness and elastic modulus using load and displacement sensing indentation experiments. *J. Mater. Res.* **1992**, *7*, 1564–1583. [CrossRef]
30. Zhuang, X.; Ma, S.; Zhao, Z. Effect of particle size, fraction and carbide banding on deformation and damage behavior of ferrite–cementite steel under tensile/shear loads. *Model. Simul. Mater. Sci. Eng.* **2016**, *25*, 015007. [CrossRef]
31. Ma, S.; Zhuang, X.; Zhao, Z. Effect of particle size and carbide band on the flow behavior of ferrite–cementite steel. *Steel Res. Int.* **2016**, *87*, 1489–1502. [CrossRef]
32. Zhao, K.; Zhuang, X.C.; Pei, X.H.; Zhao, Z. Influence of Geometrical Imperfection on Failure Mode of DP780 Steel Utilizing Damage Models Embedded RVE Technique. *Key Eng. Mater.* **2017**, *725*, 471–476. [CrossRef]
33. Viloria, A.; Nova, D.M.; Salinas, D.A.G.; Barbosa, W.; Espinosa, C.C.P.; Toledo, F.A.R.; Ballesteros, D.Y.P.; Rodriguez, J.G.D. Microhardness Profile and Residual Stresses Evaluation in a Shot Peened SAE 5160H Steel. *Rev. UIS Ing.* **2024**, *23*, 103–114.
34. Bhadeshia, H. Cementite. *Int. Mater. Rev.* **2020**, *65*, 1–27. [CrossRef]
35. Hu, X.; Van Houtte, P.; Liebeherr, M.; Walentek, A.; Seefeldt, M.; Vandekinderen, H. Modeling work hardening of pearlitic steels by phenomenological and Taylor-type micromechanical models. *Acta Mater.* **2006**, *54*, 1029–1040. [CrossRef]
36. Wang, H.; Wang, F.; Qian, D.; Chen, F.; Dong, Z.; Hua, L. Investigation of damage mechanisms related to microstructural features of ferrite-cementite steels via experiments and multiscale simulations. *Int. J. Plast.* **2023**, *170*, 103745. [CrossRef]

Disclaimer/Publisher's Note: The statements, opinions and data contained in all publications are solely those of the individual author(s) and contributor(s) and not of MDPI and/or the editor(s). MDPI and/or the editor(s) disclaim responsibility for any injury to people or property resulting from any ideas, methods, instructions or products referred to in the content.

Article

The Influence of Temperature in the Wire Drawing Process on the Wear of Drawing Dies

Maciej Suliga [1,*], Piotr Szota [1], Monika Gwoździk [1], Joanna Kulasa [2] and Anna Brudny [2]

1. Faculty of Production Engineering and Materials Technology, Czestochowa University of Technology, Armii Krajowej 19, 42-201 Czestochowa, Poland; piotr.szota@pcz.pl (P.S.); monika.gwozdzik@pcz.pl (M.G.)
2. Łukasiewicz Research Network–Institute of Non-Ferrous Metals, Sowińskiego 5, 44-100 Gliwice, Poland; joanna.kulasa@imn.lukasiewicz.gov.pl (J.K.); anna.brudny@imn.lukasiewicz.gov.pl (A.B.)
* Correspondence: maciej.suliga@pcz.pl; Tel.: +48-343250786

Abstract: This paper presents a wear analysis of tungsten carbide drawing dies in the process of steel wire drawing. The finite element method (FEM) analysis showed a significant correlation between drawing die geometry, single reduction size and drawing speed on the rate of drawing die wear. It has been shown that in steel wire drawing at higher drawing speeds, intense heating of the drawing die occurs due to friction at the wire/drawing die interface, leading to premature wear. Tribological tests on the material for the drawing die cores (94%WC+6%Co) confirmed the gradual abrasion of the steel and carbide sample surfaces with the "products" of abrasion sticking to their surfaces. The increase in temperature increases the coefficient of friction, translating into accelerated wear of the drawing dies.

Keywords: wire drawing; steel; die wear; FEM; tungsten carbide; tribology; friction

Citation: Suliga, M.; Szota, P.; Gwoździk, M.; Kulasa, J.; Brudny, A. The Influence of Temperature in the Wire Drawing Process on the Wear of Drawing Dies. *Materials* **2024**, *17*, 4949. https://doi.org/10.3390/ma17204949

Academic Editor: Frank Czerwinski

Received: 13 September 2024
Revised: 8 October 2024
Accepted: 9 October 2024
Published: 10 October 2024

Copyright: © 2024 by the authors. Licensee MDPI, Basel, Switzerland. This article is an open access article distributed under the terms and conditions of the Creative Commons Attribution (CC BY) license (https://creativecommons.org/licenses/by/4.0/).

1. Introduction

The drawing die is the basic tool of the drawing process, and it has a decisive influence on the process parameters and the properties of the steel wires [1,2]. Cemented carbides, synthetic diamonds, and cemented carbides with a ceramic coating are used to manufacture drawing dies [3–5]. Currently, millions of tons of wire and wire products (ropes, springs, screws, anchors, nails, fasteners, etc.) are produced in the world every year using the drawing method. These products are used in many branches of the economy (including construction, machine industry, and mining).

Drawing technology is the determining factor for using a particular type of drawing dies. The drawing die is built with a steel casing to protect the drawing die core from mechanical damage. In steel wire wet drawing, drawing dies with tungsten carbide or diamond inserts are used, while special ceramic coatings are also applied to the tungsten carbide surfaces to increase their durability. On the other hand, in dry wire drawing, with wire diameters over 1 mm, tungsten carbide inserts are usually used for financial reasons. Drawing dies used in wire manufacturing are also classified according to the shape of the drawing part, i.e., concave, convex sigmoid, curved and conical dies. Drawing dies can also be divided according to the drawing method, i.e., conventional, hydrodynamic [6,7], and roller dies [8,9]. Hence, when designing the drawing process, a specific type of drawing die is selected based on the surface treatment of the wire (mechanical descaling, pickling) [10–12], the application of lubricant layers, the type of wire drawing machine (dry drawing, wet drawing), the wire diameter, and the number of draws [13–15].

Depending on the type of drawing die, its lifespan varies from a few to more than 100 tonnes of processed material. The most common type of drawing die in the wire industry is the conical die, which consists of a steel casing and a tungsten carbide core. High pressures in the contact area between the material and the drawing die cause the drawing die to wear, cracks on its surface, and changes in geometry. The wear rate of drawing dies depends largely on the drawing technology. The basic parameters of the drawing process are drawing geometry, distribution of single and total reduction, degree of wire strengthening, steel grade, lubrication conditions [16,17], and drawing speed [18]. The drawing speed is one of the main factors that can affect the wear rate of drawing dies. In multi-stage drawing, an increase in drawing speed generates a large amount of heat at the wire/drawing die interface, leading to a significant rise in temperature, which, on the one hand, contributes to a deterioration in lubrication conditions and an increase in friction and, on the other hand, reduces the wear resistance of the carbide. Hence, in industrial practice, drawing die wear is common after drawing 1 tonne of wire.

Today, an experimental approach can be used to analyse the wear of drawing dies by measuring the wear and the change in geometry of the drawing dies after the drawing process depending on the amount of material drawn and the drawing conditions [19,20], tribological studies of the materials used for drawing dies and computer simulations of the drawing process (FEM) [21–25]. The design of metal forming tools is a complex, time-consuming, and resource-intensive issue. Hence, the combined experimental and numerical analysis methods are currently used for optimisation.

This paper presents an analysis of the influence of drawing process parameters, i.e., drawing speed and angle, single reduction size, on wire heating, stress distribution and drawing die wear. On the other hand, tribological studies made it possible to determine the relationship between temperature and the friction coefficient and wear of the steel/carbide friction pair.

2. Materials and Methods

2.1. Computer Simulations of the Wire Drawing Process

The study involved numerical modelling of the wire drawing process using the Simufact 2023.2 commercial computer program (Hexagon, Hamburg, Germany). The program uses the finite element method to solve the computational problem. The wire drawing process was axisymmetric in a plane deformation condition. A total of 3600 flat tetrahedral elements with an average mesh edge size of 0.15 mm were used for the calculations.

The MARC solver developed based on the displacement method was used for the calculations. The MARC methodology is based on the stiffness of the system and is based on the force–displacement relationship (1):

$$K \cdot u = f \tag{1}$$

where K is the system stiffness matrix, u is the nodal displacement and f is the force vector.

Once the displacement vector u is determined, the deformations in each element can be calculated based on the displacement (2):

$$\varepsilon_{el} = \beta \cdot u_{el} \tag{2}$$

The stress in an element is determined based on the stress–deformation relationship (3):

$$\sigma_{el} = L \cdot \varepsilon_{el} \tag{3}$$

where σ_{el} is the stress and ε_{el} is the deformation in the elements, and u_{el} is the displacement vector associated with the element nodal points; and β and L are the deformation–displacement and stress–deformation relations, respectively.

The hardening curve taken from the Simulfact 2023.2 program database, equation $\sigma(\varepsilon, \dot{\varepsilon}, T)$ (4), was used to simulate the drawing.

$$\sigma = C_1 \cdot e^{(C_2 \cdot T)} \cdot \varepsilon^{(n_1 \cdot T + n_2)} \cdot e^{\left(\frac{l_1 \cdot T + l_2}{\varepsilon}\right)} \cdot \dot{\varepsilon}^{(m_1 \cdot T + m_2)} \tag{4}$$

where T—temperature, ε—true strain, and $\dot{\varepsilon}$—strain rate.

The model adopts an elastic–plastic model of the deformed body (wire), while rigid models with the possibility of calculating heat distribution were adopted for the insert, insert housing, and water (Figure 1).

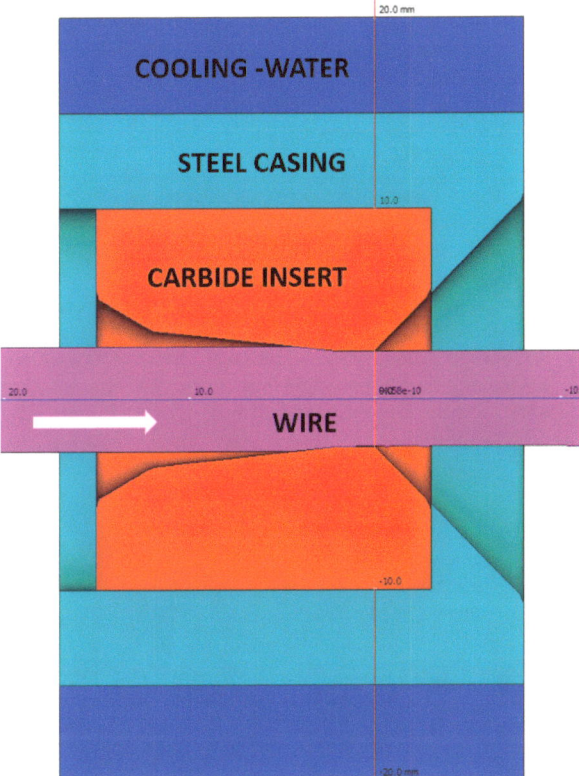

Figure 1. The drawing process tools.

Determining the wear of the drawing die insert during the drawing process requires the determination of the working conditions of the drawing die as well as the conditions prevailing on the drawing die surface in the area of the crushing cone and the calibrating part.

The following initial parameters and boundary conditions were adopted for the numerical modelling:

- Initial temperature: water 20 °C, drawing die housing 20 °C, insert 20 °C, wire 100 °C;
- Thermal conductivity coefficient: wire insert: 1000 W/(m²-K); housing insert: 20,000 W/(m²-K); housing water: 10,000 W/(m²-K);
- Friction coefficient: 0.07;
- C45 steel properties, Table 1.

Table 1. Yield stress function coefficients (4) for C45 steel/MPa.

				Coefficient Values				
C_1	C_2	m_1	m_2	l_1	l_2	n_1	n_2
692.2	−0.00064	0.000028	0.000709	−0.0000473	−0.01224	−0.00031	0.162026

For the numerical modelling, it was necessary to determine the remaining physical–mechanical properties of the material:

- Young's modulus of C45 steel—temperature-dependent;
- Poisson's number of C45 steel: 0.283;
- Thermal conductivity of C45 steel: temperature-dependent;
- The dependence of K10 carbide hardness on temperature was taken from the paper [26];
- Carbide thermal conductivity: 80 W/(m·K);
- Carbide heat capacity: 0.2 J/(g·K);
- Carbide density: 14,700 kg/m^3.

The purpose of the numerical calculatons was to determine the temperature distribution in a drawing set consisting of a tungsten carbide insert and a C45 steel housing, which is cooled with water on the side of the cylindrical surface. A constant water temperature, 20 °C, was assumed in the calculations due to its circulation in the cooling system. The drawing speed was assumed to be 5 m/s. As a result of the performed modelling, the temperature distribution in the elements of the wire drawing set with an initial diameter of 5.35 mm to a diameter of 5.0 mm was obtained. The drawing process was carried out for a wire length of 10 m and process time of 2 s. Numerical calculations made it possible to determine the temperature distribution in the drawing die and the tool heating rate as a function of drawing time; see Figures 2–4.

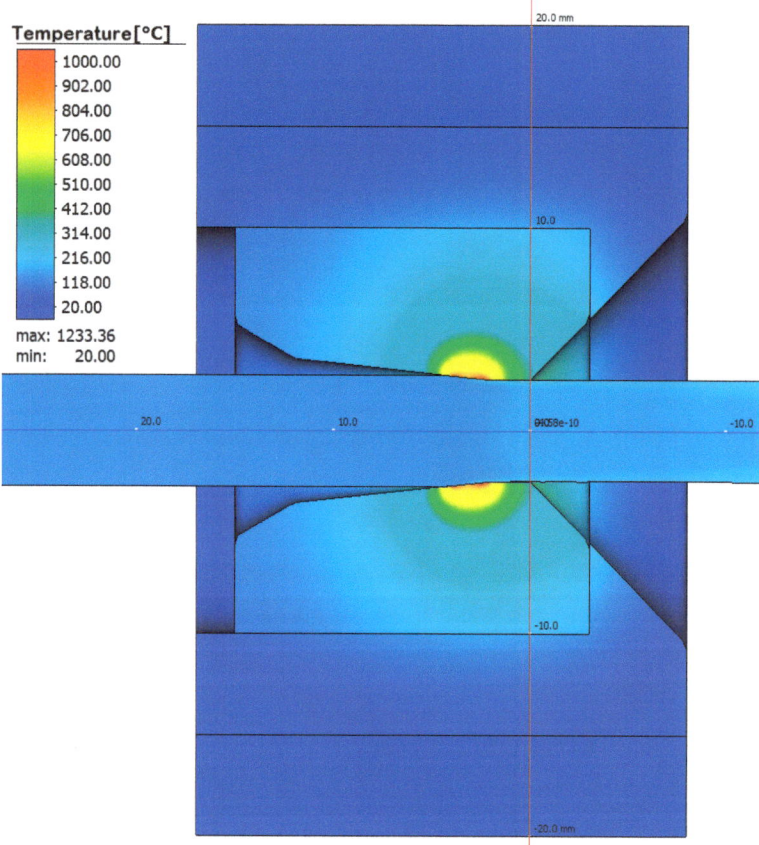

Figure 2. Temperature distribution in the tools during wire drawing at 5 m/s.

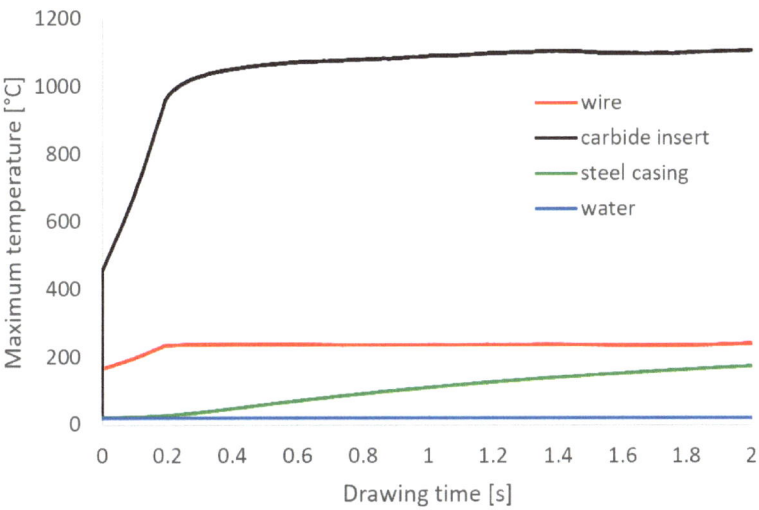

Figure 3. Maximum temperature of drawing die elements after drawing 10 m of wire.

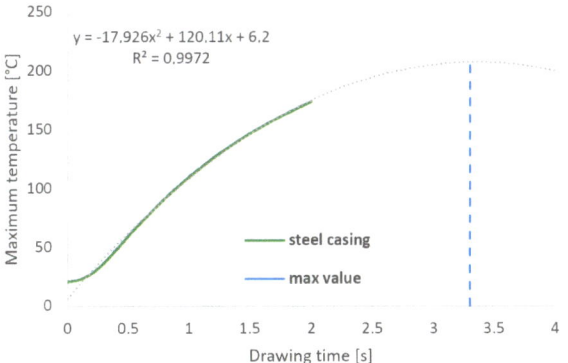

Figure 4. Maximum temperature of steel casing after drawing 10 m of wire.

Based on the numerical modelling results shown in Figure 3, a clear increase in temperature on the working surface of the drawing die insert is noticeable. In the area 0.06 mm from the working surface of the drawing die, the temperature reaches a value from 950 to 1100 °C. The temperature settles within 1 s and does not increase further in the process. The heat is dissipated through the insert and transferred to the drawing die housing. The drawing die housing is heated by the insert and cooled by water at the same time. Figures 3 and 4 show the temperature rise of the housing. The curve in Figure 4 represents the increase in temperature of the housing up to 2 s of the process. The insert heating curve up to 2 s was extrapolated based on the data. The extrapolation shows that the housing should reach thermal equilibrium at 208 °C within a further 1.4 s. The exact determination of the temperature equilibrium requires further time-consuming calculations. Based on the obtained temperature distribution, it was possible to determine the unit work of friction forces, which makes it possible to determine the quantitative wear of the drawing die. The precise determination of quantitative wear requires determining the wear coefficient. The data shown in Figure 5 are the result of calculations using the Archard model, which describes the abrasive wear of surfaces with a significant difference in speed between the surfaces in contact. That model can be expressed by Equation (5), which is described in the integral form used to solve an FEM-based algorithm.

$$wear = k_{zn} \int_0^t \frac{\sigma_n v_s}{H(T)} dt, [mm] \quad (5)$$

where v_s—tangential sliding velocity of the metal on the tool surface, σ_n—normal stresses, t—time, $H(T)$—tool hardness at a specified temperature, and k_{zn}—tool wear coefficient.

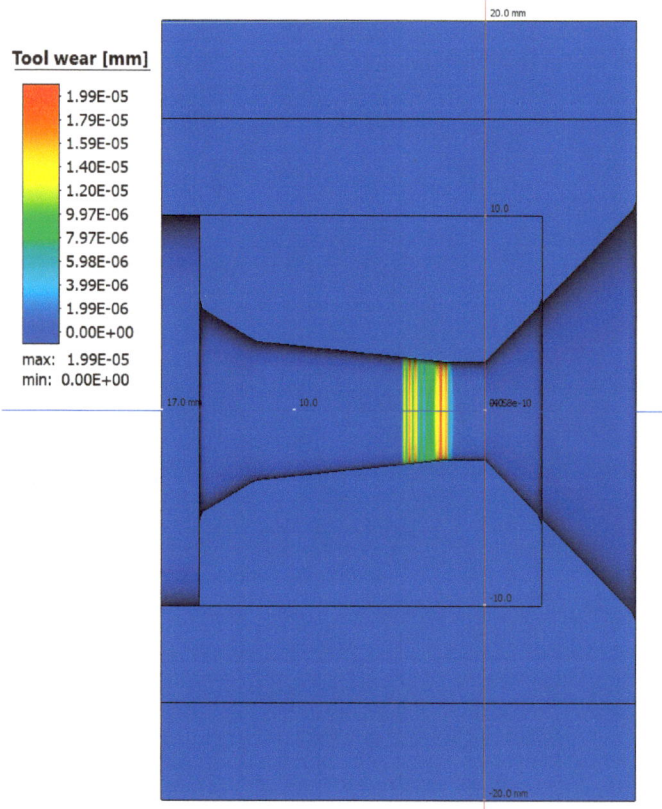

Figure 5. Drawing die wear after drawing 10 m of wire.

Preliminary computer simulations have shown that for a correct representation of the drawing process and the influence of drawing technology (e.g., single and total reduction distributions, tool geometry, drawing speed) on the drawing process and tool wear, it is necessary to assume higher initial tool and material temperatures of 500 °C and 50 °C, respectively. Simulations of wire drawing in a single sequence with the application of different values of a single reduction (Gp = 5.4; 17.4; 25.4; 33.1%) were performed for two drawing speeds of 1 and 5 m/s. The drawing process was carried out using drawing dies with angles 2α = 6°, 10°, and 14°.

2.2. Tribological Tests

Tribological tests were carried out as part of this study to determine the influence of temperature on friction conditions and wear of tungsten carbide. Tribological tests were conducted in dry conditions without lubrication. A high-temperature CSM tribometer operating in the Pin–Ball-on-Disk system was used to assess carbide wear (Anton Paar

GmbH's, Graz, Austria). It enables the measurement of tribological properties in the range from ambient temperature to 1000 °C. In the industry, the most commonly used carbide for the manufacture of drawing dies is grade K10 carbide (94%WC+6%Co) [1]. It is used in the drawing of steel wires with carbon content ranging from 0.08%C (low-carbon steel) to 0.9%C (high-carbon steel). Hence, steel samples of grade C45 (the same steel grade as in the computer simulations—Section 2.1) were prepared for tribological testing. The performed carbide abrasive wear tests in a sense reproduce the actual drawing process, during which the wire passing through the drawing die causes its heating and gradual wear. Table 2 presents the parameters of the tribological tests.

Table 2. Tribological test parameters.

Parameter	Variant I	Variant II	Variant III	Variant IV
Counter-sample			WC	
Sampling frequency			60 Hz	
Abrasion radius			3 mm	
Load			20 N	
Linear velocity			50 cm/s	
Friction path			13,000 m	
Time			26,000 s	
Process temperature	ambient	200 °C	400 °C	600 °C

Microscopic analyses were carried out using an Olympus GX41 light microscope (Olympus, Tokyo, Japan), a VHX-7000 digital microscope and a scanning electron microscope (SEM). Chemical composition was analysed using scanning electron microscopy SEM+EDS (JEOL Ltd., Tokyo, Japan).

Surface roughness was assessed using a VHX microscope with a Gaussian filter. Based on the performed tests, the following parameters were determined: Sa (arithmetic mean height), Sz (maximum height), Sp (height of highest elevation), Sv (depth of lowest indentation), and Sq (mean squared height).

3. Results

3.1. Numerical Analysis of the Drawing Process

Temperature is the main parameter influencing the wear rate of drawing dies in the drawing process. The performed computer simulations made it possible to determine the correlation between the single reduction size, the drawing die angle, the drawing speed, and the temperature of the drawing dies and wire in contact zone; see Figure 6.

The wire and drawing die temperatures depend on the amount of generated heat. In the drawing process, the source of heat is the work of deformation and the work of friction forces. Based on Figure 6, it can be noticed that the temperature of the drawing die increases with an increase in single reduction, it reaches a maximum value, and then it decreases slightly. However, in the case of wire, there is a continuous increase in temperature over the entire range of the discussed single reduction values. Another fundamental parameter in the drawing process influencing the heating of wire and tools is the geometry of the drawing dies. Computer simulations showed that increasing the drawing die angle from 6 to 14° resulted in an increase in the temperature of the drawing dies, while the wire temperature decreased. The described phenomenon can be explained as follows. The drawing angle influences the duration of contact of the wire and the drawing die; the higher the angle value, the lower the friction forces resulting in the heating of the drawing dies. At the same time, the contact duration of the wire and the drawing die decreases. In such conditions, a larger amount of the generated heat is absorbed by the wire and a smaller amount is absorbed by the drawing die. An additional factor causing heat accumulation in the wire-drawing die tribological system is the drawing speed. Increasing the drawing speed from 1 to 5 m/s resulted in a five-fold increase in deformation work per unit of time. Section 2.1 shows that the drawing die is water-cooled during its operation, which increases the heat transfer from the tool. The drawing die housing is heated by the insert and cooled by water at the same time. At a given drawing speed, which depends on the reduction size, tool geometry and wire grade, the temperature of the drawing dies rises sharply, as the heat flow generated by plastic deformation and friction significantly exceeds the heat flow absorbed by the water cooling the drawing die housing. An approximately five-fold increase in drawing speed results in an approximately two-fold increase in the temperature of the drawing dies and wire. An increase in wire temperature has a negative impact on the lubrication conditions, leading to even greater heating of the drawing dies and, in extreme cases, to damage to the drawing die surface and premature wear.

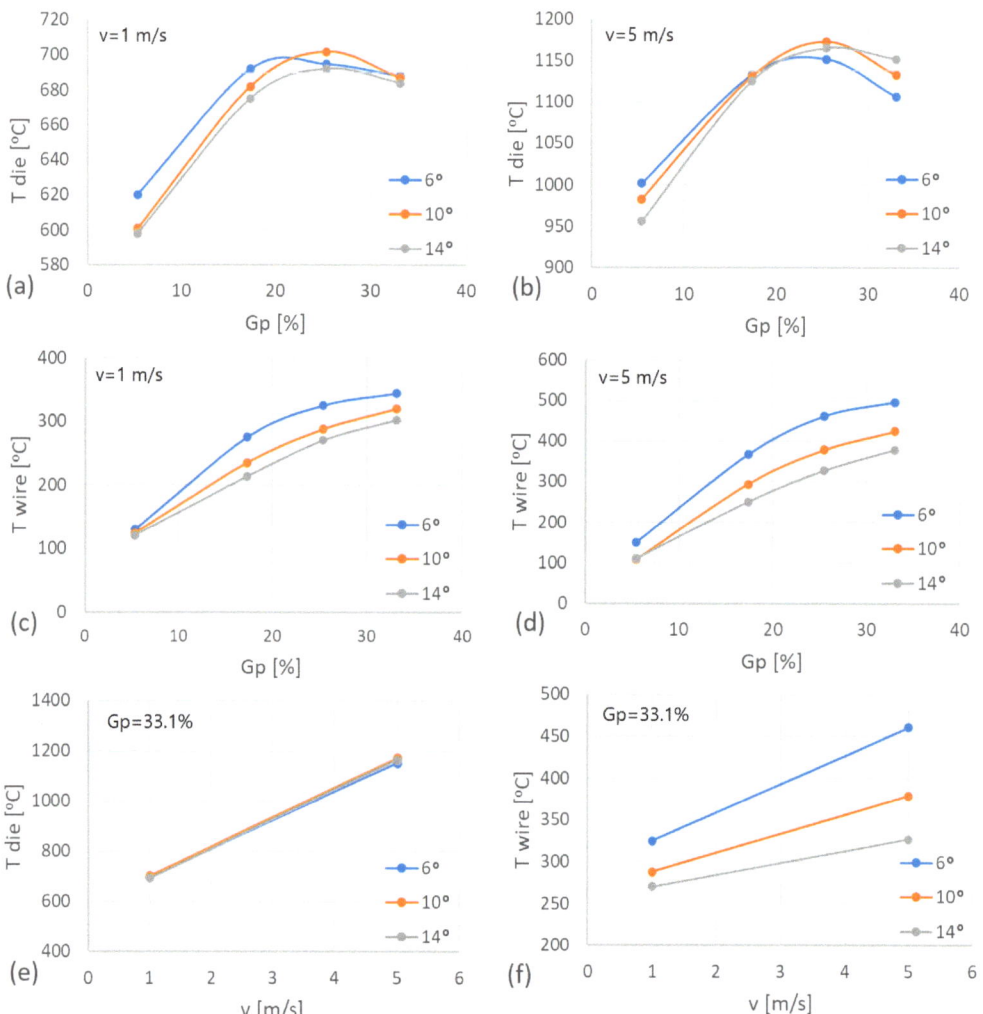

Figure 6. Influence of drawing parameters on the heating of drawing dies and wire: (**a**) die temperature in single reduction function for v = 1 m/s; (**b**) die temperature in single reduction function for v = 5 m/s; (**c**) wire temperature in single reduction function for v = 1 m/s; (**d**) wire temperature in single reduction function for v = 5 m/s; (**e**) die temperature in drawing speed function; (**f**) wire temperature in drawing speed function; where: v—drawing speed; Gp—single reduction; T—temperature; $2\alpha = 6°, 10°, 14°$.

A parameter that significantly influences tool wear in metal-forming processes is unit pressure. Figure 7 shows the distribution of unit pressure along the wire/tool contact area in the crushing part of the drawing die for angles 2α = 6, 10, 14° (drawing from 5.5 to 4.5 mm).

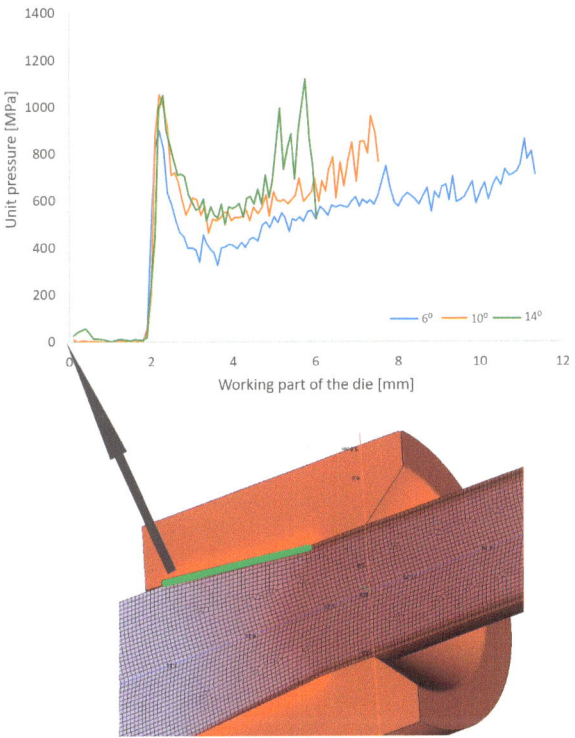

Figure 7. Distribution of unit pressure along the wire/tool contact area in the working part of the drawing die, where 2α = 6°, 10°, 14°.

Based on the data in Figure 7, it can be seen that the unit pressure is not a linear function with the highest values at the wire entry into the drawing die and near the wire transition from the crushing part of the drawing die to the calibrating part. Increasing the angle of the drawing die from 6° to 14° increased the unit pressure, depending on the location, by approximately 50%, on average. The unit pressure influences the rate of tool wear. Thus, it can be assumed that a large increase in the pressure in the drawing die will result in its accelerated wear and, in extreme cases, even its damage. Comparing Figures 6 and 7, a certain correlation can be observed. An increase in the drawing angle results in a decrease in the temperature of the wire with a simultaneous increase in the unit pressure. A large drawing angle means lower friction (reduced contact between the rubbing surfaces), decrease in wire temperature and a simultaneous increase in deformation resistance and unit pressure. Depending on which parameter is decisive, the applied particular drawing die and single reduction geometry may result in an increase or decrease in drawing die wear. Therefore, the wear rate of drawing dies after drawing 10,000 kg of wire was determined in the study, and the calculation results are presented in Figures 8 and 9.

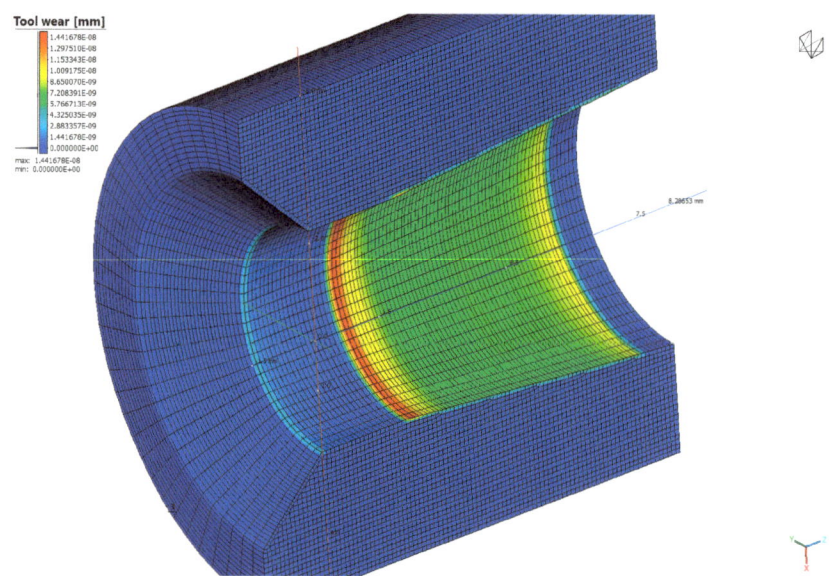

Figure 8. Example of drawing die wear in wire drawing process for drawing die angle 2α = 10°, single reduction Gp = 33%, drawing speed v = 1 m/s.

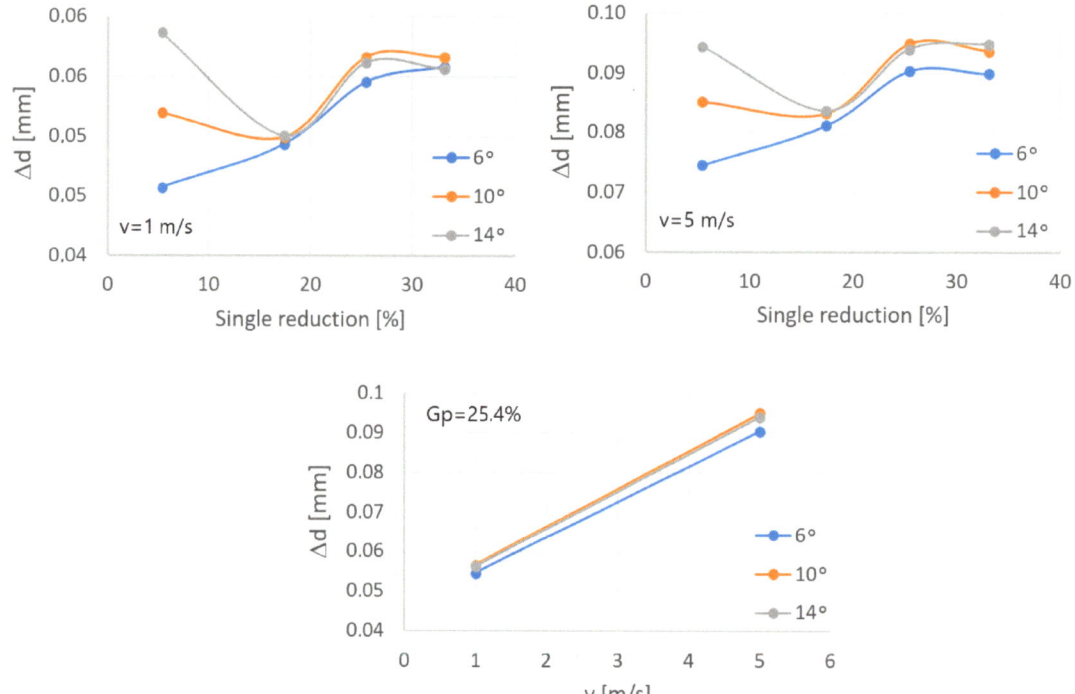

Figure 9. Change in the diameter of the calibrating part of the drawing die after the drawing process, where Δd—drawing die wear expressed in mm; 2α = 6°, 10°, 14°.

Based on the performed analysis of drawing die wear, the optimum drawing angle depends on the single reduction size. The application of large drawing die angles with unit deformations below 10% results in a significant increase in drawing die wear (the difference between the angle of 6 and 14° was over 30%). However, with larger single deformations, the differences in drawing die wear between the analysed variants decrease. According to the authors, this is related to the occurrence of opposing phenomena during drawing, i.e., friction, which depends on the duration of contact of the material and tool, the wire and drawing die temperatures and the pressure on the drawing die. Thus, the degree of drawing tool wear in the wire drawing process is the result of the above-mentioned factors. Industrial practice shows that when designing single reduction distributions to optimise drawing die wear, the applied principle is that the optimal angle of the working part of the drawing die depends on the degree of wire reinforcement and the single reduction size. Hence, in multi-stage drawing, as the wire goes through successive drawing dies, the single reduction value and the angle of the working part of the drawing die decrease.

3.2. Results of Tribological Tests

The performed tribological tests confirmed the significant influence of temperature on the friction conditions and wear of the carbide/steel friction pair. Based on the data in Table 3, it can be seen that while an initial increase in temperature to 200 °C does not result in a significant difference in friction coefficient values (a difference of approximately 3%), a considerable deterioration in lubrication conditions can be expected at temperatures above 400 °C. The differences in the values of the average friction coefficient between variants I and IV were over 30%.

Table 3. Mean friction coefficient with mass loss of the rubbing pair.

Material	Mean Friction Coefficient		Mass Loss/Gain, g	
	Mean Friction Coefficient	Standard Deviation	Samples	Counter-Samples
Variant I	0.5642	0.0282	0.0141	0.0027
Variant II	0.5466	0.1101	0.0115	0.0041
Variant III	0.6729	0.0541	+0.0051	0.0013
Variant IV	0.7572	0.0542	+0.0372	0.0020

The analysis of the change in the masses of the samples and counter-samples shown in Table 3 is a complex issue. In the performed tests, there was a gradual abrasion of the steel and carbide sample surfaces with the "products" of abrasion sticking to their surfaces. Unlike in the case of tribological tests performed at ambient temperature, when testing at higher temperatures, an additional factor influencing the condition of the surface layer is the phenomenon of scale formation on the surface of the test samples. Hence, among the analysed variants, the highest total change in the sample and counter-sample mass was recorded for variant IV (T = 600 °C). During the testing, pieces of scale are deposited on the rubbing surfaces. Consequently, depending on the resultant of the above-mentioned factors, the value of the friction coefficient is not constant and changes periodically, as shown in Figure 10. The obtained results confirm the hypothesis contained in Section 2 regarding the influence of temperature on the accelerated wear of drawing dies. An increase in the friction value in the process causes an increase in the deformation resistance at the wire/tool interface and accelerates the wear of the drawing dies. Therefore, in the actual drawing process, lubricants are used to reduce friction and to protect the surface of the drawing die from abrasion when the wire is pulled through it. An increase in drawing die temperature during drawing causes a deterioration in the lubricant's rheological properties, e.g., the dynamic viscosity coefficient, as well as a reduction in the lubricant film thickness in the contact zone. This leads to an even faster wear of the drawing die. Obviously, the tribological tests presented in this paper do not ideally reflect the frictional conditions of the drawing process, especially in industrial drawing (a shorter friction path and lower pressures in the contact zone were implemented in the laboratory tests). Nevertheless, they confirm the role of temperature in the formation of the carbide surface layer and its resistance to abrasion. Hence, as part of the study, metallographic tests of the sample and counter-samples were performed after tribological testing.

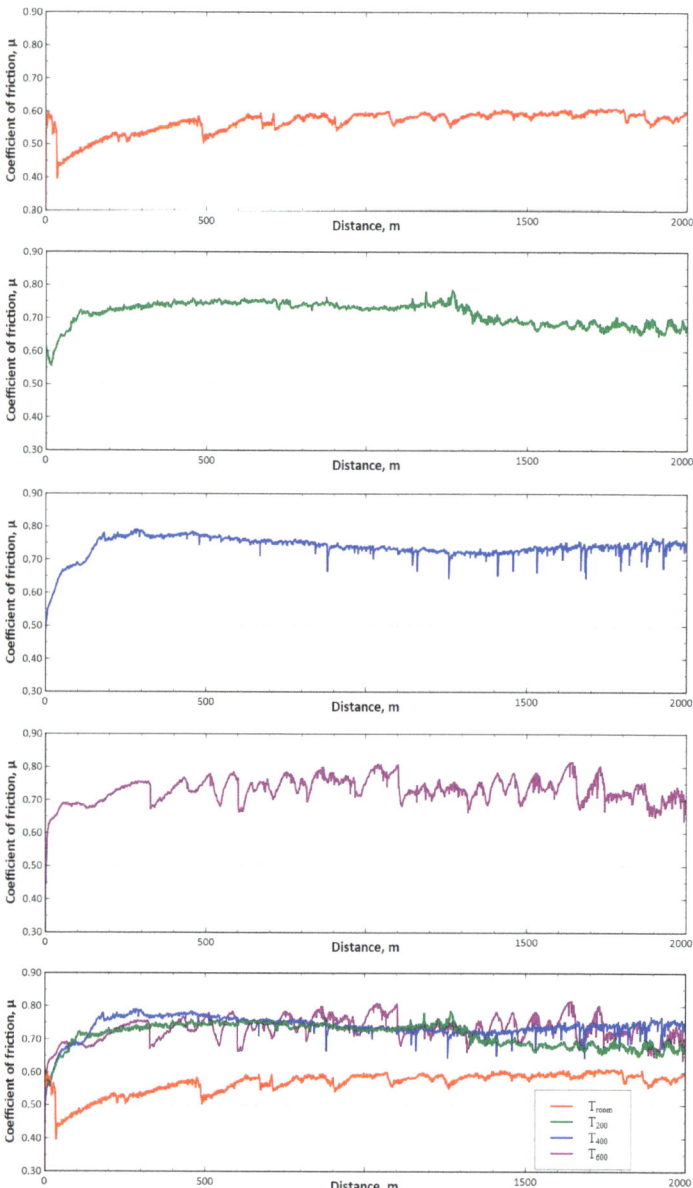

Figure 10. Dependence of the friction coefficient on the friction path (distance) for variants I–IV.

The analysis of the surface layer, after tribological testing, consisted of assessing changes in the structure and properties of C45 steel (sample) and WC carbide (counter-sample). In the first stage, the influence of temperature on the structural changes of the sample, outside the abrasion area, was determined.

The obtained microscopic test results, shown in Figure 11, confirm that regardless of the process temperature, the structure of the C45 steel does not change.

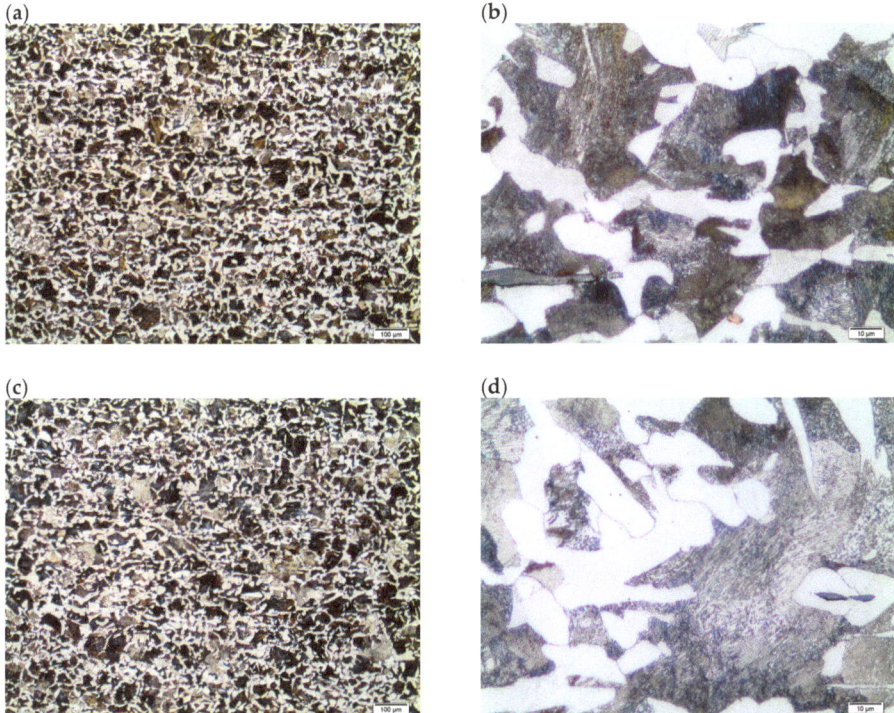

Figure 11. Structure of C45 steel, LM: (**a**) variant I, 100× magnification; (**b**) variant I, 1000× magnification; (**c**) variant IV, 100× magnification; (**d**) variant IV, 1000× magnification.

For variants I, II, III and IV, a ferritic–perlitic structure with different grain sizes was observed. On the majority of the surface, the perlite was surrounded by a lattice of ferrite (Figure 11a,d). At 1000× magnification, the perlite clearly revealed its lamellar structure, which contained both thicker and thinner lamellae (Figure 11b,d).

The analysis of the surface of the test samples showed that the temperature in the tribological tests contributed to the formation of oxides/deposits on the C45 steel. The appearance of the surface of C45 steel next to the abrasion area is shown in Figure 12. Based on the analysis of the EDS chemical composition, in the case of variant I, only the elements included in C45 steel were determined, i.e., carbon, manganese and iron. In the other cases, in addition to the above-mentioned components, oxygen was also present, which confirms the formation of iron oxides on the steel surface (variants II, III, IV). As part of the microscopic analysis, a significant difference in surface morphology was observed for variant IV compared to variants I, II and III. In this case, the structure has a coniferous form; see Figure 12e.

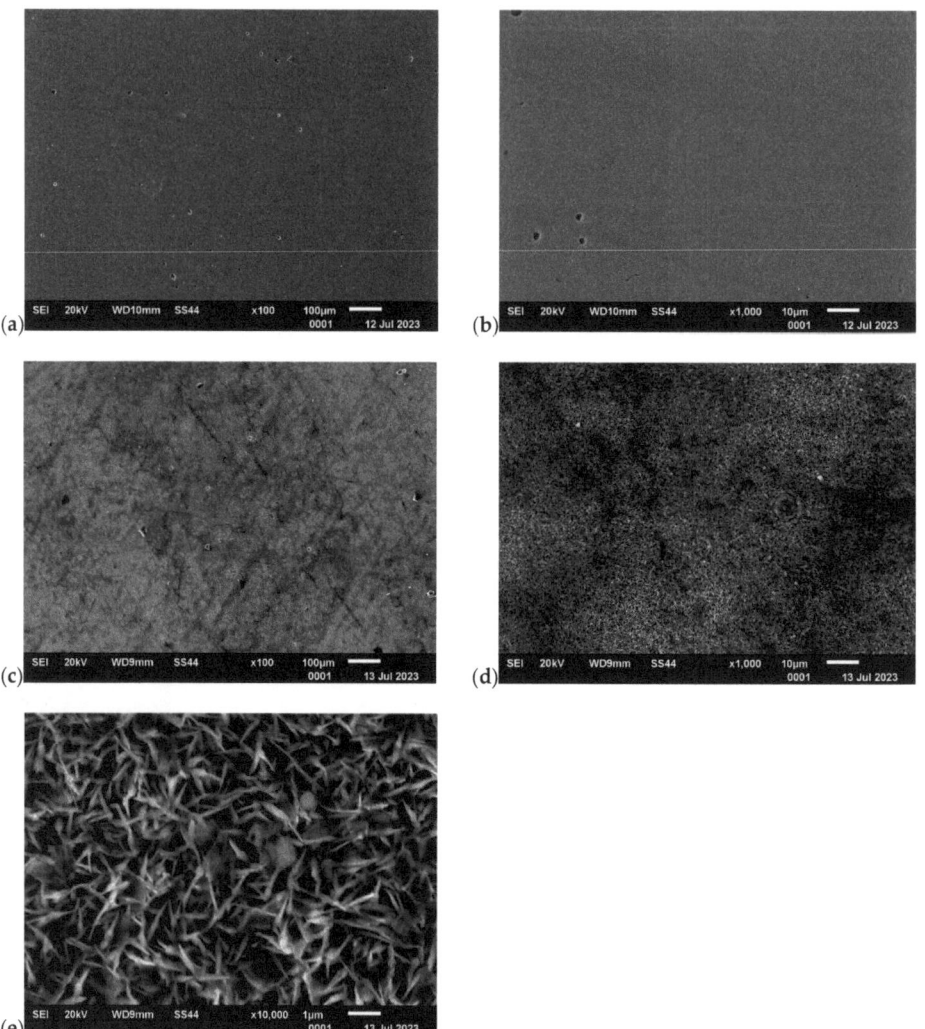

Figure 12. Surface next to the abrasion area, C45 steel, SEM: (**a**) variant I, 100× magnification; (**b**) variant I, 1000× magnification; (**c**) variant IV, 100× magnification; (**d**) variant IV, 1000×; (**e**) variant IV, 10,000× magnification.

The changes in the surface layer of the steel samples indicate that the temperature during tribological testing has a significant influence on the abrasion area. The results of the microscopic analysis and the chemical composition at the abrasion area for the different variants are presented in Figure 13, Figure 14, Figure 15, Figure 16, and Figure 17, respectively.

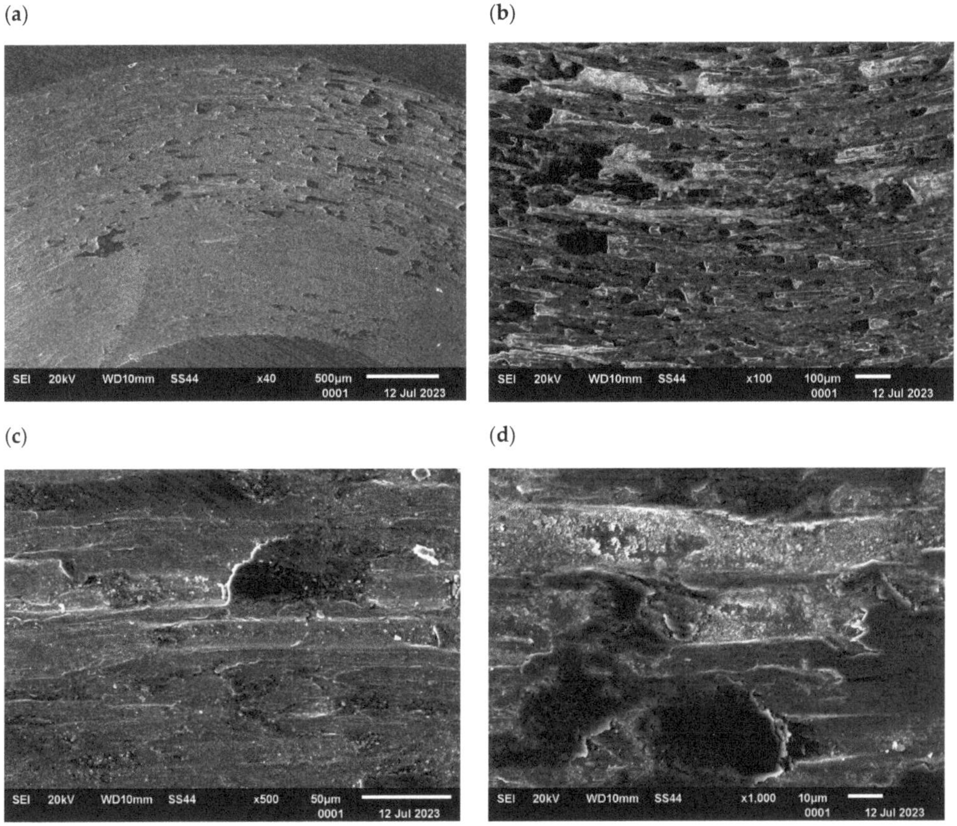

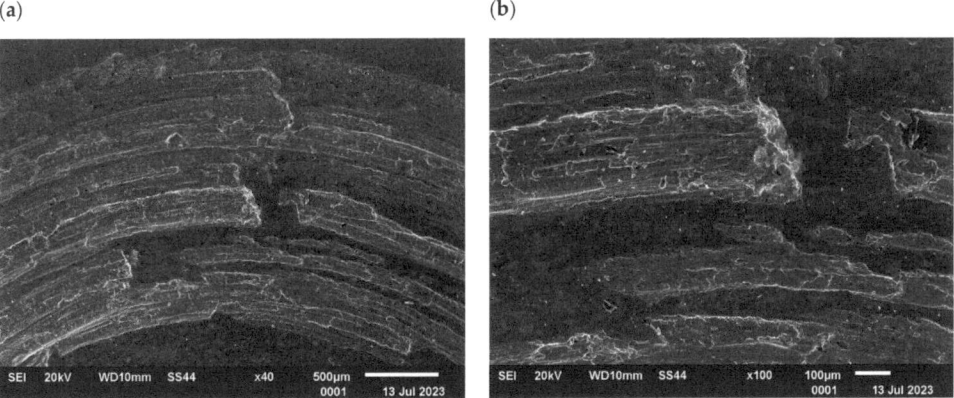

Figure 13. Variant I, C45 steel surface, SEM: (**a**) abrasion area, 40× magnification; (**b**) abrasion area, 100× magnification; (**c**) abrasion area, 500× magnification, (**d**) abrasion area, 1000× magnification.

Figure 14. *Cont.*

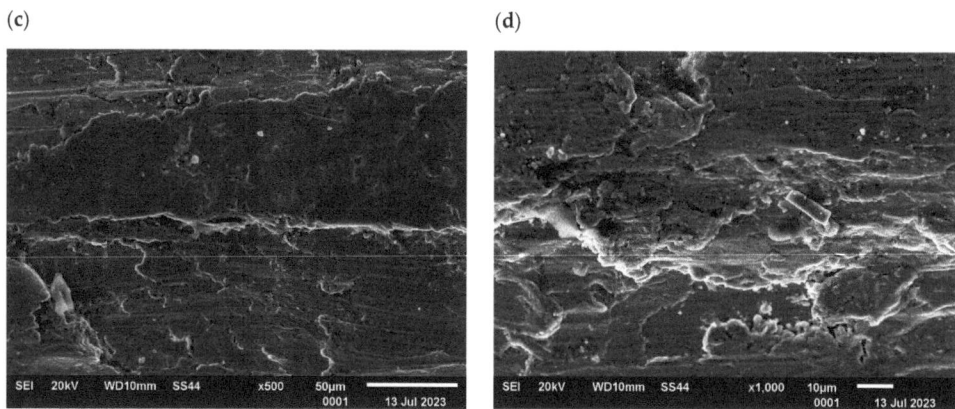

Figure 14. Variant II, C45 steel surface, SEM: (**a**) abrasion area, 40× magnification; (**b**) abrasion area, 100× magnification; (**c**) abrasion area, 500× magnification, (**d**) abrasion area, 1000× magnification.

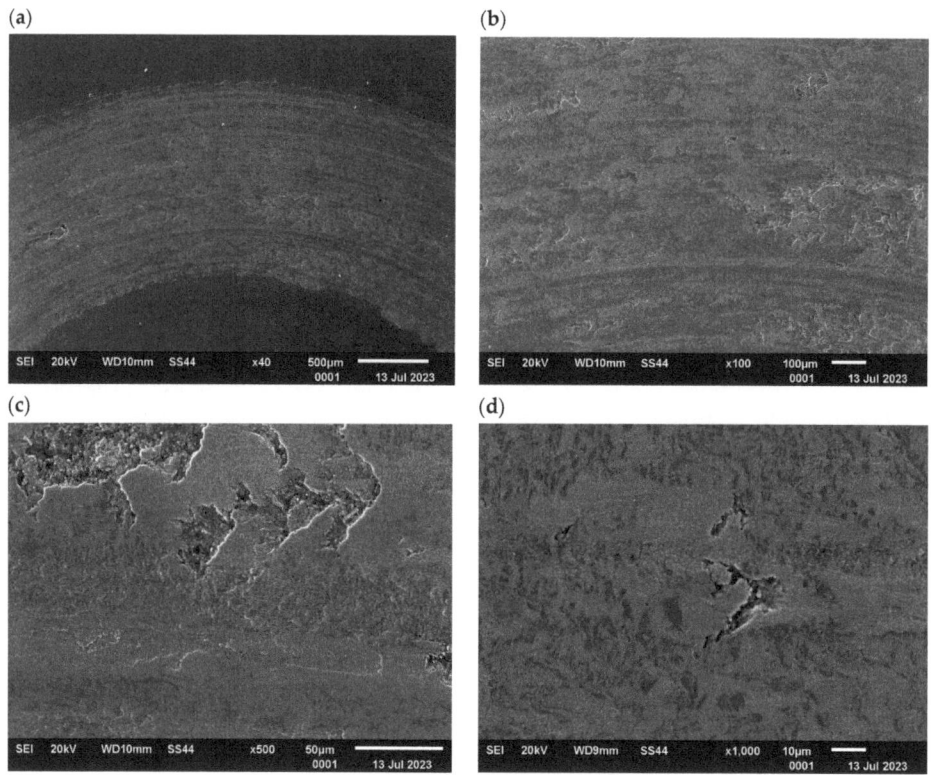

Figure 15. Variant III, C45 steel surface, SEM: (**a**) abrasion area, 40× magnification; (**b**) abrasion area, 100× magnification; (**c**) abrasion area, 500× magnification, (**d**) abrasion area, 1000× magnification.

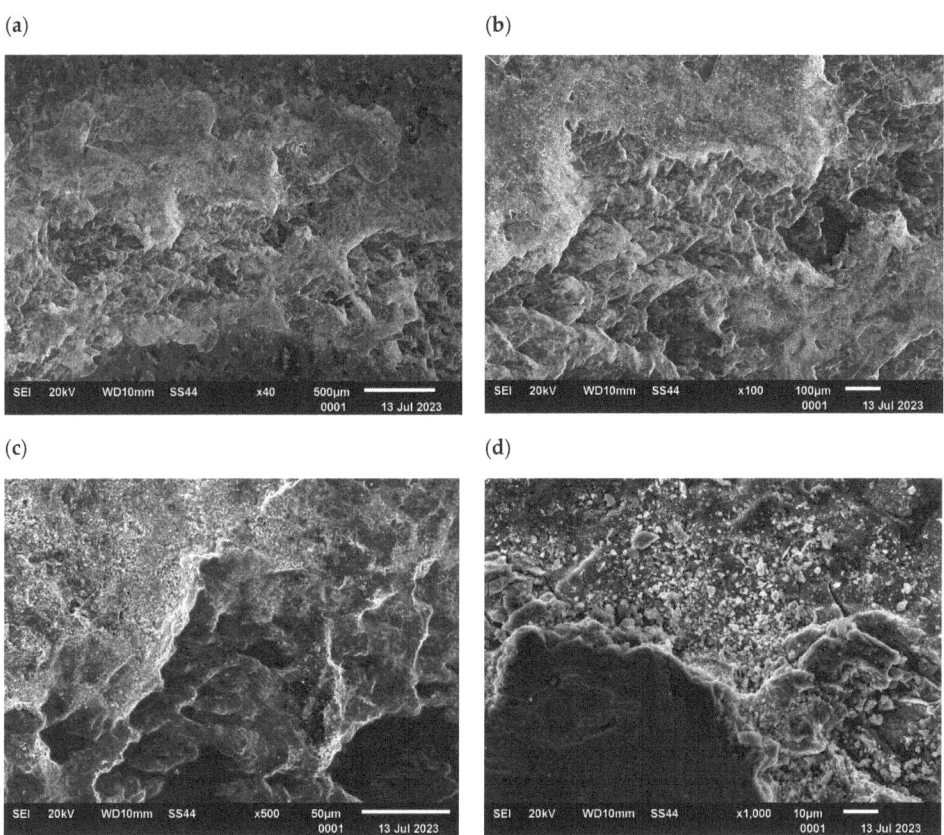

Figure 16. Variant IV, C45 steel surface, SEM: (**a**) abrasion area, 40× magnification; (**b**) abrasion area, 100× magnification; (**c**) abrasion area, 500× magnification, (**d**) abrasion area, 1000× magnification.

(**a**)

Element	Weight %	Atomic %
C K	11.13	32.34
O K	8.48	18.50
Mn K	0.45	0.28
Fe K	77.44	48.39
W M	2.50	0.48
Total	100.00	

(**b**)

Element	Weight %	Atomic %
C K	12.63	36.38
O K	6.44	13.92
Mn K	0.62	0.39
Fe K	79.22	49.10
W M	1.09	0.21
Total	100.00	

Figure 17. *Cont.*

(c)

Element	Weight %	Atomic %
C K	6.66	23.40
O K	16.50	43.50
Mn K	0.24	0.18
Fe K	26.18	19.77
Co K	3.26	2.33
W M	47.15	10.81
Total	100.00	

(d)

Element	Weight %	Atomic %
C K	3.48	12.36
O K	20.21	53.86
Fe K	26.65	20.35
Co K	3.88	2.81
W M	45.77	10.62
Total	100.00	

Figure 17. EDS analysis, C45 steel, abrasion area: (**a**) variant I, (**b**) variant II, (**c**) variant III, (**d**) variant IV.

The study showed significant differences in the abrasion area between the variants. In the case of variant I, there is more damage on the outer side of the abrasion area. Significant material loss is noticeable in that place. Variant II, in turn, is characterised by more uniform abrasion along the entire length of the abrasion radius. In the case of sample III, pitting occurs locally in the abrasion area, while variant IV is characterised by the formation of deposits on the surface, which is probably due to the higher process temperature. Analyses of the chemical composition in the abrasion area showed (Figure 17) that in the case of processes I and II, oxygen and tungsten were present in addition to the elements that made up C45 steel, such as C, Mn and Fe. In those variants, the amount of observed tungsten did not exceed 2.5% by weight. However, for variants III and IV, its proportion increased dramatically and exceeded 45% by weight. Moreover, the presence of cobalt—in the amount of 3% by weight—was recorded in those variants. The presence of tungsten and cobalt in the abrasion area was due to the fact that material from the counter-sample (ball) was deposited onto the steel ("rubbed into"). Further analysis showed significant differences in abrasion width and depth for the individual variants (Figures 18 and 19, Table 4). Microscopic observations enabled noticing irregularities in abrasion marks. This was most visible for variant I. The largest abrasion mark was also observed in this case (the measurement was taken at the widest point).

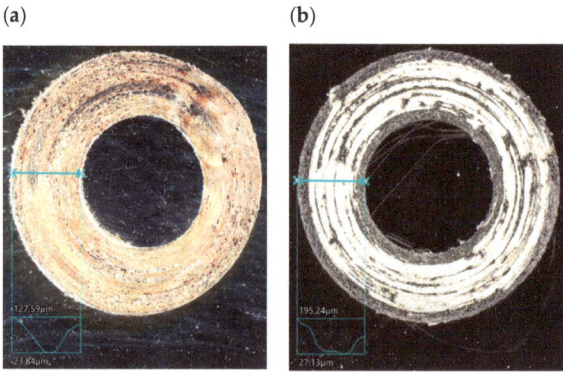

Figure 18. *Cont.*

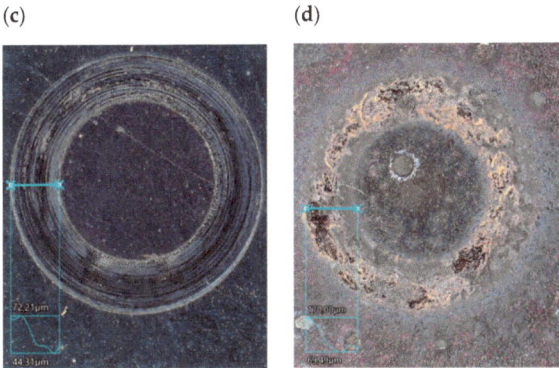

Figure 18. Image of C45 steel abrasion area, digital microscope: (**a**) variant I, (**b**) variant II, (**c**) variant III, (**d**) variant IV.

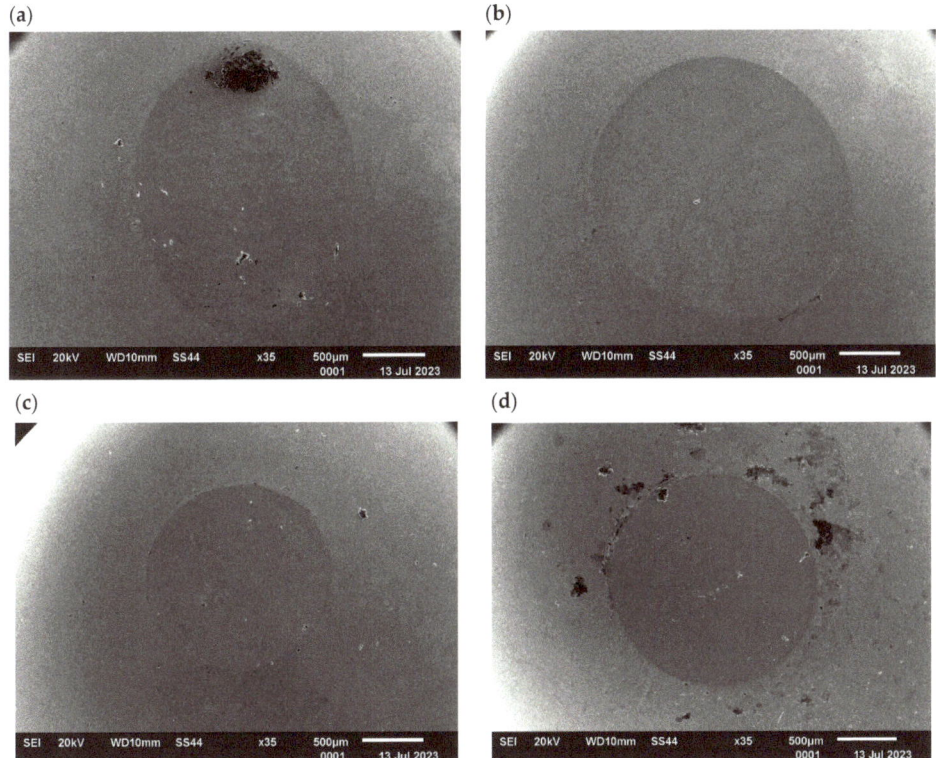

Figure 19. Image of ball abrasion area, SEM: (**a**) variant I, (**b**) variant II, (**c**) variant III, (**d**) variant IV.

Table 4. Results of C45 steel and ball interaction; W—abrasion width, D—abrasion depth.

Variant I			Variant II			Variant III			Variant IV		
Steel		Ball	Steel		Ball	Steel		Ball	Steel		Ball
W, µm	D, µm	D, µm	W, µm	D, µm	D, µm	W, µm	D, µm	D, µm	W, µm	D, µm	D, µm
2215	103	2249	2193	168	2197	1499	27	1498	1708	100	1716

For variant II, the abrasion area width was approximately 22 units smaller than for variant I, while the depth was over 64 units greater in this case. For variant III, the width and depth of the abrasion area dropped to approximately 1500 µm and 28 µm, respectively. For variant IV, in turn, the width and depth of the abrasion area increased (~1708 units—first parameter, ~100 units—second parameter).

The analysis of the surface of the counter-samples (balls) confirmed the presence of typical abrasion marks; however, no surface cracks were found. The performed tests showed that in the contact area of the sample/counter-sample during tribological testing, the phenomenon of sample material deposition on the surface of the counter-sample occurs; see Figure 20. This confirms the possibility of the steel "sticking" to the surface of the drawing dies during the drawing process.

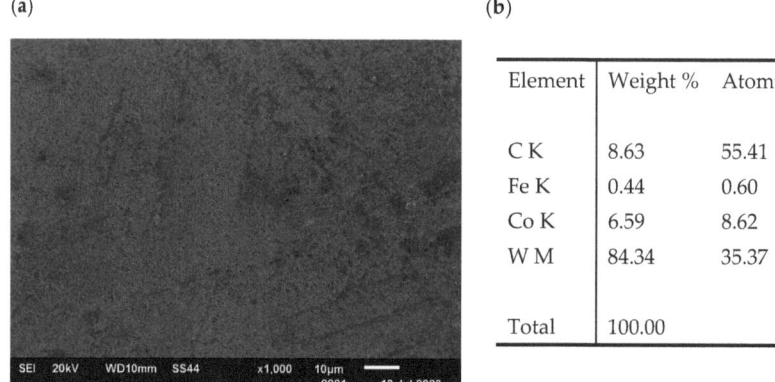

Figure 20. Image of the ball abrasion area (**a**) and chemical composition (**b**) for the variant III sample.

The parameter that influences the rate of tool wear is the surface topography. The following parameters were selected to be analysed: arithmetic mean surface roughness Sa, root mean square roughness Sq, ten-point height of irregularities surface Sz (mean absolute height of the five highest and five lowest vertex cavities), the maximum height of the elevation surface Sp (the distance from the highest point of the mean plane), and the maximum depth of the cavity surface Sv (the distance from the lowest point of the mean plane). Hence, the influence of temperature on the surface topography of steel samples was determined in this study. The measurement results are shown in Figure 21 and Table 5.

Table 5. Roughness measurement results.

Parameter	Variant I	Variant II	Variant III	Variant IV
Sa, µm	2.49	3.40	3.21	3.91
Sq, µm	13.03	3.99	3.89	4.81
Sz, µm	15.96	16.41	19.89	30.72
Sp, µm	5.82	6.20	8.66	18.85
Sv, µm	10.14	10.21	11.24	11.87

The performed tests proved the influence of the test temperature on the surface topography of the steel samples, indicating that the roughness profile deviations increase. There are discrepancies in the values of the Sa, Sq, Sz, and Sp, and based on the data shown in Figure 21, as the temperature increases, the height parameters and Sv parameters between variants I and IV were approximately 100%. The increase in surface roughness, especially for variant IV, is associated with the oxidation processes of the steel surface during tribological testing. Surface roughness has a strong influence on friction conditions. The final stage in the manufacture of drawing tools is polishing, which is performed to obtain a smooth surface on the drawing dies. As more wire is drawn through, the surface of the drawing dies becomes increasingly rough, which leads to increased friction and accelerated wear. This is confirmed by surface topography tests on steel samples, which prove that there is a correlation between surface roughness and friction coefficient (Figures 10 and 21).

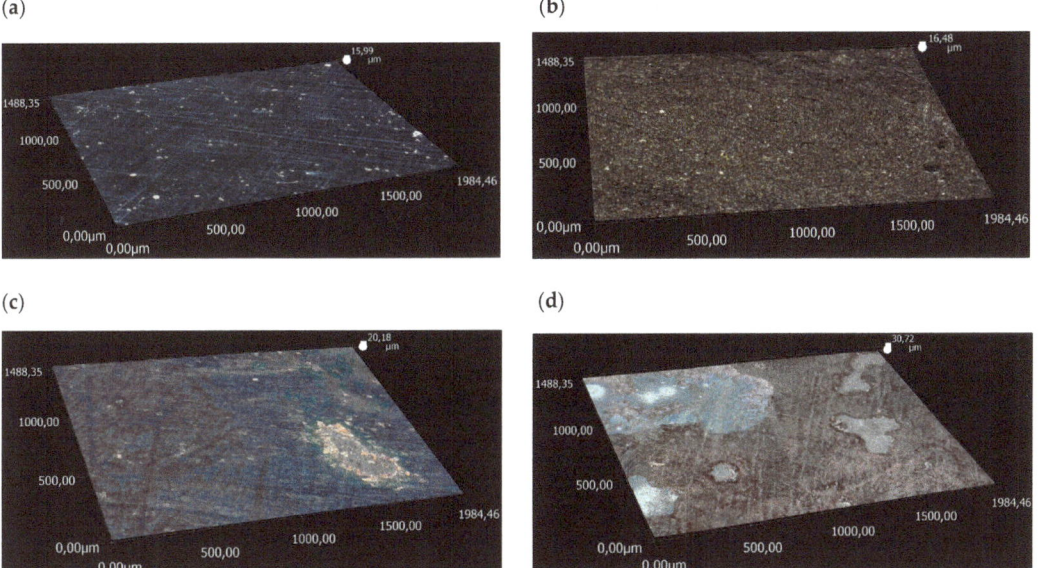

Figure 21. Three-dimensional (3D) image steel simples after tribological tests for (**a**) variant I, (**b**) variant II, (**c**) variant III, and (**d**) variant IV.

4. Conclusions

Numerical analysis has shown that after drawing approximately 10 m of wire, the temperature of the tungsten carbide insert stabilises, and its value depends on the drawing process parameters, i.e., single reduction, drawing die angle, and drawing speed.

An increase in drawing angle results in a decrease in wire temperature with a corresponding increase in unit pressure. A large drawing angle means lower friction (reduced contact between the rubbing surfaces), decrease in wire temperature, and a simultaneous increase in deformation resistance and unit pressure. Depending on which parameter is decisive, the applied particular drawing die and single reduction geometry may result in an increase or decrease in drawing die wear.

Based on the performed analysis of drawing die wear, the optimum drawing angle depends on the single reduction size. The application of large drawing die angles with unit deformations below 10% results in a significant increase in drawing die wear (the difference between the angle of 6 and 14° was over 30%).

The wear rate of drawing dies in wire drawing is the result of friction, which depends on the duration of contact of the material and tool, the wire and drawing die temperatures and the pressure on the drawing die.

It has been shown that during tribological testing, there was a gradual abrasion of the steel and carbide sample surfaces with the "products" of abrasion sticking to their surfaces. Analysis of the chemical composition at the steel abrasion area showed that the share of tungsten and cobalt increases with increasing temperature. At 400 °C, it is more than 45% for tungsten and 3% for cobalt. At higher temperatures, of approximately 600 °C, an increase of about 30% in the friction coefficient was observed.

The analysis of the surface of the counter-samples (tungsten carbide) confirmed that in the sample/counter-sample contact area during tribological testing, the phenomenon of sticking of the sample material (steel) to the surface of the counter-sample (tungsten carbide) occurs, which confirms the possibility of sticking of the steel to the surface of the drawing die during the drawing process.

The research conducted in this work has shown a significant effect of temperature on the friction conditions and abrasive wear of WC+Co sintered carbide. Hence, in the next works, the authors will focus on searching for new materials for drawing cores. The abrasion resistance of dies made of different materials will be compared.

Author Contributions: Conceptualization, M.S., P.S., M.G., J.K. and A.B; methodology, M.S., P.S., MG., J.K. and A.B.; software, M.S. and P.S.; validation, M.S., P.S., M.G. and A.B.; formal analysis, M.S., P.S. and J.K.; investigation, M.S., P.S., M.G. and A.B.; resources, M.S., P.S. and J.K.; data curation, M.S., P.S. and M.G.; writing—original draft preparation, M.S., P.S. and A.B.; writing—review and editing, M.S., M.G. and J.K.; visualization, M.S. and P.S.; supervision, P.S.; project administration, M.S. All authors have read and agreed to the published version of the manuscript.

Funding: This research received no external funding.

Institutional Review Board Statement: Not applicable.

Informed Consent Statement: Not applicable.

Data Availability Statement: The original contributions presented in the study are included in the article, further inquiries can be directed to the corresponding author.

Conflicts of Interest: The authors declare no conflicts of interest. The funders had no role in the design of the study; in the collection, analyses, or interpretation of data; in the writing of the manuscript; or in the decision to publish the results.

References

1. Enghag, P. *Steel Wire Technology*; Materialteknik HB: Orebro, Sweden, 2002.
2. Wright, R.N. *Wire Technology. Process Engineering and Metallurgy*; Butterworth-Heinemann: Oxford, UK, 2011.
3. Pirso, J.; Viljus, M.; Juhani, K.; Kuningas, M. Three-body abrasive wear of TiC-NiMo cermets. *Tribol. Int.* **2010**, *43*, 340–346. [CrossRef]
4. Klaasen, H.; Kübarsepp, J. Abrasive wear performance of carbide composites. *Wear* **2006**, *261*, 520–526. [CrossRef]
5. Bonny, K.; De Baets, P.; Vleugels, J.; Huang, S.; Van der Biest, O.; Lauwers, B. Impact of Cr3C2/VC addition on the dry sliding friction and wear response of WC-Co cemented carbides. *Wear* **2009**, *267*, 1642–1652. [CrossRef]
6. Hollinger, S.; Georges, J.M.; Mazuyer, D.; Lorentz, G.; Aguerre, O.; Du, N. High-pressure lubrication with lamellar structures in aqueous lubricant. *Tribol. Lett.* **2001**, *9*, 143–151. [CrossRef]
7. Suliga, M.; Wartacz, R.; Hawryluk, M. The Multi-Stage Drawing Process of Zinc-Coated Medium-Carbon Steel Wires in Conventional and Hydrodynamic Dies. *Materials* **2020**, *13*, 4871. [CrossRef] [PubMed]
8. Sas-Boca, I.M.; Tintelecan, M.; Pop, M.; Iluţiu-Varvara, D.A. Severe reduction in section of copper wire using the technology of wire cold cassette roller dies. *Procedia Manuf.* **2018**, *22*, 73–78. [CrossRef]
9. El Amine, K.; Larsson, J.; Pejryd, L. Experimental comparison of roller die and conventional wire drawing. *J. Mater. Process. Technol.* **2018**, *257*, 7–14. [CrossRef]
10. Gillström, P.; Jarl, M. Mechanical descaling of wire rod using reverse bending and brushing. *J. Mater. Process. Technol.* **2006**, *172*, 332–340. [CrossRef]
11. Smolarczyk, Z. Effect of descaling methods of wire rod on surface texture of drawn wires. *Wire J. Int.* **2001**, *34*, 76–81.
12. Oehr, K.H.; Roth, R.C.A. Cleaning steel wire without acid. *Wire J. Int.* **1996**, *29*, 100–103.
13. Lee, S.K.; Ko, D.C.; Kim, B.M. Pass schedule of wire drawing process to prevent delamination for high strength steel cord wire. *Mater. Des.* **2009**, *30*, 2919–2927. [CrossRef]
14. Lee, S.K.; Lee, S.B.; Kim, B.M. Process design of multi-stage wet wire drawing for improving the drawing speed for 0.72 wt% C steel wire. *J. Mater. Process. Technol.* **2010**, *210*, 776–783. [CrossRef]
15. Verma, S.; Sudhakar Rao, P. Design and analysis of process parameters on multistage wire drawing process—A review. *Int. J. Mech. Prod. Eng. Res. Dev.* **2018**, *9*, 403–412. [CrossRef]
16. Wright, R.N. Physical conditions in the lubricant layer. *Proc. Annu. Conv. Wire Assoc. Int.* **1996**, *30*, 145–149.
17. Saied, E.K.; Elzeiny, N.I.; Elmetwally, H.T.; Abd-Eltwab, A.A. An experimental study of lubricant effect on wire drawing process. *Int. J. Adv. Sci. Technol.* **2020**, *29*, 560–568.

18. Suliga, M. Analysis of the heating of steel wires during high speed multipass drawing process. *Arch. Metall. Mater.* **2014**, *59*, 1475–1480. [CrossRef]
19. Hollinger, S.; Depraetere, E.; Giroux, O. Wear mechanism of tungsten carbide dies during wet drawing of steel tyre cords. *Wear* **2003**, *255*, 1291–1299. [CrossRef]
20. Gillström, P.; Jarl, M. Wear of die after drawing of pickled or reverse bent wire rod. *Wear* **2007**, *262*, 858–867. [CrossRef]
21. Nowotyńska, I.; Kut, S. The influence of die shape and back tension force on its wear in the process of wire drawing. *Arch. Metall. Mater.* **2019**, *64*, 1131–1137. [CrossRef]
22. Stoyanov, P.; Romero, P.A.; Järvi, T.T.; Pastewka, L.; Scherge, M.; Stemmer, P.; Fischer, A.; Dienwiebel, M.; Moseler, M. Experimental and numerical atomistic investigation of the third body formation process in dry tungsten/tungsten-carbide tribo couples. *Tribol. Lett.* **2013**, *50*, 67–80. [CrossRef]
23. Stoyanov, P.; Stemmer, P.; Järvi, T.T.; Merz, R.; Romero, P.A.; Scherge, M.; Kopnarski, M.; Moseler, M.; Fischer, A.; Dienwiebel, M. Friction and wear mechanisms of tungsten-carbon systems: A comparison of dry and lubricated conditions. *ACS Appl. Mater. Interfaces* **2013**, *5*, 6123–6135. [CrossRef] [PubMed]
24. Magazzù, A.; Marcuello, C. Investigation of Soft Matter Nanomechanics by Atomic Force Microscopy and Optical Tweezers: A Comprehensive Review. *Nanomaterials* **2023**, *13*, 963. [CrossRef] [PubMed]
25. Kumar, R.; Singh, S.; Aggarwal, V.; Singh, S.; Pimenov, D.Y.; Giasin, K.; Nadolny, K. Hand and Abrasive Flow Polished Tungsten Carbide Die: Optimization of Surface Roughness, Polishing Time and Comparative Analysis in Wire Drawing. *Materials* **2022**, *15*, 1287. [CrossRef] [PubMed]
26. Dobrzański, L.A. *Podstawy Nauki o Materiałach*; Wydawnictwo Politechniki Śląskiej: Gliwice, Poland, 2012.

Disclaimer/Publisher's Note: The statements, opinions and data contained in all publications are solely those of the individual author(s) and contributor(s) and not of MDPI and/or the editor(s). MDPI and/or the editor(s) disclaim responsibility for any injury to people or property resulting from any ideas, methods, instructions or products referred to in the content.

Article

Structure and Mechanical Properties of AlMgSi(Cu) Extrudates Straightened with Dynamic Deformation

Dariusz Leśniak [1,*], Józef Zasadziński [1], Wojciech Libura [1], Beata Leszczyńska-Madej [1], Marek Bogusz [1], Tomasz Latos [1] and Bartłomiej Płonka [2]

[1] Faculty of Non-Ferrous Metals, AGH University of Krakow, 30-059 Kraków, Poland; zas@agh.edu.pl (J.Z.); libura@agh.edu.pl (W.L.); bleszcz@agh.edu.pl (B.L.-M.) bogusz@agh.edu.pl (M.B.); tomaszlatos@interia.pl (T.L.)
[2] Łukasiewicz Research Network—Institute of Non-Ferrous Metals, 32-050 Skawina, Poland; bartlomiej.plonka@imn.lukasiewicz.gov.pl
* Correspondence: dlesniak@agh.edu.pl; Tel.: +48-12-617-31-96

Abstract: Before artificial ageing, extruded aluminium profiles are subjected to stretching with a small cold deformation in the range of 0.5–2%. This deformation improves the geometrical stability of the extruded product and causes changes in the microstructure of the profile, which leads to the strain hardening of the material after artificial ageing. The work has resulted in the creation of the prototype of an original device, which is unique in the world, for the dynamic stretching of the extruded profiles after quenching. The semi-industrial unit is equipped with a hydraulic system for stretching and a pneumatic system for cold dynamic deformation. The aim of this research paper is to produce advantageous microstructural changes and increase the strength properties of the extruded material. The solution of the dynamic stretching of the profiles after extrusion is a great challenge and an innovation not yet practised. The paper presents the results of microstructural and mechanical investigations carried out on extruded AlMgSi(Cu) alloys quenched on the run-out table of the press, dynamically stretched under different conditions, and artificially aged for T5 temper. Different stretching conditions were applied: a static deformation of 0.5% at a speed of 0.02 m/s, and dynamic deformation of 0.25%, 0.5%, 1%, and 1.5% at speeds of 0.05 and 2 m/s. After the thermomechanical treatment of the profiles, microstructural observations were carried out using an optical microscope (OM) and a scanning electron microscope (SEM). A tensile test was also carried out on the specimens stretched under different conditions. In all the cases, the dynamically stretched profiles showed higher strength properties, especially those deformed at a higher speed of 2 m/s, where the increase in UTS was observed in the range of 7–18% compared to the classical (static) stretching. The microstructure of the dynamically stretched profiles is more homogeneous with a high proportion of fine dispersoids.

Keywords: AlMgSi(Cu) alloys; extrusion; straightening; dynamic deformation; microstructure; mechanical properties

Citation: Leśniak, D.; Zasadziński, J.; Libura, W.; Leszczyńska-Madej, B.; Bogusz, M.; Latos, T.; Płonka, B. Structure and Mechanical Properties of AlMgSi(Cu) Extrudates Straightened with Dynamic Deformation. *Materials* **2024**, *17*, 3983. https://doi.org/10.3390/ma17163983

Received: 29 May 2024
Revised: 31 July 2024
Accepted: 7 August 2024
Published: 10 August 2024

Copyright: © 2024 by the authors. Licensee MDPI, Basel, Switzerland. This article is an open access article distributed under the terms and conditions of the Creative Commons Attribution (CC BY) license (https://creativecommons.org/licenses/by/4.0/).

1. Introduction

After hot extrusion and quenching, aluminium alloy profiles are usually subjected to a stretching process on the run-out table of the press. The stretching process is carried out at room temperature when it is safe to handle the material.

In fact, stretching is part of the thermomechanical processing of the extruded profiles, which consists of three steps: quenching on the run-out table of the press, cold tension in the stretcher, and ageing. Cold forming, e.g., by stretching, plays an important role in this process. A small static tensile deformation in the range of 0.5–2% is applied to straighten the profile, which reduces the residual stress level within the product and causes additional changes within the material microstructure. These changes enhance the subsequent ageing process, and ultimately result in the strengthening of the profile.

In his work [1], Kazanowski investigated the effect of various process and material parameters on the stretching of a profile. He also evaluated the theory of material deformation at the different stages of the stretching process.

Many papers deal with the influence of stretching, also referred to as pre-stretching, on the behaviour of aluminium alloys during thermomechanical treatment. Furu et al. [2] investigated the effect of pre-stretching on the aluminium alloys AA6005, AA6060, and AA6082. Rectangular profiles were extruded and pre-stretched with the plastic strains of 1.0%, 5.0%, and 10.0% before artificial ageing. For all three alloys, the experimental results showed a difference in yield strength for different levels of pre-stretching. For AA6005 pre-stretched 10.0%, the yield strength was approximately twice that of the unstretched sample in the underaged condition.

The effect of pre-stretching on plasticity and fracture was investigated by Abi-Akl and Mohr for AA6451 aluminium sheets [3]. The selected specimens were pre-stretched at 2.0% and 5.0%, followed by artificial ageing at 180 °C for 20 min for all the specimens. Uniaxial tensile tests showed that a higher degree of pre-stretching resulted in a higher yield strength.

In [4], the effect of pre-stretching on the mechanical behaviour of extruded rectangular hollow profiles made from three aluminium alloys, namely AA6063, AA6061, and AA6110, was investigated. Either 0.5% or 4.0% pre-stretch was applied after extrusion and before artificial ageing to temper T6. The uniaxial tensile tests showed that the 4.0% pre-stretched alloys exhibited significantly better ductility than the 0.5% pre-stretched alloys. The improved ductility was also obtained in the crush tests of the profiles, where the 4.0% pre-stretched alloys showed fewer cracks than the 5.0% pre-stretched alloys.

Qvale et al. [5] investigated the effect of pre-stretching on the microstructure and mechanical behaviour of the extruded profiles of the aluminium alloys AA6063 and AA6082. The profiles were pre-stretched at 0.5% and 4.0% after extrusion and prior to artificial ageing to T6 temper. The uniaxial tensile tests showed a higher yield strength for AA6063 and a lower yield strength for AA6082 in the 4.0% pre-stretched condition compared to the 0.5% pre-stretched alloys.

Wang et al. [6] investigated a novel thermomechanical treatment that improved the strength of an AA6061 aluminium alloy sheet. The main steps in the process included under-ageing, cold rolling, and re-ageing, where they were able to achieve a yield strength of 542 MPa and a UTS of 560 MPa for the alloy. The increased strength was the result of the large amount of cold work applied (75% reduction in sheet thickness). However, this level of cold work is not achievable for extrusions that have been stretched prior to artificial ageing.

Kolar et al. [7] studied alloy 6060 subjected to various degrees of pre-stretching between 0 and 10% followed by artificial ageing at 190 °C for between 10 and 300 min. The results showed that the yield stress and UTS improved with the increasing levels of pre-stretching, with the greatest effect at the shorter ageing times.

Ma and Robson [8] investigated a strategy to achieve ultra-high strength for the 7075 alloy by combining work hardening and precipitation strengthening. To achieve this, they carried out experiments consisting of solution heat treatment, pre-ageing, deformation, and second ageing. The results showed that uniaxial deformation to a 10% strain had little effect on precipitates but introduced sufficient dislocations to cause significant work hardening resulting in strength greater than that obtained with a T6 temper.

The influence of plastic deformation prior to artificial ageing on the microstructure evolution and mechanical properties of a novel Al–Li–Cu–X alloy was investigated in [9,10]. The plastic deformation ranged from non-stretched to 8% stretch, with intermediate stretches of 2%, 4%, and 6%. The pre-age deformation improves the ageing kinetics, number density, and strength of fine precipitates by introducing heterogeneous matrix nucleation sites. Increasing the pre-strain level to 15% resulted in an increase in the T8 yield strength to ∼670 MPa with a reduction in ductility from ∼11 to 7.5% [10].

Zuo et al. [11] studied the effect of creep strain, mechanical properties, and microstructure of the 7055 alloy under different pre-stretch conditions. The results show that the range of pre-stretch from 1.6% to 3.3% is suitable for the creep-aged 7055-T6 alloy to obtain better mechanical properties.

Yang et al. [12] investigated the effects of pre-deformation on the creep strain, mechanical properties, and microstructures of the AA2219 aluminium alloy. The pre-deformation can prolong the duration of the primary creep stage and facilitate the creep strain. The effects of 0~9% stretching on the microstructure and mechanical properties were investigated in Al-Li-Cu-Mg alloys [13]. Stretching made T(AlCuLi) precipitates finer and more uniform. Stretching at 6% improved the yield strength of the aged alloy from 328~342 MPa to 466~488 MPa, but reduced the elongation from 9.7~10.4% to 5.7%.

In [14], the influence of pre-straining and pre-ageing on the precipitation behaviour and age-hardening response for Al-Mg-Si alloys was investigated. The authors found that the dislocations introduced by the pre-strain treatment enhanced the inhibition of natural ageing and hardening at room temperature. The dislocations provided heterogeneous nucleation for the precipitates that formed the β'' strengthening phase during the bake hardening treatment. The age-hardening response of the Al-Mg-Si alloys improved with the formation of the denser and larger β'' strengthening phase.

In [15], the effects of pre-stretching and various ageing treatments on the tensile properties and fracture toughness of the 7050 aluminium alloy were studied. The results showed that the peak-aged alloy had higher strength but poor fracture toughness.

Quan et al. [16] investigated the effect of pre-stretching after solution treatment on the hardness and microstructure of aged 2524 aluminium alloy. The results showed that compared with the unstretched samples, the hardness increased and the time to reach the peak hardness decreased with increasing pre-strain. The number density of S (Al_2CuMg) phases was increased and the length was shortened in the pre-stretched alloy.

Yu et al. [17] presented a new pre-deformation treatment method for the 2324 aluminium alloy. The treatment method combined cold rolling deformation and pre-stretching deformation which increased the crystal defects. A large number of dislocation networks and tangles significantly improved the strength of the alloy and its microstructure.

Similar results were obtained for the 2219 alloy by Liu et al. in [18]. They investigated the influence of 2% pre-deformation on the ageing process.

The influence of pre-stretching on the mechanical properties of 2219 Al alloy sheets was investigated in [19]. The introduction of pre-stretching resulted in an increase in yield strength. The peak yield strengths of 387.5 and 376.8 MPa are obtained when the specimens pre-stretched by 10% are aged at 150 and 170 °C, respectively, which are higher than those obtained for the unstretched specimens (319.2 MPa).

Taichman et al. [20] studied the effect of deformation prior to ageing on the precipitate microstructure and precipitate types in an undeformed and a 10 pct pre-deformed condition of the commercial AA6060 alloy. The tensile tests showed that the yield strength was higher with pre-deformation for different ageing times.

The above publications show that the stretching process followed by artificial ageing has a complex effect on the microstructure. Pre-stretching creates dislocations which become heterogeneous nucleation sites for precipitates during the subsequent ageing.

Łatkowski carried out comprehensive investigations on the thermomechanical treatment of aluminium alloys [21]. He applied plastic strains of 0%, 5%, 10, and 15% to alloy 6060 before artificial ageing. The hardness increased to about 77 HB as the pre-ageing deformation increased up to 15%. For the 2017 alloy, after 24 h of natural ageing and 8 h of artificial ageing, he obtained the yield strength and UTS of about 650 MPa and 670 MPa, respectively.

Research on the addition of various other elements, such as Ni, Co, Au, and Cd to 6xxx alloys is reported in the review paper [22] to summarise the influence of these elements on the evolution of microstructure and its correlation with mechanical properties. Cu alters the precipitation sequences in these alloys and produces a variety of precipitates, resulting in the improved strength properties of the alloys.

It is clear from the papers cited that static pre-deformation prior to ageing improves the strength properties of the alloys tested. On the other hand, some studies reported in the literature indicate a very promising possibility of using the dynamic deformation of the material prior to ageing. Such deformation of about 1–1.5% produces a dislocation microstructure which improves the strength properties of the alloys after ageing.

The yield strengths obtained from dynamic loading at a strain rate of 4000 s^{-1} are 17% and 19% higher than those obtained from quasi-static compression for the T6 and HT specimens of 6061 alloy, respectively [23]. The higher initial dislocation density in T6 specimens hinders the movement of newly formed dislocations and therefore increases the strength of the material.

The influence of a novel thermomechanical process route on the microstructural evolution and dynamic tensile deformation behaviour of two aluminium alloys, i.e., AA6082 and AA7075, was investigated by Sharifi et al. [24]. Dynamic tensile tests were carried out at strain rates of 40, 200, and 400 s^{-1}. As the strain rate increased, the strength and elongation to failure of the alloys investigated increased for all the conditions.

The work of Tong et al. [25], which is not directly applicable to the extrusion process, explains to some extent the influence of dynamic stretching on the strengthening effect in the aluminium alloys subjected to thermomechanical treatment. During dynamic stretching, the deformation time is short and there is not enough time for dislocations to undergo annihilation and rearrangement processes. The dynamic recovery process is strongly inhibited as the higher density of dislocations is maintained and the uniformity of the dislocation distributions is improved. As the strain rate increases, the speed of dislocation movement also increases, resulting in a significant increase in peak strength.

Based on the data on the influence of dynamic deformation on the microstructure and mechanical properties of aluminium alloys, it can be concluded that even a small deformation of 1–1.5%, as usually used in the stretching of an extruded profile, can produce a microstructural effect equivalent to about 10% of the static deformation.

In the present work, the authors have built a prototype device for the dynamic stretching of extruded profiles after quenching. The semi-industrial device is equipped with a hydraulic system for stretching and a pneumatic system for cold dynamic deformation. The aim of this project is to produce favourable microstructural changes and increase the strength properties of the extruded material.

The microstructure and mechanical properties of the profiles were studied on extruded AlMgSi(Cu) alloys, which were water-quenched on the run-out table of the press, and subjected to static and dynamic stretching before artificial ageing to T5 temper.

2. Materials and Methods
2.1. Characterisation of AlMgSi(Cu) Alloys

Billets with the chemical composition shown in Table 1 and a diameter of 100 mm were direct-chill (DC) cast under semi-industrial conditions. Three alloys within the AlMgSi(Cu) grade were investigated. In all the alloys studied, the low-melting microstructural components were dissolved during homogenisation soaking to a degree sufficient for practical use—no incipient melting peaks were observed on the differential scanning calorimetry (DSC) curves. As a result, a significant increase in the solidus temperature was achieved, and the values obtained ranged from 574.6 °C for alloy 3 to 596.1 °C for alloy 1 (Table 2).

Table 1. Chemical composition of the AlMgSi(Cu) alloys under study, mass per cent.

Alloy Denotation	Si	Fe	Cu	Mg	Cr	Zn	Ti	Zr
AlMgSi(Cu) alloy 1	1.04	0.05	0.61	0.68	0.25	0.01	0.02	0.15
AlMgSi(Cu) alloy 2	1.20	0.04	0.81	0.79	0.23	0.01	0.02	0.15
AlMgSi(Cu) alloy 3	1.21	0.06	1.22	0.80	0.41	0.01	0.02	0.15

Table 2. DSC test results of as-cast and homogenised AlMgSi(Cu) alloys.

Alloy	Solidus Temperature, °C	Incipient Melting Heat, J/g
AlMgSi(Cu) alloy 1	544.0	0.61
AlMgSi(Cu) alloy 2	542.8	1.54
AlMgSi(Cu) alloy 3	509.2	0.21
AlMgSi(Cu) alloy 1 (homogenised)	596.1	0.13
AlMgSi(Cu) alloy 2 (homogenised)	584.3	0.32
AlMgSi(Cu) alloy 3 (homogenised)	574.6	1.00

2.2. Device for Dynamic Straightening

In order to add the effect of dynamic deformation to the static stretching of the extruded aluminium profiles, a semi-industrial device was designed and equipped with a dedicated dynamic system. The system stretches the profile statically by means of a hydraulic cylinder and simultaneously applies the dynamic effect in the form of cyclic impacts of the special dynamic hammers. The maximum stretching force is assumed to be as high as 40–60 kN. The structure of the device is based on a solid assembly table and two heads for the profiles (left and right). The two heads equipped with jaws are movable with respect to the table, one of them being able to move by pulling force only with a constant force parameter mediated by an elastomer suppression system, while the other one moves with the increase in the pulling force and with the deformation of the profile. Figure 1 shows the profile view of the entire workstation; Figure 2 shows the view from the left and right sides (back and front). Figure 3 shows the right side of the set, where a system for maintaining the constant tension of the profile is located. The system is based on the application of a special elastic element between the two steel plates that close when the pulling force is applied. The two plates move on the four sliding columns positioned vertically with respect to the assembly table. On the leftmost plate, there is a jaw for clamping the profile, which is equipped with a hydraulic cylinder to generate the correct clamping force. The hardness of the elastic element must be calculated on the basis of the section and condition of the profile being tested in order to ensure that the system operates within its optimum force parameters.

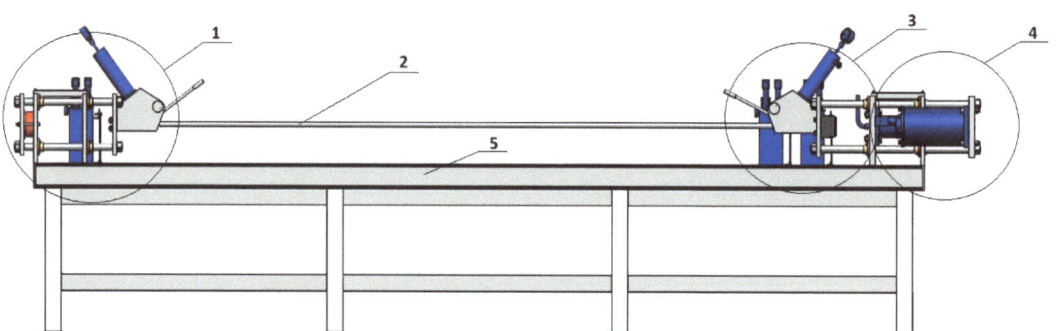

Figure 1. Device for the cold dynamic stretching of the extruded profiles from the AlMgSi(Cu) alloys, view from the face: 1—clamp jaw with elastomer; 2—aluminium profile; 3—clamp jaw on the right side; 4—dynamic force system; 5—pedestal.

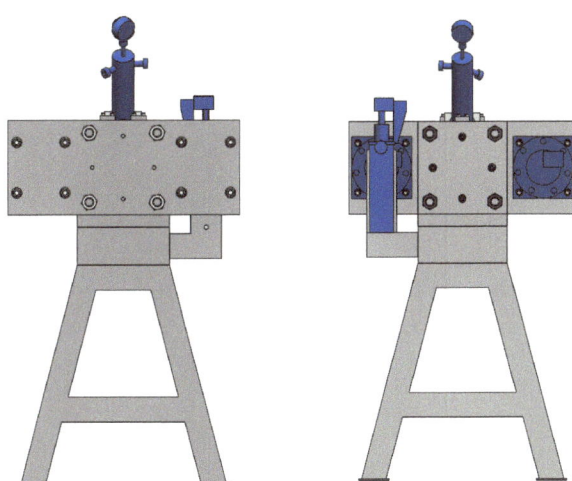

Figure 2. Device for the cold dynamic stretching of the extruded profiles from the AlMgSiCu) alloys, view from the back and front.

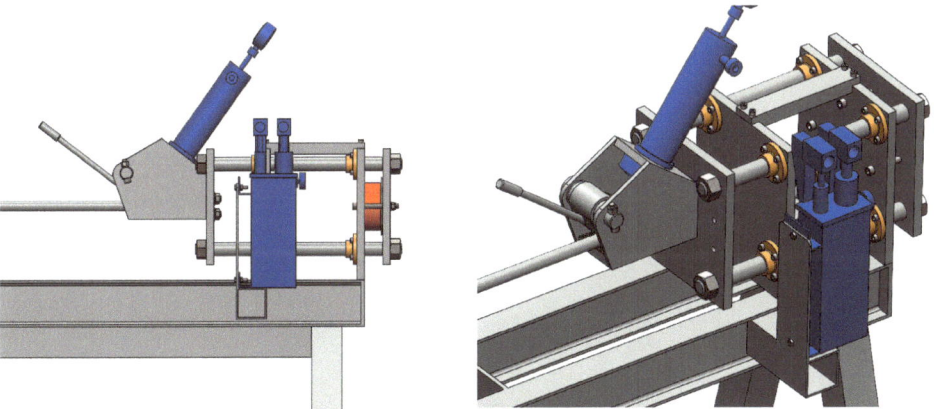

Figure 3. Device for the cold dynamic stretching of the extruded profiles from the AlMgSi(Cu) alloys, view from the right side of the system.

Figure 4 shows the left side of the dynamic system, where the hydraulic cylinder is mounted to generate the static force and the set of two synchronised pneumatic hammers to generate the additional dynamic forces. The two moving plates are also located on this side. These two plates can be moved on the four sliding columns that are perpendicular to the assembly table. The jaw is mounted on the lower plate to hold the profile, similar to the one on the right side of the system. The dynamic set is mounted on the larger plate. In addition, the larger plate is pushed by the 50 T hydraulic cylinder on the vertical axis to generate the stretching force. The number of cycles and the duration of the impact are programmable and must be adapted to the given profile on the basis of the experimental data and its initial characteristics.

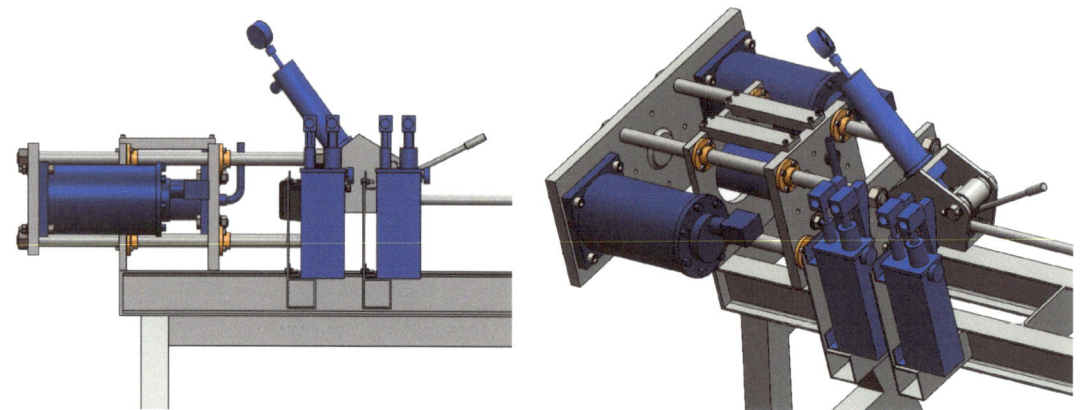

Figure 4. Device for the cold dynamic stretching of the extruded profiles from the AlMgSi(Cu) alloys, view from the left side.

2.3. Extrusion Trials and Straightening Process

The extrusion trials for the 60 × 40 × 2 mm (alloy 1 and alloy 2) and 50 × 30 × 3 mm (alloy 3) from the AlMgSi(Cu) alloys using the porthole dies were carried out on a 5 NM hydraulic direct press equipped with a 4″ diameter container and water wave installation (Figure 5a). The following process parameters were recorded during the trials: a metal exit speed of approximately 10 m/min, extrusion force of approximately 4.5 MN, and profile temperature of approximately 540 °C using the data acquisition system. The surface quality was checked online for cracks or tarnishing. The maximum extrusion speed applied causes cracks to appear on the surface of the extrudates. The samples taken from the profiles extruded under different process conditions were subjected to optical scanning to check their dimensional accuracy. Similar optical scanning was carried out on the used dies and their dimensions were compared with those of the new dies.

Figure 5. The 5 MN 4-inch extrusion press run-out table with water wave installation (**a**) and the profile from AlMgSi(Cu) alloy after extrusion with cooling by water on the run-out table (**b**).

The dynamic stretching tests were carried out with the prototype device on the water-quenched profiles of the following dimensions: 60 × 40 × 2 mm (alloy 1A and 2A) and 50 × 30 × 3 mm (alloy 3A). The dynamic deformation during the test was increased to the level of 0.25%, 0.5%, 1%, 1.5%, and 2% at stretching speeds of 0.05 and 2 m/s. For comparison, static stretching was also performed for a deformation of 0.5% and at a stretching speed of 0.02 m/s. Figure 6 shows the photographs of the test using the prototype device. After the static and dynamic stretching, the profiles were subjected to artificial ageing at a temperature of T = 175 °C and for an ageing time of 8 h.

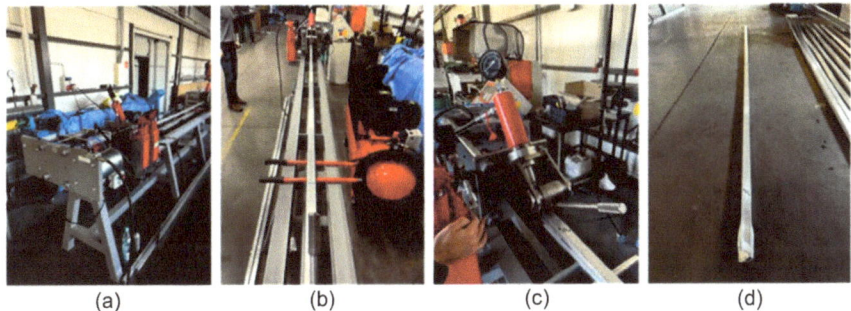

Figure 6. The prototype device for the dynamic stretching of the extruded profiles from the AlMgSi(Cu) alloys—the photos from the experimental tests: (**a**) side view, (**b**) front view, (**c**) the close-up view of the clamp jaw and dynamic force system and (**d**) the sample of the extruded profile after the dynamic straightening.

The aluminium profiles used in the tests were 3600 mm long. The profiles are mounted on both sides of the system in the hydraulic jaws of 25 T capacity. After clamping, an initial tension of up to 50% of the force required for the plastic deformation of the profile is applied. The profile is then stretched by means of a hydraulic cylinder with a maximum force of 50 T to achieve plastic deformation. At the same time, the pneumatic hammers synchronously strike the main plate on the left side to generate an additional deformation by the dynamic shock wave. The number of impacts and repeatable cycles depends on the test assumptions and the final result for the given profile. During the impact of the pneumatic hammers, the elongation of the profile occurs, and the constant force maintenance system keeps the device under tension all the time. When the correct deformation is obtained, as read on the straightedge, the profile is removed from the jaws. The device is now ready for the next test.

All the conditions of the extrusion process of the profiles, extrudate stretching, and ageing were shown in Table 3.

Table 3. The parameters of the extrusion process, extrudate stretching, and ageing for the AlMgSi(Cu) alloys of different chemical composition.

Alloy	AlMgSi(Cu) alloy 1, 2 and 3
Billet dimensions	Ø100 × 300 mm
Billet temperature	490 °C
Container/Die temperature	500 °C
Metal exit speed	10 m/min
Extrudates temperature	540 °C
Extrudates length	3600 mm
Stretching strain static	0.5%
Stretching strain dynamic	0.25%, 0.5%, 1%, 1.5% and 2%
Stretching speed	0.05 and 2 m/s
Ageing conditions	175 °C/8 h

2.4. Methodology of the Microstructural and Mechanical Examination

The samples for microscopic examination were mounted in resin, mechanically ground with sandpaper of appropriate grit, and then mechanically polished in two stages using a diamond paste suspension and a colloidal silica oxide suspension for the final polishing. In order to reveal the microstructure of the samples for observation by light microscopy, the samples were anodised in a Barker reagent with a composition of 100 mL H_2O + 2 mL HBF_4. The microstructure of the samples was examined by light microscopy (OLYMPUS GX51 microscope, Tokyo, Japan) and scanning electron microscopy (Hitachi SU 70 microscope, Tokyo, Japan). In addition, the chemical composition within the micro-areas was analysed by energy dispersive X-ray spectroscopy (EDS) (Thermo Fisher Scientific, Waltham, MA, USA). An analysis of the chemical composition within the grains was carried out to determine the content of the individual alloying elements. A minimum of 20 spot analyses were performed in each case. The test was performed at an acceleration voltage of 15 kV. For the selected variants ions, the microstructure of the thin films was examined using a scanning electron microscope equipped with a thin film observation attachment.

The basic mechanical properties: yield strength (YS), ultimate tensile strength (UTS), and percentage elongation (A, %) were determined using an INSPECT 100 tensile testing machine (with a maximum tensile strength of 100 kN). All the tests were performed at least three times.

2.5. Optical Scanning of Extruded Tubes and Dies

A GOM Atos Core 200 scanner (Figure 7) was used to scan the extruded elements on the inner and outer surfaces of the samples to measure the wall thickness deviations. The surface was cleaned of impurities prior to the scanning. The coloured map of the deviations was also obtained from the CAD model and the scanned element.

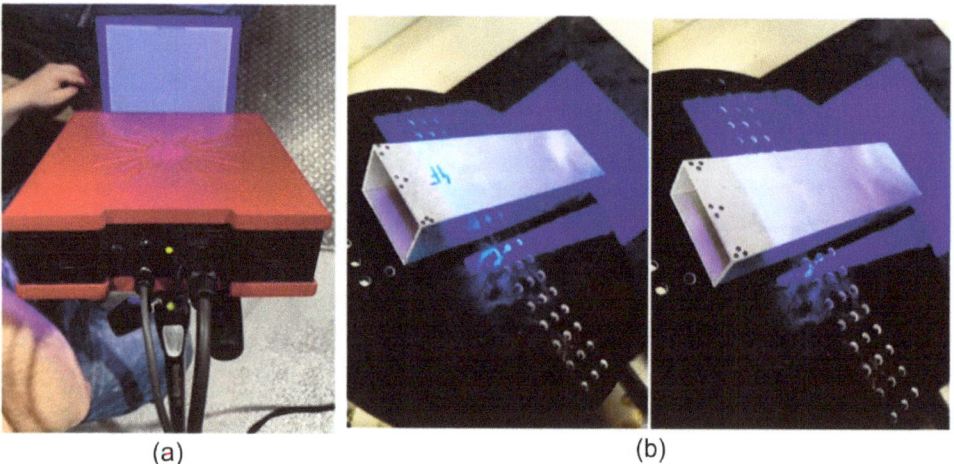

(a)   (b)

Figure 7. The scanner GOM Atos Core 200 for optically measuring the geometry of the extruded profiles (**a**) and the scanned extruded profiles (**b**).

3. Results

3.1. Microstructural Examination

The thermomechanical treatment consisted of straightening the profiles after extrusion, combined with supersaturation during the press run and prior to artificial ageing. The aim of this treatment was to improve the mechanical properties of the profiles. Following the thermomechanical treatment carried out under different conditions (deformation, speed), the ageing process took place at T = 175 °C for t = 8 h. Subsequently, microstructural observations and uniaxial tensile tests were carried out. The representative results of the microstructure tests are shown in Figures 8–16 below. These results were obtained after water-cooled extrusion on the press run and the subsequent static or dynamic straightening using the original device designed by the authors. The data are presented as a function of stretch speed and straightening speed.

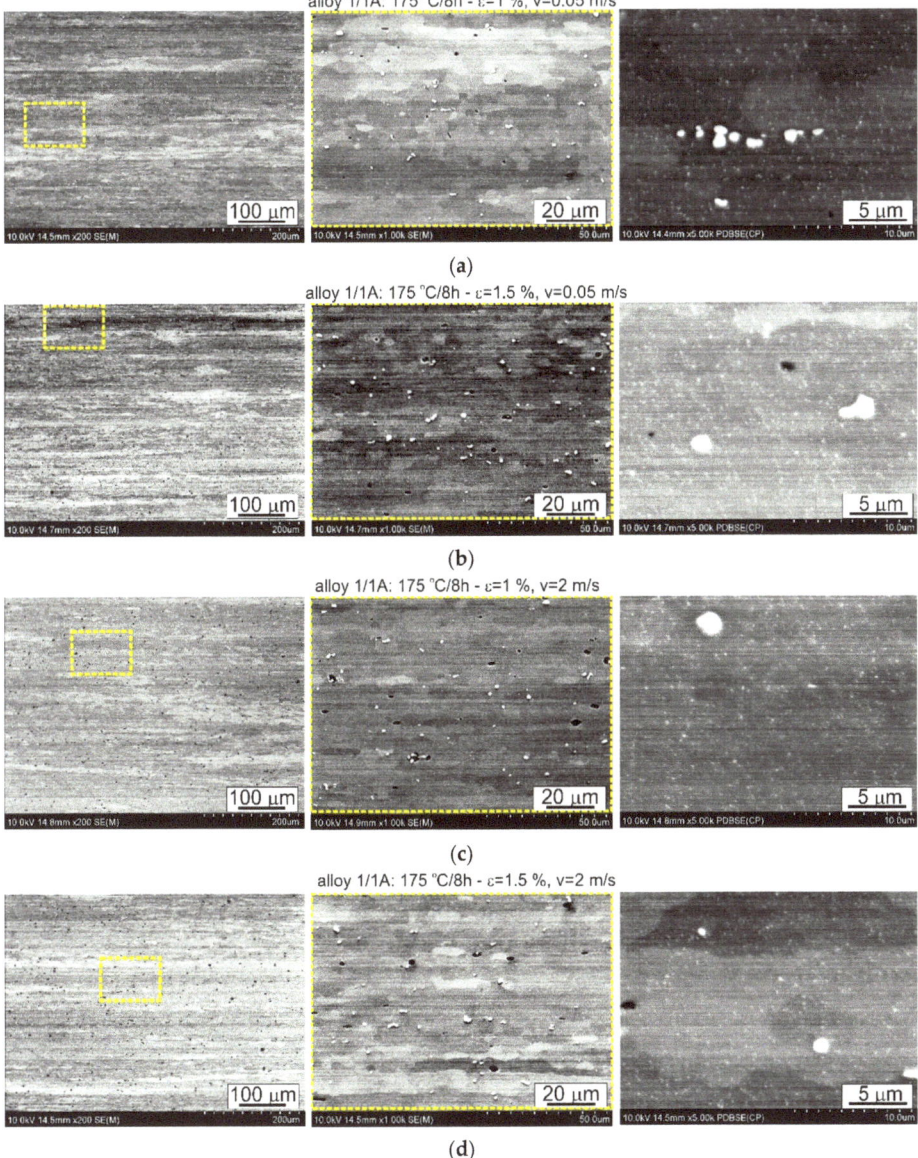

Figure 8. Microstructure of profiles extruded from alloy 1/1A and dynamically straightened at (**a**) $\varepsilon = 1\%$, $v = 0.05$ m/s; (**b**) $\varepsilon = 1.5$, $v = 0.05$ m/s; (**c**) $\varepsilon = 1\%$, $v = 2$ m/s; (**d**) $\varepsilon = 1.5\%$, $v = 2$ m/s; SEM (The middle pictures are enlargements of the area marked by the yellow box in the pictures on the left).

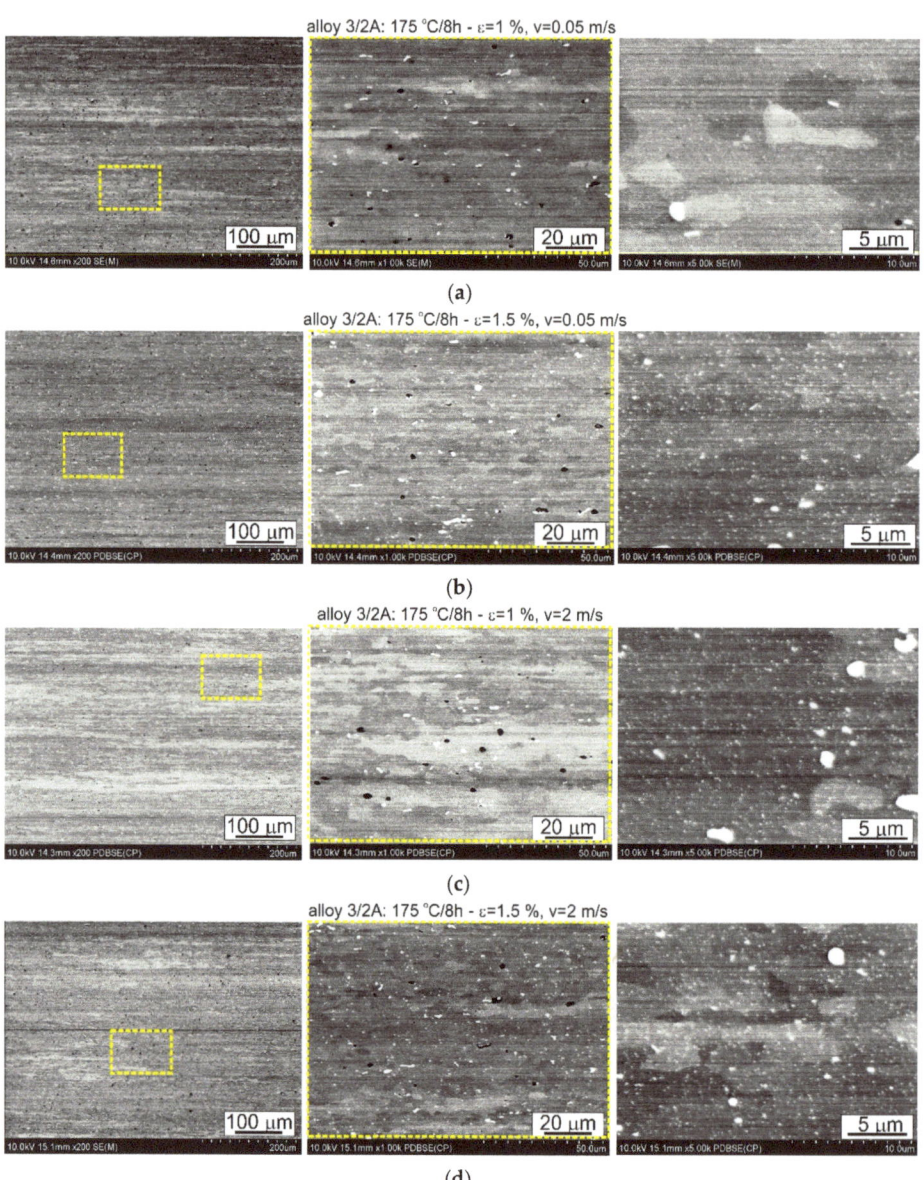

Figure 9. Microstructure of profiles extruded from alloy 3/2A and dynamically straightened at (**a**) $\varepsilon = 1\%$, $v = 0.05$ m/s; (**b**) $\varepsilon = 1.5\%$, $v = 0.05$ m/s; (**c**) $\varepsilon = 1\%$, $v = 2$ m/s; (**d**) $\varepsilon = 1.5$, $v = 2$ m/s; SEM (The middle pictures are enlargements of the area marked by the yellow box in the pictures on the left).

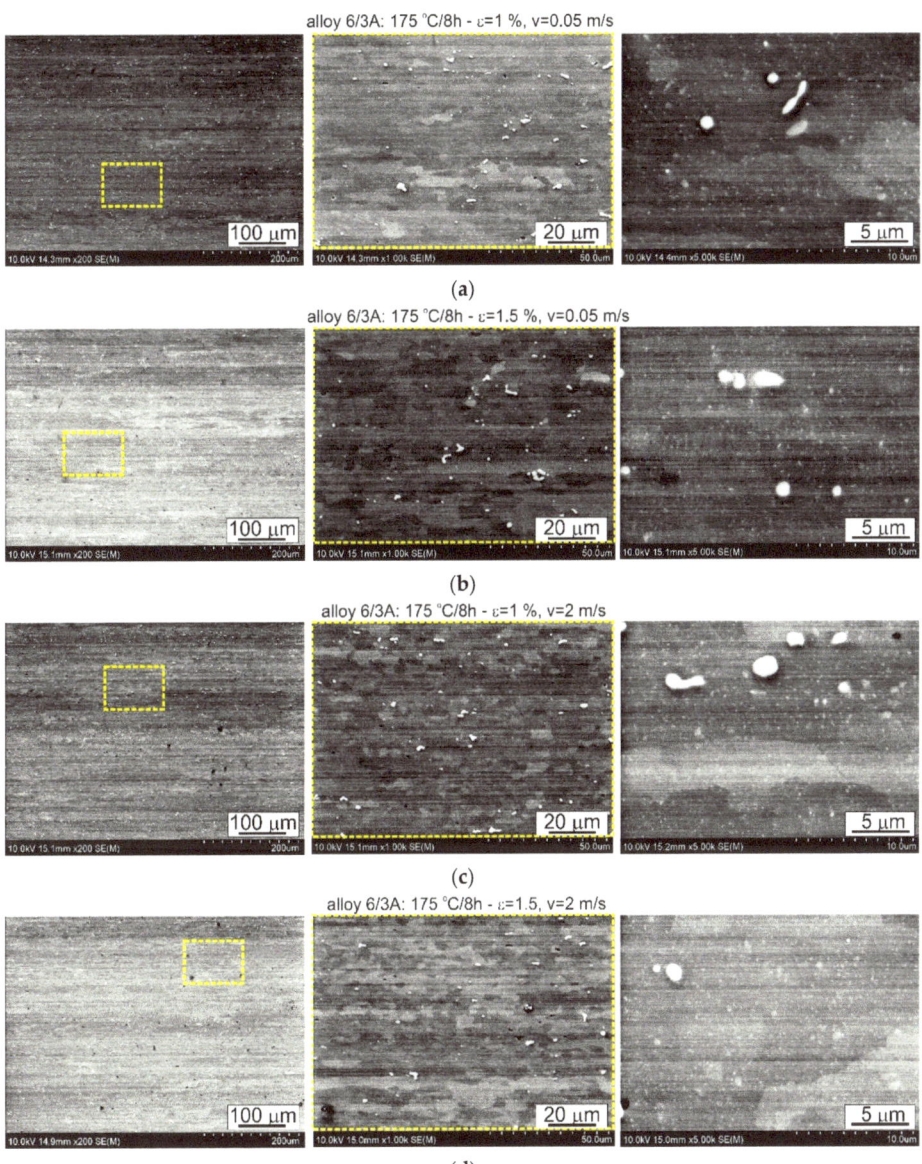

Figure 10. Microstructure of profiles extruded from alloy 6/3a and dynamically straightened at (**a**) $\varepsilon = 1\%$, $v = 0.05$ m/s; (**b**) $\varepsilon = 1.5\%$, $v = 0.05$ m/s; (**c**) $\varepsilon = 1\%$, $v = 2$ m/s; (**d**) $\varepsilon = 1.5\%$, $v = 2$ m/s; SEM (The middle pictures are enlargements of the area marked by the yellow box in the pictures on the left).

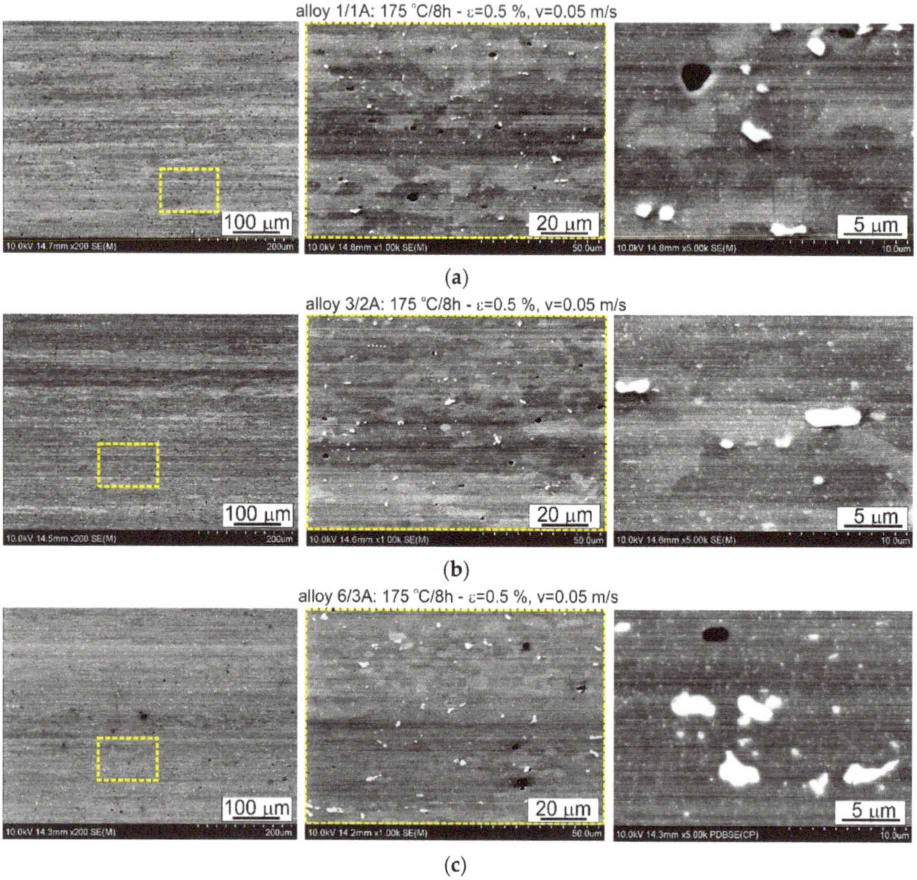

Figure 11. Microstructure of profiles statically straightened at ε = 0.5%, v = 0.05 m/s: (**a**) alloy 1/1A, (**b**) alloy 3/2A, and (**c**) alloy 6/3A; SEM (The middle pictures are enlargements of the area marked by the yellow box in the pictures on the left).

alloy 3/2A: 175 °C/8h - ε=1 %, v=2 m/s

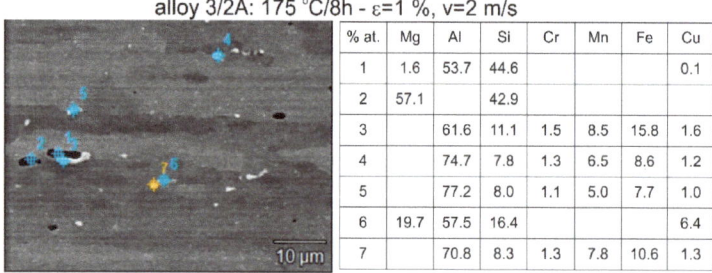

% at.	Mg	Al	Si	Cr	Mn	Fe	Cu
1	1.6	53.7	44.6				0.1
2	57.1		42.9				
3		61.6	11.1	1.5	8.5	15.8	1.6
4		74.7	7.8	1.3	6.5	8.6	1.2
5		77.2	8.0	1.1	5.0	7.7	1.0
6	19.7	57.5	16.4				6.4
7		70.8	8.3	1.3	7.8	10.6	1.3

Figure 12. Illustrative outcomes of the chemical analysis depicting the particles within the microstructure of the alloy 3/2A-extruded profiles supersaturated during the press run. Noteworthy particles identified include β-Mg_2Si, Q-$Al_5Cu_2Mg_8Si_6$, and a phase comprising Al, Si, Fe, and Mn.

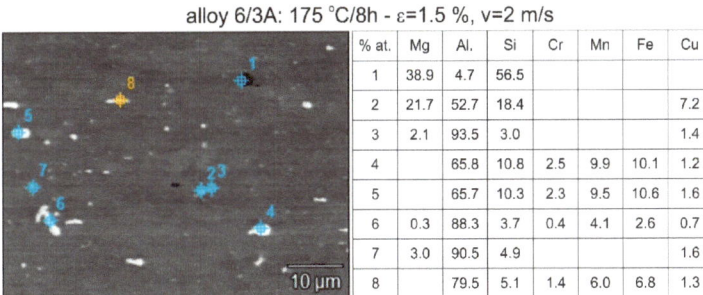

alloy 6/3A: 175 °C/8h - ε=1.5 %, v=2 m/s

% at.	Mg	Al	Si	Cr	Mn	Fe	Cu
1	38.9	4.7	56.5				
2	21.7	52.7	18.4				7.2
3	2.1	93.5	3.0				1.4
4		65.8	10.8	2.5	9.9	10.1	1.2
5		65.7	10.3	2.3	9.5	10.6	1.6
6	0.3	88.3	3.7	0.4	4.1	2.6	0.7
7	3.0	90.5	4.9				1.6
8		79.5	5.1	1.4	6.0	6.8	1.3

Figure 13. Illustrative outcomes of the chemical analysis depicting the particles within the microstructure of the alloy 6/3A-extruded profiles supersaturated during the press run. Noteworthy particles identified include β-Mg_2Si, Q-$Al_5Cu_2Mg_8Si_6$, and a phase comprising Al, Si, Fe, and Mn.

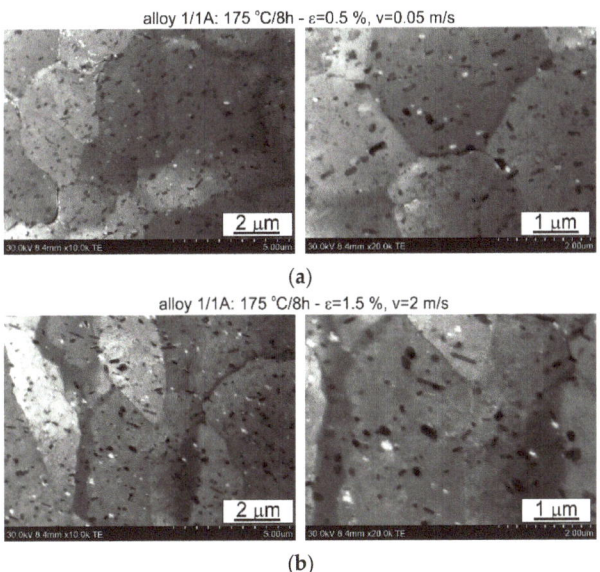

Figure 14. Microstructure of profiles extruded from alloy 1/1A: (a) statically straightened at ε = 0.5, v = 0.05 m/s; (b) dynamically straightened at ε = 1.5, v = 2 m/s; STEM.

Differences in the microstructure of the profiles which were tested are revealed by the results of the post-strain microstructure analysis. The microstructure of these profiles typically shows elongated grains which flatten with wall thickness. The character of the grain boundaries varies, taking on a more bulged appearance or remaining flat depending on the specific deformation process variant and alloy chemical composition. Detailed observations at high magnification reveal the presence of near-axial fine grains within the elongated grains, forming bands. The number, shape, and size of these fine grains depend on the parameters of the deformation process, including the deformation size, speed, and whether static or dynamic strain is applied (see Figures 8–11). The microstructure of the investigated profiles also contains numerous precipitates. The chemical composition studies using SEM/EDS microarrays confirm the presence of β-Mg_2Si phase particles up to several μm in size and Q-$Al_5Cu_2Mg_8Si_6$ phase particles (see Figures 12 and 13). In addition, advantageous dispersoids are present which improve the mechanical properties of the extruded profiles (see Figures 8–11).

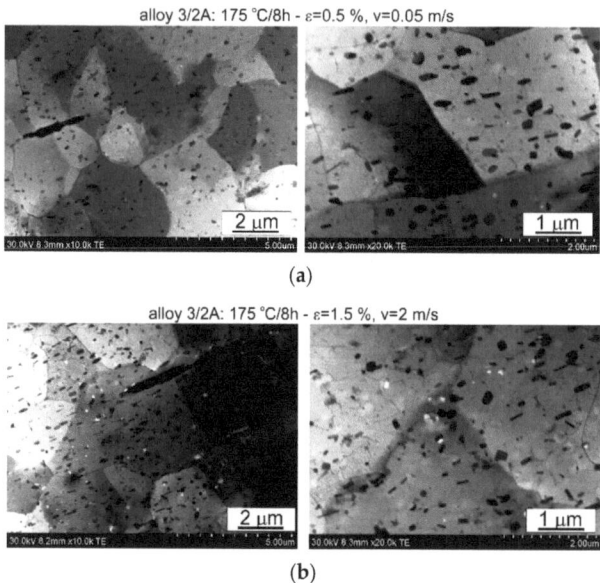

Figure 15. Microstructure of profiles extruded from alloy 3/2A: (**a**) statically straightened at $\varepsilon = 0.5$, $v = 0.05$ m/s; (**b**) dynamically straightened at $\varepsilon = 1.5$, $v = 2$ m/s; STEM.

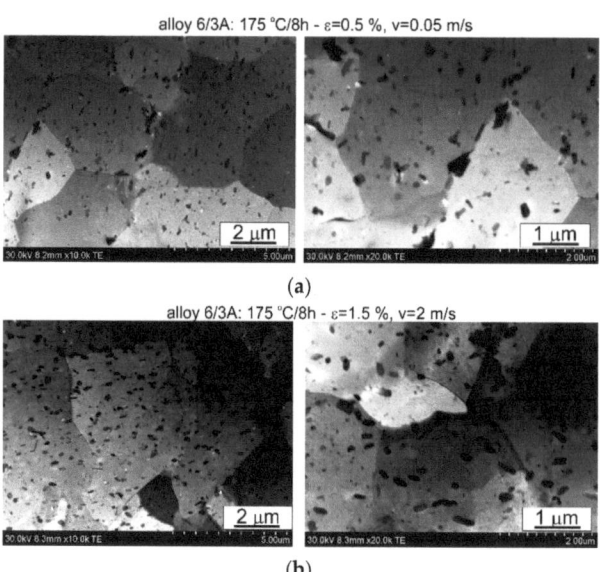

Figure 16. Microstructure of profiles extruded from alloy 6/3A: (**a**) statically straightened at $\varepsilon = 0.5$, $v = 0.05$ m/s; (**b**) dynamically straightened at $\varepsilon = 1.5$, $v = 2$ m/s; STEM.

For the selected variations, the microstructure of the thin films was examined using a scanning electron microscope equipped with a thin film observation attachment. The representative microstructural images are presented in Figures 14–16. The analysis of the thin films revealed that irrespective of the extrusion process parameters applied to the profiles, their microstructure consisted of grains with localised low-energy dislocation systems and a significant presence of precipitates. The quantity and size of these precipitates depend on the alloy composition, with higher alloying elements corresponding to an increase in the quantity and size of the precipitate, as confirmed by the statistical analysis presented later in the article. The low dislocation content is attributed to the presence of microstructural renewal processes during the plastic deformation process. Moreover, the ageing temperature used (175 °C) is sufficient to facilitate microstructural renewal processes. The rearrangement and annihilation of dislocations may have occurred prior to the precipitation of the strengthening phases during the ageing process.

3.2. Mechanical Properties

Figure 17 shows the stress/strain curves recorded during the static tensile test of the samples from the alloys 1/1A, 3/2A, and 6/3A-extruded, statically and dynamically stretched and artificially aged. Figure 18 shows the correlation between the achieved tensile strength (UTS) of the profiles made from the alloys 1/1A, 3/2A, and 6/3A. The profiles were extrusion-cooled in the press run and then subjected to static or dynamic straightening on the original prototype device. They were then artificially aged (175 °C/8 h). The relationship between the amount of cold deformation and the straightening speed is shown. In particular, there is a significant increase in the tensile strength for the samples subjected to dynamic straightening compared to those straightened statically—approximately an 18% increase for the alloys 1/1A and 3/2A and a 7% increase for alloy 6/3A. In general, slightly higher tensile strength values were observed for the higher stretching speed variant, i.e., 2 m/s, compared to 0.05 m/s. For alloy 1/1A which has the lowest Cu content (0.61%), the highest tensile strength (UTS) was recorded at 2% dynamic deformation, reaching approximately 387 MPa. This contrasts with 326 MPa at 0.5% static deformation. In the case of alloy 3/2A with an intermediate Cu content (0.81%), the maximum tensile strength (UTS) was observed at 1% dynamic deformation which was approximately 432 MPa (compared to 360 MPa for 0.5% static deformation). Similarly, for alloy 6/3A which has the highest Cu content (1.22%), the maximum tensile strength (UTS) was attained with a dynamic deformation of 2%, measuring approximately 441 MPa (as opposed to 412 MPa for a static deformation of 0.5%).

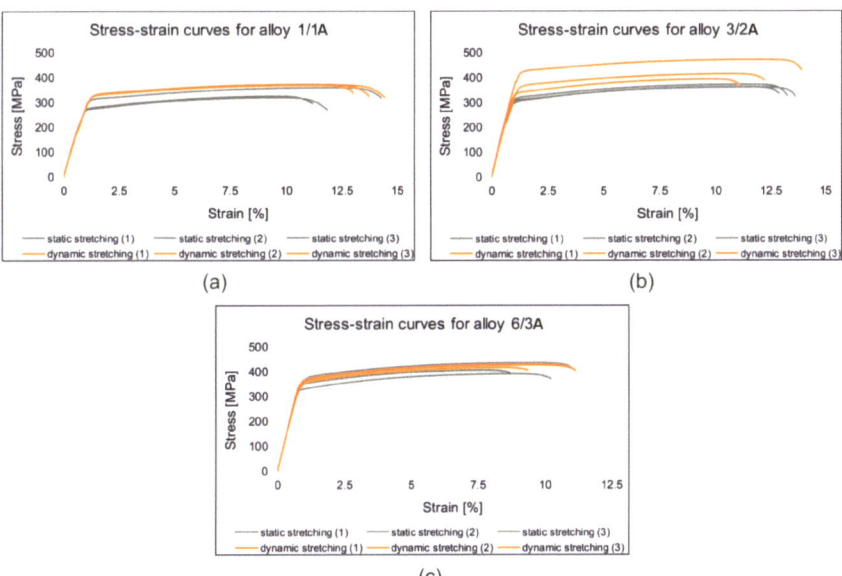

Figure 17. Stress/strain curves recorded during static tensile test of samples from alloys 1/1A (**a**), 3/2A (**b**), and 6/3A-extruded (**c**), statically and dynamically stretched and artificially aged.

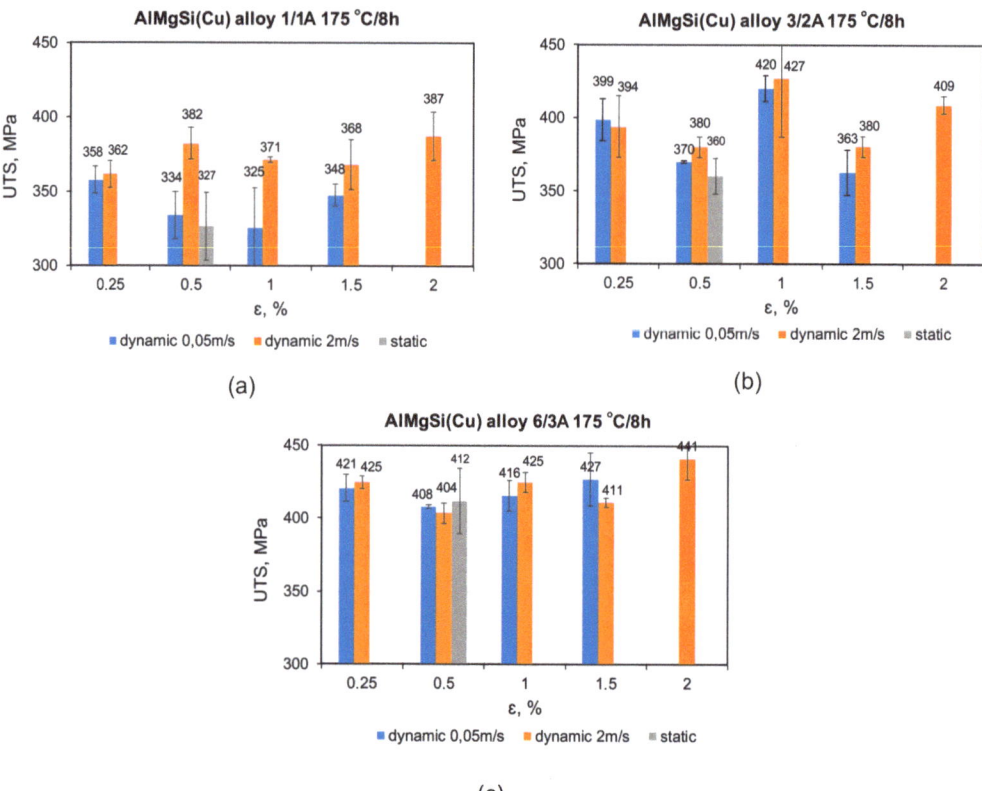

Figure 18. Dependence of tensile strength (UTS) for extruded profiles of alloys (**a**) 1/1A, (**b**) 3/2A, and (**c**) 6/3A, subjected to press run cooling, static or dynamic straightening, and subsequent artificial ageing at 175 °C for 8 h.

The elongation value (A, %) for most variants is comparable for the alloys with low and medium Cu content (alloy 1/1A and alloy 3/2A) averaging around 11–13% (Figure 19a,b). For the alloy with the highest Cu content (6/3A), the elongation is lower, at around 9–10% for the samples after the dynamic straightening and 8% for the samples after the static straightening (Figure 19c). These results correlate well with the level of tensile strength, as this alloy has the highest level of mechanical properties.

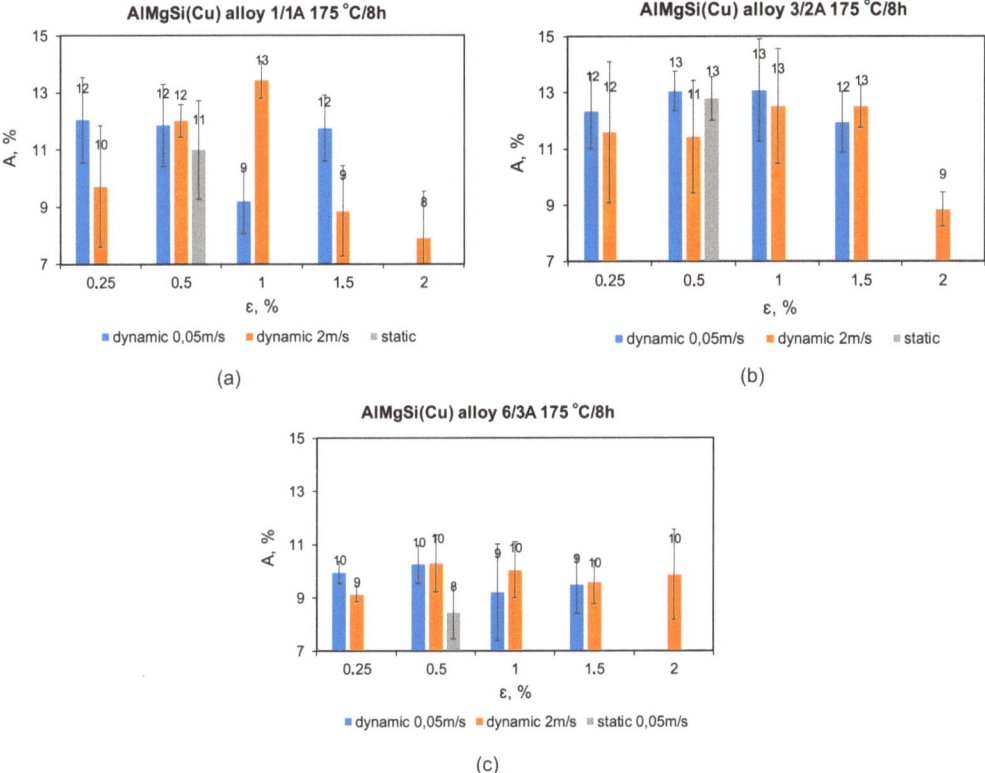

Figure 19. Dependence of elongation (A) for extruded profiles of alloys (**a**) 1/1A, (**b**) 3/2A, and (**c**) 6/3A, subjected to press run cooling, static or dynamic straightening, and subsequent artificial ageing at 175 °C for 8 h.

3.3. Geometrical Inspection

Figures 20–22 show the results of the 3D optical scanning of the profiles extruded from the AlMgSi(Cu) alloys and dynamically stretched with a 2% deformation at a speed of v = 2 m/s: the 60 × 40 × 2 mm 1/1A alloy profile (Figure 20), the 60 × 40 × 2 mm 3/2A alloy profile (Figure 21), and the 50 × 30 × 3 mm 6/3A alloy profile (Figure 22). In all the cases, the upper figure shows the coloured map of the dimensional deviations, while the lower figure on the left shows the dimensional deviations of the wall thickness of the profile, and, on the right, the dimensional deviations of the width and height of the profile after extrusion and dynamic stretching. The coloured maps show the relatively large deviations on the longitudinal edge and on the side and bottom walls of the 60 × 40 × 2 mm 1/1A alloy profile (Figure 20). Relatively higher deviations are observed at the ends of the analysed profile, which may be due to the deformation of the ends in the jaws during stretching. However, the most important conclusion concerns the wall thickness, width, and height of the profiles, which are generally within the permissible range of the relevant standard [26]. For all three profiles studied, the acceptable limits of wall thickness deviations according to the standard are in the range of ±0.35 mm and are not exceeded in any case. In the case of the height/width dimensions for the first two profiles, the acceptable limits of deviation are in the range of ±1.00 mm and are also not exceeded, similarly for the 3A alloy, where the acceptable limits of deviation are within the range of ±0.80 mm.

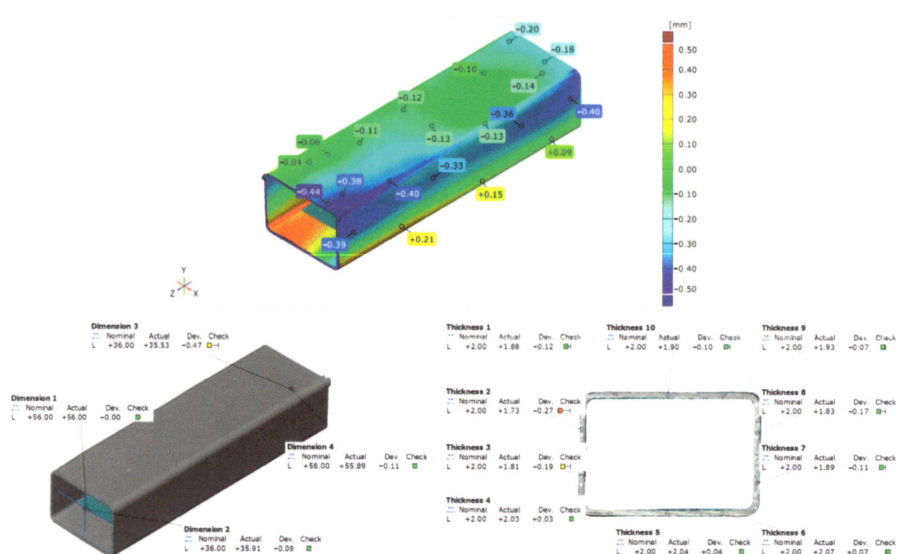

Figure 20. Results of 3D optical scanning of profile of 60 × 40 × 2 mm extruded from alloy 1/1A and dynamically straightened at ε = 2.0, v = 2 m/s.

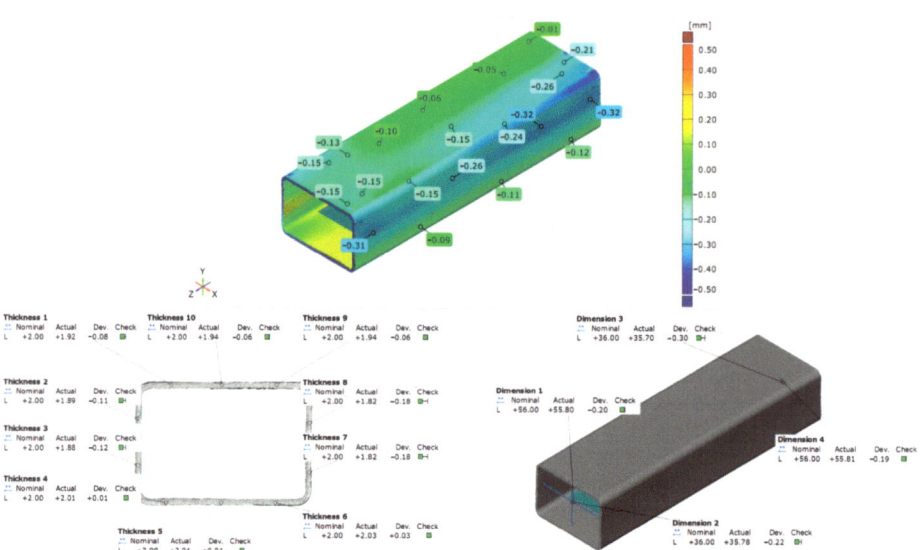

Figure 21. Results of 3D optical scanning of profile of 60 × 40 × 2 mm extruded from alloy 3/2A and dynamically straightened at ε = 2.0, v = 2 m/s.

The analysis of the dimensional deviations of the profiles studied shows that they are very accurate. The higher the strength of the alloy (from alloy 1A to alloy 3A), the lower the dimensional deviations of the wall thickness and height/width.

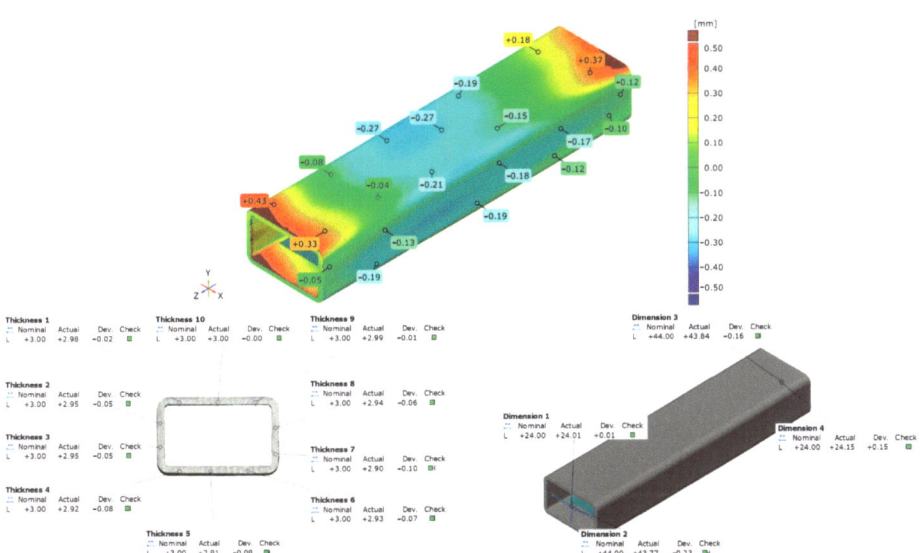

Figure 22. Results of 3D optical scanning of profile of 50 × 30 × 3 mm extruded from alloy 6/3A and dynamically straightened at ε = 2.0, v = 2 m/s.

4. Discussion

The laboratory dynamic stretching device is relatively simple to operate and maintain. It requires at least two operators working together, who must be trained to carry out the tests. The first operator controls the clamping jaws, the actuator feed, and the tension of the profile for static deformation. The second operator controls the work of the dynamic system. The accuracy of the force measurements is sufficient; for the parameters tested, the load of up to approximately 100 kG on a maximum scale of 50 T represents only about 0.2%. The displacement recorded by a digital straightedge with a reading accuracy of 0.02 mm is sufficient to measure the displacement of the pulling plate and to calculate the deformation. The mechanical system used to clamp the profile in the jaws is sufficient to immobilise it. The set of pneumatic hammers is sufficient to generate the dynamic deformations for the specified cross-section of the profiles. When designing industrial equipment, it is recommended to take into account the digital measurement of static force on hydraulic valves and the measurement of displacement with an accuracy of 0.02 mm. For a higher frequency of impacts, multiple pneumatic hammers and their alternate cyclic work should be applied. If very high deformation and high energy of impacts are required, pneumatic hammers should be replaced by hydraulic cylinders. The cradle for profiles over 15 mm in length should be replaced by the independent traveller with the dynamic set. It is necessary to have a system for damping vibrations and maintaining the initial tension of the profile. A gas spring kit or elastomers can be used. It is recommended to apply the programmer to set the cycle time and the number of repetitions of the hydraulic system. However, it should be remembered that the application of changes and modifications depends on the final design of the device and the assumptions for its use.

The introduction of dynamic stretching facilitates the occurrence of dynamic recrystallisation, a phenomenon which is particularly pronounced in the microstructure of dynamically stretched profiles at ε = 1% and ε = 1.5% with stretching speeds of v = 2 m/s. Typically, fine, equiaxial grains are organised in bands, and occasionally, the protrusions of the original grain boundaries are locally visible. Yu et al. [27] observed a similar phenomenon during the porthole die extrusion process of aluminium alloy 6063. Equiaxed grains enter the weld zone and elongate along the extrusion direction under the influence of compression and shear. Subsequently, grain boundaries migrate due to interfacial energy, resulting in grain entrapment and the growth of elongated coarse grains. These elongated coarse grains undergo continuous dynamic recrystallisation and geometric dynamic recrystallisation at high temperatures and pressures, resulting in the formation of fine equiaxed grains.

The profiles subjected to dynamic deformation exhibit superior mechanical properties, especially at higher speeds such as 2 m/s. Consider, for example, the alloy 3/2A profile, where the highest ultimate tensile strength (UTS) is observed at $\varepsilon = 1\%$. Compared to both the dynamically and statically stretched profiles, its microstructure appears more homogeneous. Dispersoids are smaller and more abundant, especially in the statically straightened profiles (see Figures 14–16). Furthermore, in several cases, particularly with dynamic deformation, the grain boundaries appear to be bulged and numerous fine grains, probably formed during recrystallisation, can be seen in their vicinity.

In order to further analyse the results which were obtained, a statistical analysis of the dispersoids was carried out for the variants from which thin films were made. The free software ImageJ 1.54i was used to determine the cross-sectional area of the dispersoids, and the results are presented graphically in Figure 23 and in Table 4. The results confirmed previous observations regarding the amount of dispersoids as a function of alloy type. The highest amount of dispersoids was found in the alloy with the highest Cu content (alloy 6/3A), occupying 8.1% and 7.7% of the analysed area for the statically and dynamically straightened profiles, respectively. The tensile strength for both the straightening variants was comparable, averaging UTS = 411–412 MPa (Table 3). It is worth noting that for the dynamically straightened alloy 6/3A, 50% of the dispersoids had a cross-sectional area below 0.005 μm^2, while for the statically straightened variant, 36% of the analysed dispersoid population fell within this range (Figure 23).

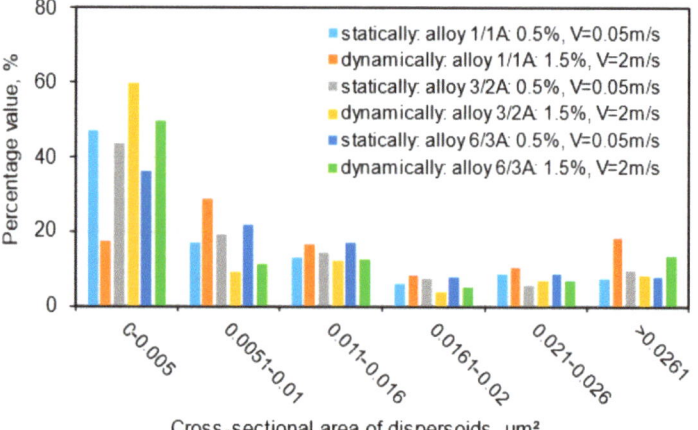

Figure 23. Statistical analysis of dispersoids for extrudates from AlMgSi(Cu) alloys with different Cu contents after static or dynamic stretching.

For both the alloys with the lowest and the medium Cu content (1/1A and 3/2A alloys, respectively), the total area occupied by dispersoids was greater in the case of static straightening (Table 4). Almost 77% of the dispersoid population measured in the statically straightened profiles of these alloys has a size below 0.016 μm^2, while for the dynamically straightened profiles, the percentages are 63% and 82%, respectively (Figure 23). The higher level of mechanical properties in the case of the dynamically straightened profiles is mainly due to the size and shape of the grain, which is finer and, depending on the straightening conditions, either elongated in the direction of extrusion or close to equiaxed (Figures 8–11).

Table 4. Results of the UTS and cross-sectional area of the dispersoid analysis depending on the alloy and stretching conditions.

	AlMgSi(Cu) Alloy 1	AlMgSi(Cu) Alloy 2	AlMgSi(Cu) Alloy 3
	Percentage value of dispersoids in the investigated area, %		
Statically stretched	7.2	7.5	8.1
Dynamically stretched	5.9	4.4	7.7
	UTS, MPa/Standard deviation, MPa		
Statically stretched	326/23	360/12	412/22
Dynamically stretched	368/17	380/7	411/3

Figure 24 shows the effects of applying the dynamic stretching ($\varepsilon = 2.0$, $v = 2$ m/s) to the extruded profiles of the AlMgSi(Cu) alloys with different alloy components, including Cu. The Cu content is used here as a parameter indicating the increase in the UTS after the dynamic stretching relative to static stretching (left) and as a parameter indicating the average deviation in the wall thickness of the extruded and stretched profiles (right). In both cases, there is a tendency towards failure from the 1/1A alloy with the lowest content of alloying elements to the 6/3A alloy with the highest content of alloying elements. The highest increase in the UTS due to dynamic deformation during the stretching of the profile was obtained for alloy 1/1A (strength factor of 1.18 means an 18% increase in the UTS after dynamic deformation compared to static deformation. Similarly, the highest mean dimensional deviation of wall thickness was observed for this alloy at the level of 0.1 mm (acceptable according to the standard [26]). This is probably due to the finest-grained and most homogeneous microstructure, numerous dispersoids present in this alloy after extrusion, dynamic stretching, and artificial ageing. This alloy, due to the lowest content of alloying elements, is characterised by the lowest yield stress, which translates into the highest dimensional deviations after stretching. For the other alloys (3/2A and 6/3A), dynamic stretching is not so advantageous as the increase in the UTS is only at the level of 13% and 7%, respectively, with respect to the statically (classically) stretched profiles. The 3/2A and 6/3A alloys are characterised by the lower average dimensional deviations of the wall thickness, 0.08 mm and 0.06 mm, respectively, due to the higher content of alloying elements and also to the higher values of the yield stress.

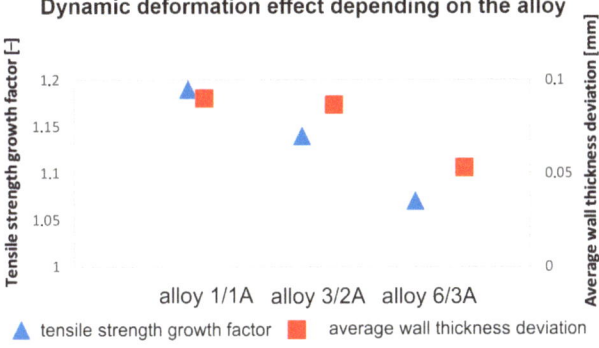

Figure 24. Dynamic deformation effect depending on the alloy for the extruded profiles of the alloys 1/1A, 3/2A, and 6/3A subjected to press run cooling, static or dynamic straightening, and subsequent artificial ageing at 175 °C for 8 h.

5. Conclusions

The following conclusions can be drawn from the results which have been obtained:

1. In this paper, the authors have built a prototype of a device for the dynamic stretching of extruded profiles after quenching. The semi-industrial device is equipped with a hydraulic system for stretching and a pneumatic system for cold dynamic deformation. The aim of this project is to induce advantageous microstructural changes and increase the strength properties of the extruded material. A guideline is formulated for the design of the industrial device in which very long profiles are stretched. This requires the intensification of the dynamic stretching. For high deformations and high impact energy, pneumatic hammers should be replaced by hydraulic actuators.
2. The highest increase in the UTS due to dynamic deformation during the stretching of the profile was obtained for alloy 1/1A (a strength factor of 1.18 means an 18% increase in the UTS after dynamic deformation compared to static deformation). This alloy contains the lowest amount of the alloying elements. For the alloys 3/2A and 6/3A, the dynamic stretching is not so favourable; the increase in the UTS is only 13% and 7%, respectively, with respect to the statically stretched profiles.
3. The application of dynamic stretching to the alloys tested makes it possible to obtain profiles with an advantageous fine-grained microstructure, containing a large number of dispersoids and a homogeneous microstructure, resulting in high strength properties.
4. Dimensional tolerances are not a critical parameter in the dynamic straightening of extruded aluminium profiles with high cold deformation reaching up to 2%. The dimensional tolerances obtained for this dynamic deformation are within the limits of the relevant standard. Therefore, it can be said that the dimensional deviations obtained are not an obstacle to the use of high dynamic deformation in the straightening of extruded aluminium profiles and are advantageous from the point of view of microstructure and mechanical properties.

Author Contributions: Conceptualisation, J.Z. and W.L.; methodology, T.L., B.L.-M. and M.B.; validation, D.L.; formal analysis, D.L.; investigation, T.L., B.L.-M., M.B. and B.P.; writing—original draft preparation, D.L.; writing—review and editing, D.L., B.L.-M. and T.L.; visualisation, T.L., B.L.-M. and M.B.; supervision, D.L.; project administration, D.L. All authors have read and agreed to the published version of the manuscript.

Funding: This research was funded by The National Centre For Research And Development, grant number TECHMATSTRATEG-III/0040/2019-00 "Development of technology for extrusion of sections from ultra-high-strength AlMgSi(Cu) alloys".

Institutional Review Board Statement: Not applicable.

Informed Consent Statement: Not applicable.

Data Availability Statement: The original contributions presented in the study are included in the article, further inquiries can be directed to the corresponding author.

Acknowledgments: The authors are grateful to Albatros Aluminium Company for the possibility of performing wide extrusion trials in industrial conditions.

Conflicts of Interest: The authors declare no conflicts of interest.

References

1. Kazanowski, P.; Dickson, R. Evaluation of Process Mechanism and Parameters for Automated Stretching Line. In Proceedings of the Tenth International Aluminum Extrusion Technology Seminar, Orlando, FL, USA, 1–4 May 2012; Volume I, pp. 327–341.
2. Furu, T.; Ryen, Ø.; Myhr, O.R. Effect of pre-deformation on age hardening kinetics in commercial 6xxx alloys. In Proceedings of the 11th International Conference on Aluminium Alloys (ICAA), Aachen, Germany, 22–26 September 2008; Wiley-VCH: Hoboken, NJ, USA, 2008; pp. 1626–1633.
3. Abi-Akl, R.; Mohr, D. Paint-bake effect on the plasticity and fracture of pre-strained aluminium 6451 sheets. *Int. J. Mech. Sci.* **2017**, *124–125*, 68–82. [CrossRef]
4. Granum, H.; Myhr, O.; Børvik, T.; Hopperstad, O.S. Effect of pre-stretching on the mechanical behaviour of three artificially aged 6xxx series aluminium alloys. *Mater. Today Commun.* **2021**, *27*, 102408. [CrossRef]

5. Qvale, K.; Hopperstad, O.S.; Reiso, O.; Tundal, U.H.; Marioara, C.D.; Børvik, T. An experimental study on pre-strained double-chamber 6000-series aluminium profiles subjected to quasi-static and dynamic axial crushing. *Thin-Walled Struct.* **2021**, *158*, 107160. [CrossRef]
6. Wang, Z.; Li, H.; Miao, F.; Fang, B.; Song, R.; Zheng, R. Improving the strength and ductility of Al-Mg-Si-Cu alloys by a novel thermo-mechanical treatment. *Mater. Sci. Eng. A* **2014**, *607*, 313–317. [CrossRef]
7. Kolar, M.; Pedersen, K.O.; Gulbrandsen-Dahl, S.; Teichmann, K.; Marthinsen, K. Effect of pre-deformation on mechanical response of an artificially aged Al-Mg-Si alloy. *Mater. Trans.* **2011**, *52*, 1356–1362. [CrossRef]
8. Ma, Z.; Robson, J.D. Understanding the effect of deformation combined with heat treatment on age hardening of Al–Zn–Mg–Cu alloy AA7075. *Mater. Sci. Eng. A* **2023**, *878*, 145212. [CrossRef]
9. Gable, B.M.; Zhu, A.W.; Csontos, A.A.; Starke, E.A., Jr. The role of plastic deformation on the competitive microstructural evolution and mechanical properties of a novel Al–Li–Cu–X alloy. *J. Light Met.* **2001**, *1*, 1–14. [CrossRef]
10. Rodgers, B.I.; Prangnell, P.B. Quantification of the influence of increased pre-stretching on microstructure-strength relationships in the Al–Cu–Li alloy AA2195. *Acta Mater.* **2016**, *108*, 55–67. [CrossRef]
11. Zuo, D.; Cao, Z.; Cao, Y.; Zheng, G. Effect of Pre-Stretching on Microstructures and Mechanical Behaviors of Creep-Aged 7055 Al Alloy and Its Constitutive Modeling. *Metals* **2019**, *9*, 584. [CrossRef]
12. Yang, Y.; Zhan, L.; Shen, R.; Yin, X.; Li, L.; Li, L.; Huang, M.; He, D. Effect of pre-deformation on creep age forming of 2219 aluminum alloy: Experimental and constitutive modelling. *Mater. Sci. Eng. A* **2017**, *683*, 227–235. [CrossRef]
13. Sin, H.S.; Jo, G.G.; Jeong, Y.H.; Sin, M.C. Effect of Stretching on the Microstructure and Mechanical Properties of Al-Li-Cu-Mg Alloys. *Korean J. Mater. Res.* **1995**, *5*, 1005–1112.
14. Gao, G.J.; Li, X.W.; Yan, L.Z. Effect of Combined Pre-Straining and Pre-Aging on the Precipitation Behavior and Age Hardening Response for Al-Mg-Si Alloys. *Mater. Sci. Forum* **2021**, *1026*, 74–83. [CrossRef]
15. Zhang, X.M.; Han, N.M.; Liu, S.D.; Ke, B.; Xin, X. Effects of pre-stretching and ageing on the strength and fracture toughness of aluminum alloy 7050. *Mater. Sci. Eng. A* **2011**, *528*, 3714–3721.
16. Quan, L.; Zhao, G.; Gao, S.; Muddle, B.C. Effect of pre-stretching on microstructure of aged 2524 aluminium alloy. *Trans. Nonferrous Met. Soc. China* **2011**, *21*, 1957–1962. [CrossRef]
17. Yu, C.; Feng, Y.; Wang, L.; Fu, L.; Kang, F.; Zhao, S.; Guo, E.; Ma, B. Effect of pre-stretching deformation treatment process on microstructure and mechanical properties of Al-Cu-Mg alloy. *Mater. Today Commun.* **2022**, *31*, 103368. [CrossRef]
18. Liu, C.; Ma, Z.; Ma, P.; Zhan, L.; Huang, M. Multiple precipitation reactions and formation of θ'-phase in a pre-deformed Al–Cu alloy. *Mater. Sci. Eng. A* **2018**, *733*, 28–38. [CrossRef]
19. Li, G.A.; Ma, Z.; Jiang, J.T.; Shao, W.Z.; Liu, W.; Zhen, L. Effect of Pre-Stretch on the Precipitation Behavior and the Mechanical Properties of 2219 Al Alloy. *Materials* **2021**, *14*, 2101. [CrossRef]
20. Teichmann, K.; Marioara, C.D.; Andersen, S.J.; Marthinsen, K. The Effect of Preaging Deformation on the Precipitation Behavior of an Al-Mg-Si Alloy. *Metall. Mater. Trans. A* **2012**, *43*, 4006–4014. [CrossRef]
21. Łatkowski, A. Obróbka cieplno-mechaniczna stopów aluminium (in Polish). *Pap. AGH Univ. Krakow Metall. Foundry* **1989**, *1272*, 128.
22. Singh, P.; Ramacharyulu, D.A.; Kumar, N.; Saxena, K.K.; Eldin, S.M. Change in the structure and mechanical properties of AlMgSi alloys caused by the addition of other elements: A comprehensive review. *J. Mater. Res. Technol.* **2023**, *27*, 1764–1796. [CrossRef]
23. Li, L.; Flores-Johnson, E.A.; Shen, L.; Proust, G. Effects of heat treatment and strain rate on the microstructure and mechanical properties of 6061 Al alloy. *Int. J. Damage Mech.* **2016**, *25*, 26–41. [CrossRef]
24. Scharifi, E.; Sajadifar, S.V.; Moeini, G.; Weidig, U.; Böhm, S.; Niendorf, T.; Steinhof, K. Dynamic Tensile Deformation of High Strength Aluminium Alloys Processed Following Novel Thermomechanical Treatment Strategies. *Adv. Eng. Mater.* **2020**, *22*, 2000193. [CrossRef]
25. Tong, M.; Jiang, F.; Wang, H.; Jiang, J.; Ye, P.; Xu, X. The evolutions of mechanical properties and microstructures of Al-Mg-Mn-Sc-Zr alloy during dynamic stretching deformation. *J. Alloys Compd.* **2021**, *889*, 161753. [CrossRef]
26. EN 755-2; Aluminium and Aluminium Alloys—Extruded Rod/Bar, Tube and Profiles—Part 8: Permissible Deviations in Size and Shape of Tubes Extruded on Porthole Dies. DIN Deutsches Institut für Normung e. V.: Berlin, Germany, 2016.
27. Yu, J.; Zhao, G.; Zhang, C.; Chen, L. Dynamic evolution of grain structure and microtexture along a welding path of aluminum alloy profiles extruded by porthole dies. *Mater. Sci. Eng. A* **2017**, *682*, 679–690. [CrossRef]

Disclaimer/Publisher's Note: The statements, opinions and data contained in all publications are solely those of the individual author(s) and contributor(s) and not of MDPI and/or the editor(s). MDPI and/or the editor(s) disclaim responsibility for any injury to people or property resulting from any ideas, methods, instructions or products referred to in the content.

Article

Ultra-Fine Bainite in Medium-Carbon High-Silicon Bainitic Steel

Xinpan Yu [1], Yong Wang [2], Huibin Wu [1,*] and Na Gong [3,*]

1. Collaborative Innovation Center of Steel Technology, University of Science and Technology Beijing, Beijing 100083, China; xinpan_yu@163.com
2. School of Materials Science and Engineering, Nanyang Technological University, Singapore 639798, Singapore
3. Institute of Materials Research and Engineering (IMRE), A*STAR (Agency for Science, Technology, and Research), 2 Fusionopolis Way, Singapore 138634, Singapore
* Correspondence: whbustb@163.com (H.W.); na_gong@imre.a-star.edu.sg (N.G.)

Abstract: The effects of austenitizing and austempering temperatures on the bainite transformation kinetics and the microstructural and mechanical properties of a medium-carbon high-silicon ultra-fine bainitic steel were investigated via dilatometric measurements, microstructural characterization and mechanical tests. It is demonstrated that the optimum austenitizing temperature exists for 0.3 wt.%C ultra-fine bainitic steel. Although the finer austenite grain at 950 °C provides more bainite nuclei site and form finer bainitic ferrite plates, the lower dislocation density in plates and the higher volume fraction of the retained austenite reduces the strength and impact toughness of ultra-fine steel. When the austenitizing temperature exceeds 1000 °C, the true thickness of bainitic ferrite plates and the volume fraction of blocky retained austenite in the bainite microstructure increase significantly with the increases in austenitizing temperature, which do harm to the plasticity and impact toughness. The effect of austempering temperature on the transformation behavior and microstructural morphology of ultra-fine bainite is greater than that of austenitizing temperature. The prior martensite, formed when the austempering temperature below M_s, can refine the bainitic ferrite plates and improve the strength and impact toughness. However, the presence of prior martensite divides the untransformed austenite and inhibits the growth of bainite sheaves, thus prolonging the finishing time of bainite transformation. In addition, prior martensite also strengthens the stability of untransformed austenite though carbon partition and enhances the volume fraction of blocky retained austenite, which reduces the plasticity of ultra-fine bainitic steel. According to the experimental results, the optimum austempering process for 0.3 wt. %C ultra-fine bainitic steel is through austenitization at 1000 °C and austempering at 340 °C.

Keywords: prior martensite; transformation behavior; retained austenite; mechanical property

1. Introduction

Ultra-fine bainite steel, also known as super bainite steel, nano-bainite steel and low-temperature bainite steel, has become a research hotspot due to its excellent combination of strength and ductility [1–3]. Ultra-fine bainitic ferrite plates, 20 nm~1 μm in thickness, within high dislocation densities provide the steel with high strength, and the film-like retained austenite embedded into the plates can absorb microcrack propagation, thereby ensuring its high toughness property [4–6]. The impressive combination of ultimate tensile strength (>2 GPa) and ductility (>15%) has been achieved latterly in medium-carbon bainitic steel, indicating that the ultra-fine bainitic steel has huge potential in the field of wear-resistant components such as gears, bearing, and so on [6–9].

The first generation of ultra-fine bainitic steel with high carbon content was obtained though austempering at low temperature (125–325 °C) from 2 to 90 days [10]. With further research about the bainite transformation mechanism and the relationship between microstructure and mechanical properties, the bainite transformation process is clearly accelerated by the following methods: (i) Through adjusting the chemical component design [11–13], the segregation of boron at prior austenite grain boundaries can retard the bainite transformation at low temperature [13]. (ii) By increasing the nucleation sites though lowering the direct austempering temperature to below Ms [14–17], the supercooled austenite is deformed [18–21] and partial bainite/athermal martensite is obtained before bainite transformation [22–26]. However, there are quite a few research studies about the effect of prior austenite on bainite transformation, and the results are controversial. On the one hand, Seok-Jae Lee [27] and Jing Zhao et al. [28] thought that the bainite transformation kinetics of medium-carbon steel increased as the austenite grain size decreased. On the other hand, F Hu et al. [29] noticed that the coarser austenite grains provide less nucleation sites, which are beneficial to bainite growth. That is to say, there is an optimum austenitizing temperature for ultra-fine bainite steels with different carbon contents.

The austempering temperature below Ms was selected to accelerate the transformation kinetics and enhance the mechanical properties of ultra-fine bainite steel though forming partial martensite in recent research [30–32]. Although the presence of prior martensite can increase the nucleation sites of bainite in the following isothermal process, the formation of the former phase also refines the untransformed austenite, and then inhibits the growth of bainitic ferrite plates [33,34]. In addition, the diffusion of carbon ejected from supersaturated martensite to austenite increases the stability of the latter phase, thereby reducing the degree of bainite transformation [35,36]. Therefore, it is necessary to contrastively analyze the effects of austempering temperatures above and below Ms on transformation kinetics and microstructural morphology to obtain the ultra-fine bainite steel with optimal mechanical properties.

In order to determine the effect of austenitizing temperature and austempering temperature on ultra-fine bainitic steel, a medium-carbon steel with 0.30 wt.%C was selected for research in the paper. A series of heat treatment processes with different austenitizing temperatures and austempering temperatures were carried out to explore the relationship between the heat treatment processes and bainite transformation kinetics, microstructural morphology, and mechanical properties of medium-carbon steel. The multiphases of martensite, bainitic ferrite with different volume fractions and dislocation densities, and the retained austenite with different morphologies and carbon concentrations along with their mechanical properties were observed and measured though scanning electronic microscopy (SEM), transmission electron microscope (TEM), X-ray diffractometer (XRD) and tensile testing.

2. Experimental Procedures

2.1. Steel Composition and Thermomechanical Treatments

The chemical composition of the tested steel used in this paper was 0.30C-1.40Si-1.50Mn-1.18Cr-1.15Al-0.49Mo-0.61Ni-0.019Nb-0.0026B (wt.%). Si can inhibit the precipitation of cementite during isothermal transformation, enhance the thermal stability of retained austenite and accelerate bainite transformation at low temperature [37,38]. The addition of Mn and Cr can increase the hardenability of medium-carbon steel [39,40]. Al can increase the driving force of austenite to ferrite ($\Delta G^{\gamma \to \alpha}$) [41,42]. The precent of Nb and Mo can refine the prior austenite grains and accelerate the following bainite transformation [11]. The boron segregation at prior austenite can affect grain size and determine the nucleation site and transformation kinetics of bainite [12]. The tested steel was smelted in a vacuum furnace and forged into a block measuring $60 \times 80 \times 100^3$ mm. Samples with a size of $\varphi 4 \times 10$ mm^3 was cut and used for measuring the transformation temperature and kinetics of the tested steel with different heat treatment processes. Samples with a size of $12 \times 60 \times 60$ mm^3 composed of sodium nitrate and potassium nitrite (1:1 in weight) were cut and used for the subsequent heat treatments in a muffle furnace and a salt bath furnace.

The Ms temperature was measured to be 328 °C by the DIL 805A thermal dilatometer (Waters, Milford, MA, USA). The heat treatment processes are given in Figure 1. As shown in Figure 1a, the samples were fully austenitized at 950–1100 °C about 30 min with a heating rate of 10 °C/s, then cooled to 340 °C for 1 h and finally quenched into room temperature to research the effect of austenitizing temperature on bainite transformation, the microstructure, and mechanical properties of the tested steel. Similarly, after austenitizing at 1000 °C for 30 min, the samples were quenched to below Ms (310 and 320 °C) and above Ms (340 and 350 °C) about 1 h to research the effect of austempering temperatures, as given in Figure 1b. The dilatation of the samples along the radial direction was recorded and used to characterize the bainite transformation kinetics.

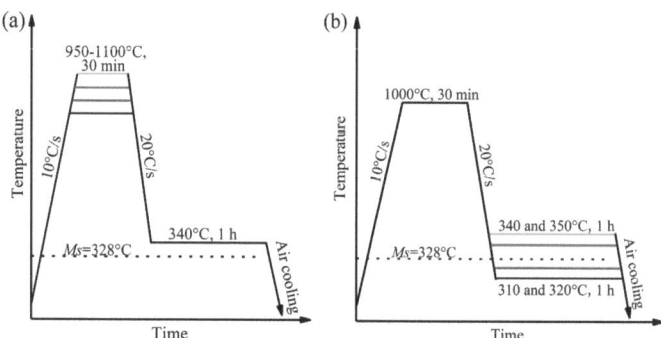

Figure 1. Heat treatment program to study the effect of austenitizing temperature (**a**) and austempering temperature (**b**) on bainite transformation.

2.2. Microstructural Characterization

SEM samples were mechanically ground, polished and etched with a 4% nital solution. The morphological characters of prior austenite grain, bainite and retained austenite were observed at low magnification though SEM (ZEISS EVO 18) (Carl Zeiss, Oberkochen, Germany). TEM samples were thinned to ~50 μm by a precision ion polishing system (Gatan 691) (Gatan, Pleasanton, CA, USA). The sizes of the bainitic ferrite were measured and counted using Tecnai F20 (FEI, Hillsboro, OR, USA), operating at 200 kV. The dislocation density within bainitic ferrite and the volume fraction and average carbon content of retained austenite were analyzed by XRD with Cu-$K\alpha$ radiation operating at 40 kV and 40 mA. The XRD pattern was drawn based on the data obtained at a scanning rate of 1°/min and 2θ range of 30–100°. The dislocation density of bainitic ferrite (ρ) was counted according to the Williamson–Hall formula, as presented in Equation (1) [43]. The volume fraction of retained austenite (V_γ) was calculated based on the integrated intensities of (200), (220) and (311) austenite peaks and (200), (211) ferrite peaks [44]. The volume fractions of prior martensite (V_M) within the samples austempered below Ms are calculated according to Equation (2). Finaly, the volume fraction of bainitic ferrite, V_B, is calculated by Equation (3). The average carbon content of retained austenite (C_γ) was determined though Equation (4) [45]. Carbon contents of retained austenite with film-like and blocky morphology were calculated by the Gaussian multi-pears fitting method, as reported in the literature [46,47].

$$\rho = 14.4\, e^2/b^2 \tag{1}$$

$$V_M = 1 - e^{-0.011(Ms-QT)} \tag{2}$$

$$V_B = 1 - V_\gamma - V_M \tag{3}$$

$$C_\gamma = (a - 3.578)/0.033 \tag{4}$$

where Ms and QT are the martensite start temperature and the directly austempering temperature, respectively, of the tested sample; e is the micro-strain measured by MDI Jade

6.0 software (6.0, Materials Data, Livermore, CA, USA); b is the Burgers vector of ferrite (0.248 nm); and a is the austenite lattice parameter in Å.

2.3. Mechanical Testing

In accordance with GB/T228.1-2021, plate samples with a parallel length of 52 mm and a cross-section of 15 mm × 4 mm were used for tensile tests, which were conducted using the MTS-CMT5105 universal testing machine (Millennium Technology Services, San Francisco, CA, USA), with a loading rate of 3 mm/min. Standard Charpy V-notch samples, with dimensions of 10 mm × 10 mm × 50 mm, were used to measure the impact toughness at room temperature using the JBDW-300D impact tester (Jinan Kehui test equipment Co., LTD, Jinan, China). For each heat treatment process, three samples were tested.

3. Results

3.1. Transformation Kinetics

Figure 2 shows the bainite transformation kinetics curves, including dilatation–temperature curves, dilatation–time curves, and transformation rate–time curves of the samples at different austenitizing temperatures. The cooling rate from austenitizing stage down to isothermal stage is large enough to avoid the formation of ferrite and pearlite, so the starting temperature of the dilatation–temperature curve of the samples was chosen to be 500 °C, as shown in Figure 2a. Temperature-independent bainite formation occurs immediately when the austempering temperature reaches 340 °C. During further cooling to room temperature, there is no obvious deviation of the dilatation–temperature curves, indicating that the blocky retained austenite does not decompose into martensite, as shown by the dashed box in Figure 2a. The dilatation–temperature curves at the austempering stage were normalized to obtain the dilatation–time curves, as shown in Figure 2b. There is no obvious transformation incubation time for all the samples at different austenitizing temperatures. The completion time of isothermal bainite transformation was determined though the tangent method. The isothermal bainite completion times of the samples at 950–1100 °C were 820.2 s, 687.0 s, 708.6 s and 802.2 s. The transformation rate–time curves were obtained by derivation of the dilatation–time curves, as shown in Figure 2c. At the beginning of the isothermal stage, the bainite transformation rate reaches the maximum value, and then decreases gradually as the isothermal time increases. It is worth mentioning that the transformation rate of the sample austenitized at 1000 °C is higher than those of the samples which were austenitized at 950 °C, 1050 °C and 1100 °C.

The bainite transformation kinetic curves of the samples austempering at different temperatures are shown in Figure 3. When the isothermal bainite transformation temperatures were 310 and 320 °C, as shown in Figure 3a, the dilatation–temperature curves of the samples deviated slightly after the temperature drops below Ms, indicating that the prior martensite formed before isothermal bainite transformation. The lower the isothermal bainite transformation temperature, the higher the volume fraction of prior martensite formation [16–19]. The dilatation–time curves and bainite transformation rate–time curves of the sample austempered at different temperatures were obtained similarly with those in the case of samples austenitized at different temperatures. The results are shown in Figure 3b,c. The transformation incubation times of the samples austempered at 310–350 °C are also not obvious like those of the samples austenitized at 950–1100 °C. Except for the samples austempered at 310 °C, the bainite transformation completion times of the samples austempered at 320 °C, 340 °C and 350 °C are not different, which are 658.8 s, 687.0 s and 678.8 s, respectively. Although a larger degree of supercooling provides higher transformation dynamic, and more α/γ interfaces provide nucleation sites for bainite transformation, the transformation completion time of the sample austempered at 310 °C is 1184.4 s, which is significantly longer than in the case of the other austempered samples. The maximum transformation rate increases with the increase in austempering temperature, and decreases after reaching the peak value at 340 °C. It is worth noting that the maximum transformation rate of the samples austempered at 350 °C appears 18 s after the beginning of the bainite transformation, while the other austempered samples have no obvious transformation incubation period, and the maximum transformation rate occurs at the beginning of bainite transformation.

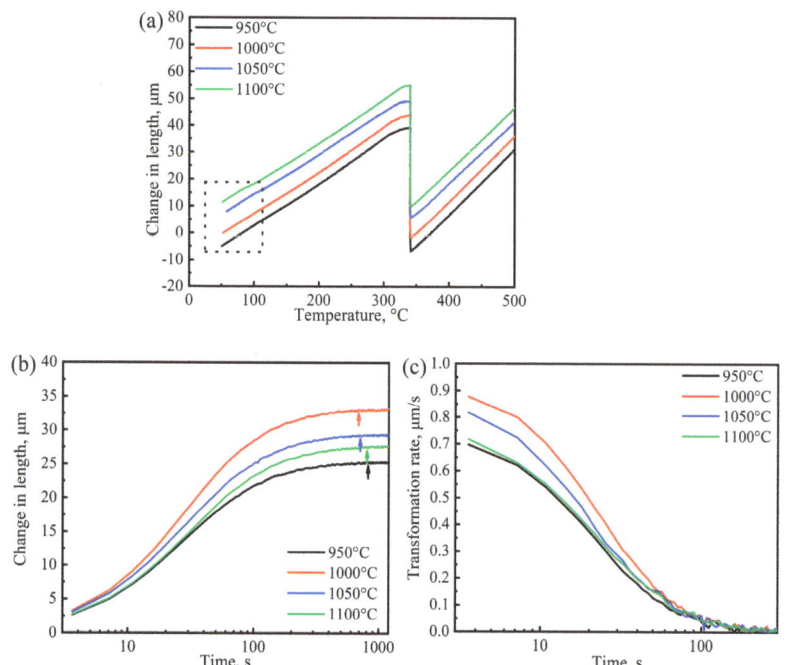

Figure 2. Dilatation–temperature (**a**), dilatation–time (**b**) and transformation rate–time (**c**) curves of the samples during holding at 340 °C directly from different austenitizing temperatures.

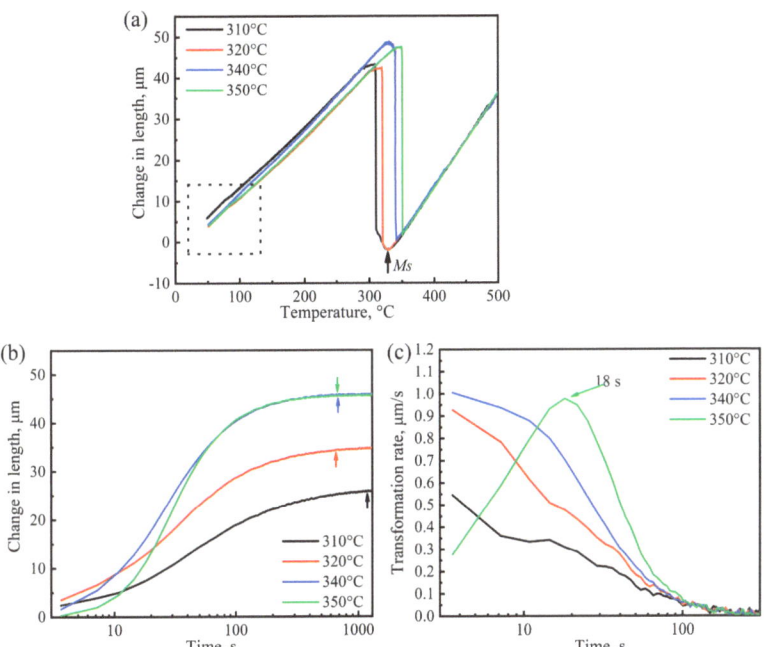

Figure 3. Dilatation–temperature (**a**), dilatation–time (**b**) and transformation rate–time (**c**) of the samples during holding at different austempering temperatures directly from 1000 °C.

3.2. Microstructure

The microstructural macroscopic morphology of ultra-fine bainite within the sample under different isothermal transformation process parameters are shown in Figures 4–7. It can be observed that the microstructure of the samples after different austempering processes is the same, which is composed of bainitic ferrite plates and retained austenite, in which the retained austenite shows film-like and blocky morphology. The microstructural composition and the morphology of ultra-fine bainite from the samples austenitized at different temperatures are shown in Figures 4 and 5, respectively. The ultra-fine bainitic ferrite plates nucleate at the prior austenite grain boundary (PAG) or α/γ interface, and grow into the grain interior. In order to measure the size of prior austenite grain size, samples were directly quenched into water from different austenitizing temperatures, and the PAG was etched using an electrochemical etching method [27]. The equivalent grain size of prior austenite in the tested sample is smaller than 30 µm after austenitizing at 950 and 1000 °C, and increases significantly when the austenitizing temperature exceeds 1000 °C (Figure 6a). It can be seen from the SEM micrographs that the size of blocky retained austenite and the length of bainitic ferrite plate (L_{BF}) increase with an increment in austenitizing temperate. To analyze quantitatively the effect of austenitizing temperature on the microstructure of ultra-fine bainite, the average length and true thickness of bainitic ferrite were measured 100 times based on SEM and TEM micrographs. The true thickness of bainitic ferrite (t_{BF}) can be obtained according the following equation:

$$t_{BF} = 2L_T/\pi \tag{5}$$

where L_T presents the mean linear intercept, which is determined along the normal direction of the bainitic ferrite plates. The statistical results are compared and listed in Figure 6b. The value of L_{BF} increases linearly with the austenitizing temperature from 950 °C to 1100 °C, but the rule does not apply to the case of t_{BF}. The t_{BF} increases slowly when the austenitizing temperature increases from 950 °C to 1000 °C. However, the t_{BF} increases significantly when the austenitizing temperature exceeds 1000 °C. The strength of ultra-fine bainite steel depends on the true thickness of bainitic ferrite plates, and the finer plates, the higher strength of the tested steel.

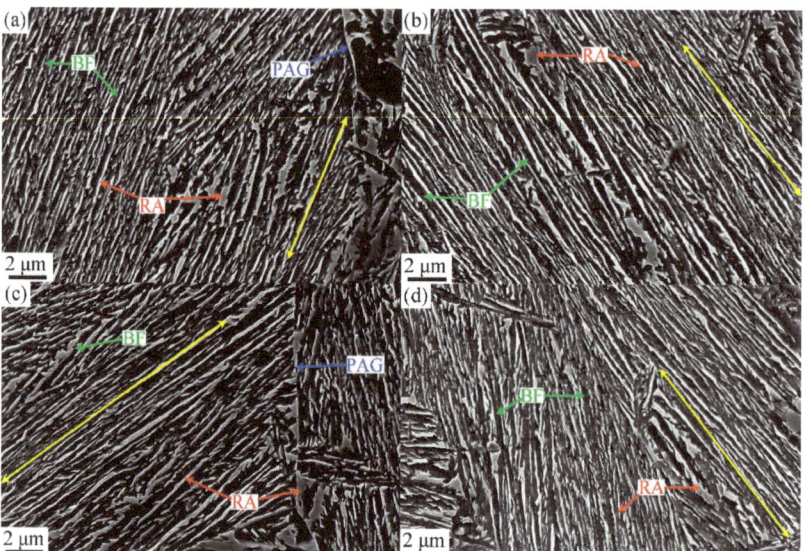

Figure 4. SEM micrographs of the samples austenitized at different temperatures: (**a**) 950 °C, (**b**) 1000 °C, (**c**) 1050 °C, (**d**) 1100 °C.

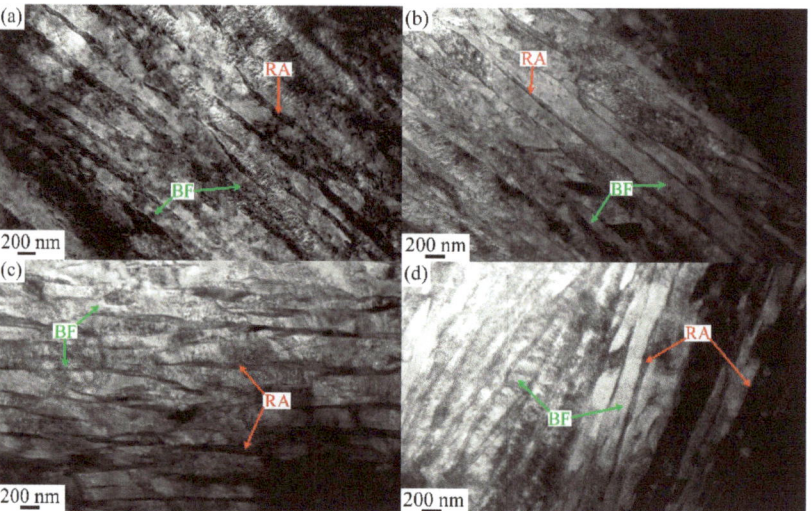

Figure 5. TEM micrographs of the samples austenitized at different temperatures: (**a**) 950 °C, (**b**) 1000 °C, (**c**) 1050 °C, (**d**) 1100 °C.

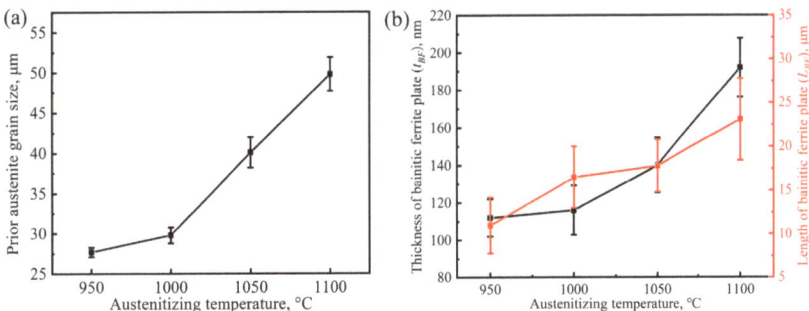

Figure 6. Grain size of prior austenite (**a**) and thickness and length of bainitic ferrite plate (**b**) changing with the austenitizing temperature.

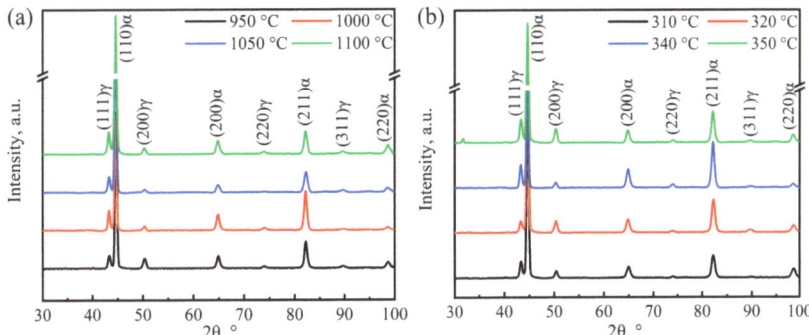

Figure 7. XRD patterns of the samples after different heat treatments: (**a**) austenitizing temperature; (**b**) austempering temperature.

The XRD patterns of the samples austenitized at different temperatures are shown in Figure 8a. the microstructure of the samples is composed of ferrite phase (α) and austenite phase (γ), and no diffraction peak of carbide. The volume fractions and average carbon contents of retained austenite were counted and are listed in Table 1. With the increase in austenitizing temperature, the volume fraction of retained austenite within the samples first decreases and then increases. The volume fraction of retained austenite reaches the lowest value of 8.8 vol.% from the sample austenitized at 1000 °C. However, the average carbon content of the retained austenite shows an opposite change, and the average carbon content of the retained austenite reaches the peak value of 1.9% when the austenitizing temperature is selected as 1000 °C. The carbon content of retained austenite is closely related to the plasticity and toughness of ultra-fine bainitic steel [48]. Based on the Gaussian multi-peak fitting method, (200) austenite peak was selected to analyze the volume fraction and carbon content of film-like and blocky retained austenite within the tested steel [6]. The corresponding calculation results are also shown in Table 1. The carbon contents of film-like retained austenite in the samples after austenitizing at 1000 and 1050 °C are significantly higher than those of other samples. However, the carbon content of blocky retained austenite has little correlation with the austenitizing temperature, which may be related to the isothermal bainite transformation process of supercooled austenite.

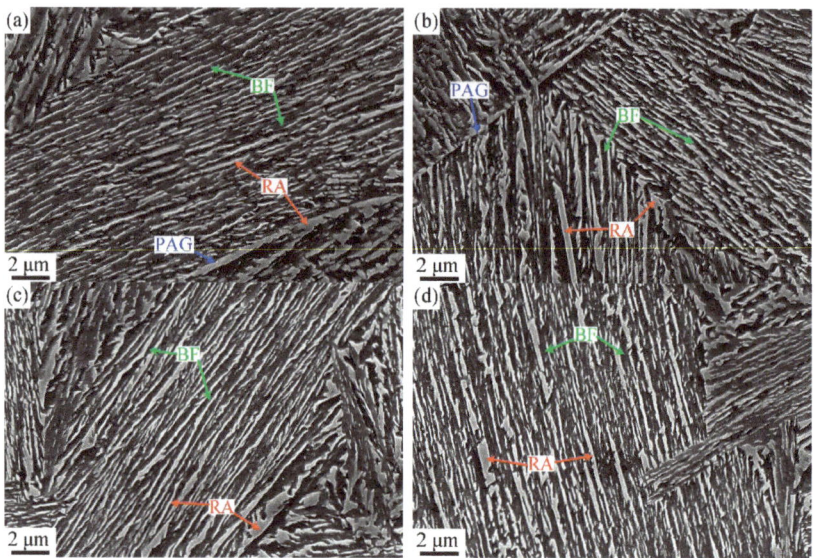

Figure 8. SEM micrographs of the samples austempered at different temperatures: (**a**) 310 °C, (**b**) 320 °C, (**c**) 340 °C, (**d**) 350 °C.

Table 1. Microstructural parameters of the tested steels at different austenitizing temperatures.

Temperature, °C	$\rho \times 10^{15}$ m^{-2}	V_B, %	V_γ, %	C_γ, wt. %	$V_{F\text{-}RA}$, %	$C_{F\text{-}RA}$, wt.%	$V_{B\text{-}RA}$, %	$C_{B\text{-}RA}$, wt.%
950	3.3 ± 0.5	83.9 ± 0.6	16.1 ± 0.4	1.3 ± 0.1	8.5	1.4	7.6	1.2
1000	5.4 ± 0.3	91.2 ± 0.5	8.8 ± 0.5	1.9 ± 0.3	3.4	2.5	5.5	1.3
1050	4.9 ± 0.7	89.6 ± 0.5	10.4 ± 0.5	1.8 ± 0.4	3.5	2.4	6.9	1.2
1100	4.1 ± 0.8	86.6 ± 0.6	13.4 ± 0.4	1.5 ± 0.3	4.5	1.6	8.9	1.4

The ultra-fine bainite microstructural morphology from the samples austempered at different temperatures is shown in Figures 8 and 9. Like the case of that austenitized at different temperatures, the microstructural parameters of the tested steel were also counted and are shown in Figure 10. When the austempered temperature is below M_s, specifically 310 °C, the true thickness of bainitic ferrite plates within the samples is significantly finer than those of the samples whose austempered temperatures are above M_s. The lower the austempered temperature, the finer the bainitic ferrite plate, which can be explained by the fact that prior martensite in supercooled austenite provides more nucleation sites for the following bainite transformation.

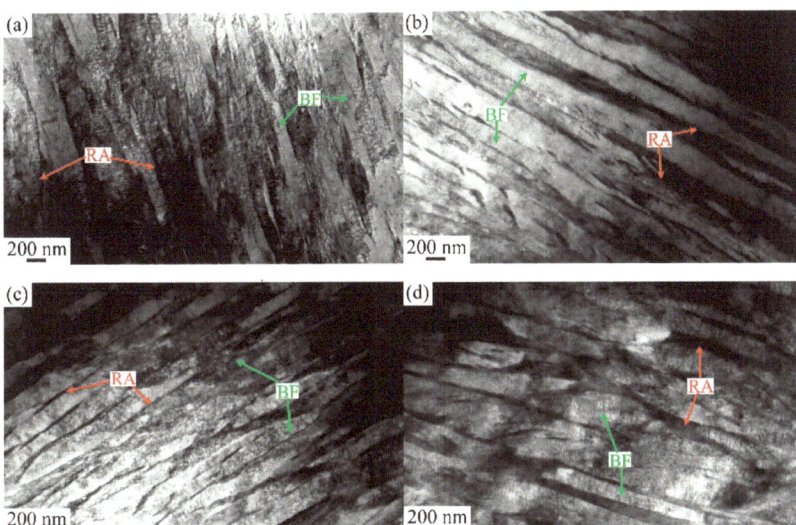

Figure 9. TEM micrographs of the samples austempered at different temperatures: (**a**) 310 °C, (**b**) 320 °C, (**c**) 340 °C, (**d**) 350 °C.

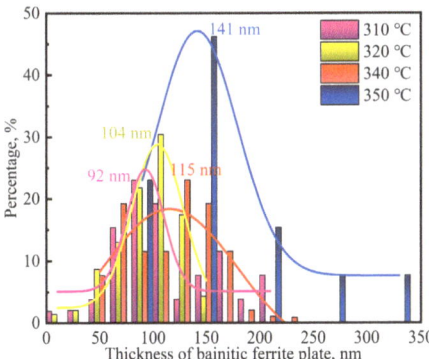

Figure 10. True thickness of bainitic ferrite plate from the samples austempered at different temperatures.

The XRD pattern of the samples austempered at different temperatures is shown in Figure 8b, and the volume fraction of prior martensite, bainitic ferrite, retained austenite and the average carbon content within the latter phase of the samples is calculated and shown in Figure 11. The constitution of the microstructures of the samples are different when the austempering temperature is below or above Ms. For the samples austempered below Ms, although the volume fraction of retained austenite changes little, the average carbon content of retained austenite decreases significantly. However, the volume fraction of retained austenite increases with an increment in temperature, and the average carbon content within it changes in the opposite way when the austempering temperature is above Ms. The calculated results of retained austenite with film-like and blocky morphology and their carbon content based on the Gaussian multi-pears fitting method are listed and compared in Table 2. On the whole, like the case of the effect of austenitizing temperature, the austempering temperature shows little effect on the carbon content of blocky retained austenite. Although the volume fraction of blocky retained austenite (V_{B-RA}) increases significantly for the sample austempered below Ms, more carbon atoms are injected into film-like retained austenite. When austempering temperature exceeds Ms, both the volume fractions of the two morphologies of the retained austenite increase with an increment in austempering temperature, but the carbon content of the film-like retained austenite decreases, as shown in Table 2.

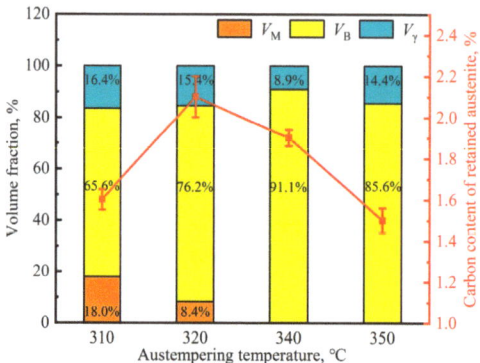

Figure 11. Volume fraction of constituting phases and carbon content of retained austenite from the samples austempered at different temperatures. VM: prior martensite, VB: bainitic ferrite, Vγ: retained austenite.

Table 2. Microstructural parameters of retained austenite from the samples austempered at different temperatures.

Temperature, °C	V_{F-RA}, %	C_{F-RA}, wt.%	V_{B-RA}, %	C_{B-RA}, wt.%
310	2.5	2.4	12.9	1.5
320	9.4	1.7	6.9	1.5
340	3.4	2.5	5.5	1.4
350	7.7	1.7	6.7	1.4

3.3. Mechanical Properties

The engineering stress–strain curves of the samples austenitized at various temperatures are shown in Figure 12a. As observed, the engineering stress–strain curves of the samples austenitized at different temperatures do not have much difference except when austenitizing at 950 °C. The value of the yield strength (YS), ultimate tensile strength (UTS), yield ratio, uniform elongation (UEL), total elongation (TEL) and impact toughness at room temperature of different austenitized samples are summarized and shown in Figure 13. The UTS and TS are higher in the samples austenitized at 1000–1100 °C (~1300 MPa and 900 MPa, respectively) compared to the sample austenitized at 950 °C (1200 MPa and 800 MPa). Although the value of t_{BF} of the 950 °C sample is close to that of the 1000 °C sample, the volume fraction of bainitic ferrite and the dislocation density within are probably due to the increase in strength. The UEL of different austenitized samples exhibit a weaker correlation with austenitizing temperature, while the TEL and impact toughness of austenitized samples increase firstly and then decrease with the increase in austenitizing temperature. The austenitized samples have lower impact toughness and show brittle behavior under room temperature; however, the 1000 °C austenitized sample exhibits an extraordinary impact toughness of approximately 25 J. By contrast, the 950 °C austenitized sample only absorbs half of this value at the same impact testing.

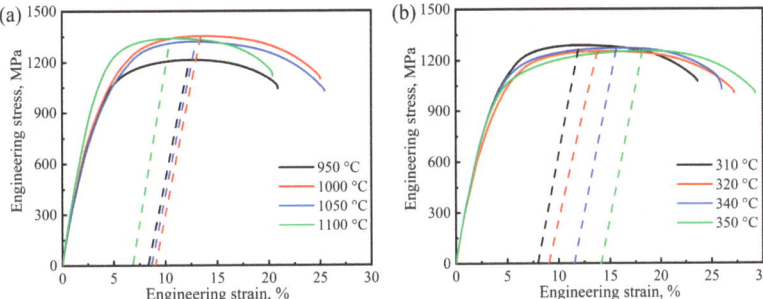

Figure 12. Engineering stress–strain curves of the tested steel after different heat treatments: (**a**) austenitizing temperature; (**b**) austempering temperature.

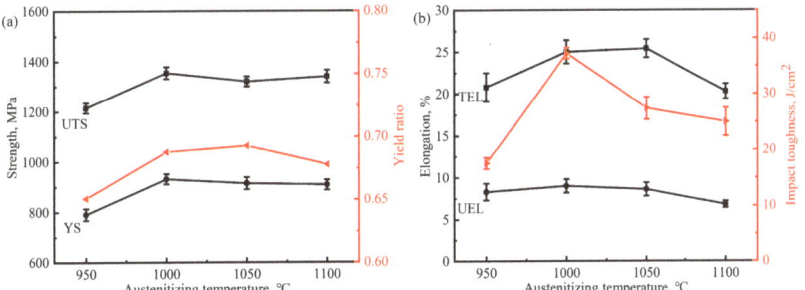

Figure 13. Mechanical property parameters of the samples after austenitizing at different temperatures: (**a**) strength and yield ratio, (**b**) elongation and impact toughness.

Figure 12b shows the engineering stress–strain curves of the samples austempered at above and below Ms. Unlike the case of austenitized samples, the tensile curves of the samples, austempered at 310 °C, 320 °C (below Ms) and 340 °C, 350 °C (above Ms), are not obviously different, but their strengths and elongation are varied. The parameters of tensile property and impact toughness of the austempered samples are summarized and compared in Figure 14. It can be seen that the austempering temperature shows little effect on the UTS of the samples. When the austempering temperature is above Ms, the value of YS and impact toughness of the samples decrease, while the elongation increase with the increase in austempering temperature. The same phenomenon also occurs when the austempering temperature is below Ms. This may be related to the volume fraction of bainite and prior martensite as well as the mechanical stability of retained austenite. The microstructure of the samples is composed of bainitic ferrite and retained austenite when austempered at 340 and 350 °C. Their mechanical properties depend on the volume fraction of bainite ferrite and the thickness of the plates as well as the mechanical stability of the residual austenite. After austempering at 310 and 320 °C, the prior martensite formed in the microstructure of the samples, and the lower the austempering temperature, the higher the volume fraction of prior martensite. The presence of prior martensite not only change the composition of the samples, but also refine the thickness of bainitic ferrite plate and decrease the average carbon content of retained austenite. The phases and corresponding microstructure parameter on the mechanical properties of the austempered samples will be discussed in Section 4.

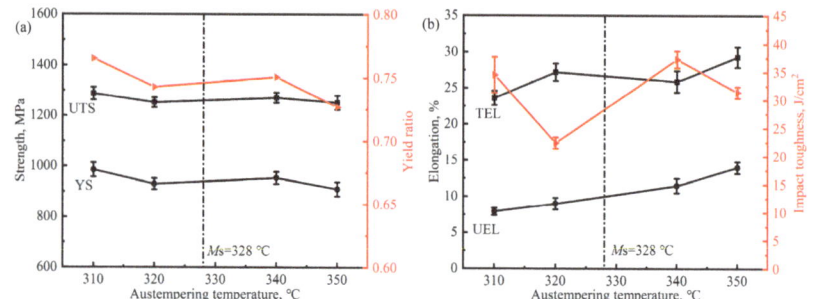

Figure 14. Mechanical property parameters of the samples after austempering at different temperatures: (**a**) strength and yield ratio, (**b**) elongation and impact toughness.

4. Discussion

In the process of manufacturing ultra-fine bainitic steel by conventional one-step austempering, the transformation kinetics, microstructural morphology and mechanical properties of bainite mainly depend on the driving force of bainite transformation and the nuclei site for bainitic ferrite plates [49]. In the present paper, the effects of transformation temperature on bainite finishing time, the thickness of bainitic ferrite plates and the density dislocation within it, the volume fraction of retained austenite and carbon content within it, and the tensile and impact properties of the tested steel at different austenitizing and austempering temperatures were all analyzed to explore the optimal one-step isothermal quenching process.

At the same austempering temperature, which is above Ms, bainitic ferrite mainly nucleates at the grain boundaries of prior austenite (PAG) or the bainite sheaves formed in the early transformation, and grows into untransformed austenite grain (Figure 1). After austenitizing at 950 °C, finer austenite grains are formed in the tested steel. Compared to that in the samples austenitized at higher temperature (e.g., 1000 °C), the proportion of PAG in the sample is higher, which means that more bainite sheaves can be formed within a single austenite grain when austempering at the same temperature. In addition, more nucleation sites also mean that the incubation period of bainite transformation is very short, which can be reflected in the transformation rate curves that the perk value of transformation rate reached in the beginning of bainite transformation (Figure 2c). Due to the formation of the prior bainite sheaves in austenite grain, the growth of the newly formed bainite lath in the untransformed austenite region is inhibited, which results in a low transformation rate and a longer transformation finish time.

Another result of the more bainite nuclei sites is that the bainitic ferrite plates is refined after the transformation is completed, which improves the strength of the tested steel. The thickness of the plates at 950 and 1000 °C is comparable, but the difference in yield strength between them is 141 MPa, which can be attributed to the dislocation density (3.3×10^{15} m^{-2} to 5.4×10^{15} m^{-2}) within the bainitic ferrite. The growth of bainite in the 950 °C austenitized sample is inhibited, resulting in retained austenite with high volume fraction in the final microstructure, especially the blocky retained austenite (8.6 vol.%). The blocky retained austenite with low carbon content tends to transform to martensite during the tensile and impact process, which is harmful to the plasticity and toughness of the tested steel. As a result, the impact toughness of the 950 °C sample is only 17.6 J, which is consistent with the results for the previous study [50].

After the austenitizing temperature exceeds 1000 °C, with the increase in austenitizing temperature, the prior austenite grain size increases obviously, but the ratio of grain boundary in tested steel decreases and then reduces the bainite nuclei sites. Nonetheless, the growth inhibition of bainitic ferrite plates is weaker, and the length of the plates increases significantly (Figure 6). After the austenitizing temperature is further increased to 1100 °C, the prior austenite grain size increases significantly, which can be reflected by the length of bainite ferrite plates after the transformation is completed. Although the highest degree of supercooling can provide the highest driving force for following bainite transformation, the proportion of PAG acted as the bainite nuclei site is lower compared with that in the 1000 °C austenitized sample, leading to the thicker bainite ferrite plates formed in the final microstructure (Figure 5d). In this case, the diffusion distance of carbon atoms from bainitic ferrite to adjacent film-like austenite increases in the isothermal process, resulting in the longer finishing time of bainite. More carbon atoms are retained in untransformed austenite after the transformation, forming more blocky retained austenite with low mechanical stability. The slight decrease in plasticity and impact toughness of the tested steel can be attributed to the increase in volume fraction of the retained austenite in the bainite microstructure.

The final microstructure of ultra-fine bainitic steel is composed of bainitic ferrite and retained austenite when the austempering temperature is above Ms. Like the case of austenitizing temperature, the transformation kinetics, microstructural size and mechanical properties of bainitic steel is determined by the degree of supercooling. Although the incubation period is also not very obvious, and the finishing time of bainite for the 350 °C austempered sample is comparable to that of the sample austempered at 340 °C, the max transformation rate for the former occurs 18 s after the beginning of the austempered process (Figure 3c). A lower degree of supercooling means a lower driving force for the following bainite transformation, leading to a lower degree completion of the final transformation. Not only did the thickness of bainitic ferrite plates increase (Figure 10), but the volume fraction of the retained austenite also increased while the average carbon content decreased (Figure 11). On the condition that the volume fraction and carbon content of blocky retained austenite in the two samples are not much different, the volume fraction of film-like retained austenite in the 350 °C austempered sample is higher and the carbon content is lower compared to the 350 °C austempered sample. The TRIP effect tends to occur in film-like retained austenite with lower carbon content during tensile test to release stress concentration and prevent microcrack propagation, which results in a higher plasticity of the 350 °C austempered sample.

It has been known that the surface or tips of the prior martensite act as the nuclei for subsequent bainitic ferrite plates and accelerate the bainite transformation [51]. However, the finishing time of bainite in the paper increased when the austempering temperature was below Ms, and the lower the austempering temperature, the longer the finishing time (Figure 3b). The phenomenon can be explained as follows. Firstly, the prior martensite divides the supercooled austenite grain and suppresses it as the bainitic ferrite grows. Secondly, more carbon atoms are ejected from prior martensite to untransformed austenite and improve the stability of the latter. And thirdly, decreasing the austempering temperature also strengthens the stability of supercooled austenite. The combined effect of all the above factors suppresses the growth of bainite, thus increasing the finishing time of bainite transformation.

As demonstrated by the aforementioned SEM and TEM micrographs, on the one hand, the introduction of prior martensite refines the bainitic ferrite plates and improves the fine grain strengthening effect. On the other hand, the presence of hard martensite grain can also improve the strength of the sample. The lower the austempering temperature, the higher the strength of the ultra-fine bainitic steel (Figure 14a). Furthermore, the size/morphology and carbon content of retained austenite are two major factors affecting the mechanical properties of ultra-fine bainitic steel: a finer grain size and a higher carbon content give rise to higher stability [52]. The prior martensite also refines the size of film-like retained austenite, and improve the carbon content within it (Table 2). The higher the carbon content of the film-like retained austenite, the higher the mechanical stability, and a more difficult TRIP effect occurs when the crack spreads to the austenite during deformation. Compared to that in the 320 °C austempered sample, more of the prior martensite formed in the 310 °C austempered sample, resulting in a lower completion degree of bainite transformation.

However, the carbon content of film-like retained austenite in the latter sample is significant, so the impact toughness of the sample clearly improved.

A competitive relationship is observed between the nuclei sites and the growth of bainite sheaves when studying the effect of transformation temperature on bainite transformation kinetics, microstructural parameters, tensile and impact properties. The nuclei sites of bainite determine the thickness of bainitic ferrite plates and thus the strength of ultra-fine bainitic steel. The degree of transformation completion, which depends on the growth of bainite sheaves, not only determines the finishing time of bainite transformation, but also determines the plasticity and impact toughness of ultra-fine bainitic steel by affecting the volume fraction of film-like and blocky retained austenite and carbon content within it. Based on the above experimental results and discussion, the optimal treatment for medium-carbon ultra-fine bainitic steel in the present paper is by austenitizing it at 1000 °C and austempering it at 340 °C.

5. Conclusions

The effects of transformation temperature, i.e., austenitizing temperature and austempering temperature, on transformation kinetics, microstructural completion and parameters and the mechanical properties of medium-carbon ultra-fine bainite steel were investigated. The main conclusions are summarized as follows:

(1) With the austenitizing temperature increase from 950 to 1000 °C, the increase in the mechanical property of medium-carbon bainitic steel can be attributed to the higher volume fraction of bainite ferrite and the higher dislocation density within it. When the austenitizing temperature exceeds 1000 °C, the volume fraction of blocky retained austenite in the final bainite microstructure increases due to the obvious coarsening of prior austenite grain, and the plasticity and impact toughness are gradually reduced.

(2) The finishing time and completion degree of bainite transformation depend on the degree of the undercooling of prior austenite when the austempering temperature is above M_s. With the increase in austempering temperature, the finishing time of bainite transformation is clearly extended, and the completion degree of bainite transformation decreases. The increment in elongation and the decrement in impact toughness for medium-carbon bainitic steel can be attributed to the coarsening of bainitic ferrite plates, the increase in the volume fraction of retained austenite with film-like morphology and the decrease in the carbon content within it.

(3) The prior martensite forms before bainite transformation when the austempering temperature is below M_s. With the decrease in austempering temperature, the higher degree of supercooling and the presence of prior martensite refines the bainitic ferrite plates and increases the carbon content of film-like retained austenite, thereby significantly improving the strength and impact toughness of medium-carbon bainitic steel. However, the introduction of prior martensite inhibits the growth of bainite sheaves, and thus reduces the completion of bainite transformation, resulting in a significant increase in the volume fraction of blocky retained austenite and a slight decrease in the elongation of medium-carbon bainitic steel. Based on the experimental results and discussion, the optimal treatment for medium-carbon ultra-fine bainitic steel in the present paper is through austenitization at 1000 °C and austempering at 340 °C.

Author Contributions: Conceptualization, Y.W. and H.W.; methodology, N.G.; software, N.G.; validation, X.Y. and N.G.; formal analysis, Y.W.; investigation, H.W.; resources, Y.W. and H.W.; data curation, X.Y. and H.W.; writing—original draft preparation, X.Y.; writing—review & editing, X.Y. and H.W.; visualization, N.G.; supervision, N.G.; project administration, Y.W.; funding acquisition, H.W. All authors have read and agreed to the published version of the manuscript.

Funding: This work was financially supported by the National Key Research and Development Project of China (No. 2022YFB3708200), Fundamental Research Funds for the Central Universities (FRF-BD-23-01) and National Natural Science Foundation of China (No. 52301165).

Institutional Review Board Statement: Not applicable.

Informed Consent Statement: Not applicable.

Data Availability Statement: Data are contained within the article.

Conflicts of Interest: The authors declare no conflict of interest.

References

1. Caballero, F.G.; Bhadeshia, H.K.D.H.; Mawella, K.J.A.; Jones, D.G.; Brown, P. Very strong low temperature bainite. *Mater. Sci. Technol.* **2002**, *18*, 279–284. [CrossRef]
2. Soliman, M.; Palkowski, H. Development of the low temperature bainite. *Arch. Civ. Mech. Eng.* **2016**, *16*, 403–412. [CrossRef]
3. Caballero, F.G.; Garcia-Mateo, C. Super-bainite. *Encycl. Mater. Met. Alloys* **2022**, *2*, 73–83.
4. Morales-Rivas, L.; Garcia-Mateo, C.; Kuntz, M.; Sourmail, T.; Caballero, F.G. Induced martensite transformation during tensile test in nanostructured bainitic steels. *Mater. Sci. Eng. A* **2016**, *662*, 169–177. [CrossRef]
5. Peet, M.J.; Fielding, L.C.D.; Hamedany, A.A.; Rawson, M.; Hill, P.; Bhadeshia, H.K.D.H. Strength and toughness of clean nanostructured bainite. *Mater. Sci. Technol.* **2017**, *33*, 1171–1179. [CrossRef]
6. Long, X.; Zhang, F.; Yan, Z.; Lv, B. Study on microstructure and properties of carbide-free and carbide-bearing bainitic steels. *Mater. Sci. Eng. A* **2018**, *715*, 10–16. [CrossRef]
7. Zhang, F.; Yang, Z. Development of and Perspective on high-performance nanostructured bainitic bearing steel. *Engineering* **2019**, *5*, 319–328. [CrossRef]
8. Zhang, C.; Li, S.; Fu, H.; Lin, Y. Microstructure evolution and wear resistance of high silicon bainitic steel after austempering. *J. Mater. Res. Technol.* **2020**, *9*, 4826–4839. [CrossRef]
9. Li, J.; Yang, Z.; Zhao, G.; Zhang, F. A simultaneously improved strength and ductility on carbide free bainite steel via novel ausrolling and twinning process based on SFE controlling. *Mater. Sci. Eng. A* **2022**, *832*, 142442. [CrossRef]
10. Garcia-Mateo, C.; Caballero, F.G.; Bhadeshia, H.K.D.H. Development of hard bainite. *ISIJ Int.* **2003**, *43*, 1238–1243. [CrossRef]
11. Hausman, K.; Krizan, D.; Spiradek-Hahn, K.; Pichler, A.; Werner, E. The influence of Nb on transformation behavior and mechanical properties of TRIP-assisted bainitic-ferritic sheet steels. *Mater. Sci. Eng. A* **2013**, *588*, 142–150. [CrossRef]
12. He, B.; Xu, W.; Huang, M. Effect of boron on bainitic transformation kinetics after ausforming in low carbon steels. *J. Mater. Sci. Technol.* **2017**, *33*, 1494–1503. [CrossRef]
13. Douguet, P.; Da Rosa, G.; Hoummada, P.M.J.D.K. Effect of boron segregation on bainite nucleation during isothermal transformation. *Scr. Mater.* **2022**, *207*, 114286. [CrossRef]
14. van Bohemen, S.M.C.; Santofima, M.J.; Sietsma, J. Experimental evidence for bainite formation below Ms in Fe-0.66C. *Scr. Mater.* **2008**, *58*, 488–491. [CrossRef]
15. Samanta, S.; Biswas, P.; Giri, S.; Singh, S.B.; Kundu, S. Formation of bainite below the Ms temperature: Kinetics and crystallography. *Acta Mater.* **2016**, *105*, 390–403. [CrossRef]
16. Feng, J.; Frankenbach, T.; Wettlaufer, M. Strengthening 42CrMo4 steel by isothermal transformation below martensite start temperature. *Mater. Sci. Eng. A* **2017**, *683*, 110–115. [CrossRef]
17. Xia, S.; Zhang, F.; Yang, Z. Microstructure and mechanical properties of 18Mn3Si2CrMo steel subjected to austempering at different temperatures below Ms. *Mater. Sci. Eng. A* **2018**, *724*, 103–111. [CrossRef]
18. Zhao, L.; Qian, L.; Zhou, Q.; Li, D.; Wang, T.; Jia, Z.; Zhang, F.; Meng, J. The combining effects of ausforming and below-Ms or above-Ms austempering on the transformation kinetics, microstructure and mechanical properties of low-carbon bainitic steel. *Mater. Des.* **2019**, *183*, 108123. [CrossRef]
19. Eres-Castellanos, A.; Caballero, F.G.; Garcia-Mateo, C. Stress or strain induced martensite and bainitic transformations during ausforming processes. *Acta Mater.* **2020**, *189*, 60–72. [CrossRef]
20. Guo, H.; Fan, Y.P.; Feng, X.Y.; Li, Q. Ultrafine bainitic steel produced through ausforming-quenching process. *J. Mater. Res. Technol.* **2020**, *9*, 3633–3659. [CrossRef]
21. Hu, H.; Tian, J.; Xu, G.; Zurob, H.S. New insights into the effects of deformation below-Ms on isothermal kinetics of bainitic transformation. *J. Mater. Res. Technol.* **2020**, *9*, 15750–15758. [CrossRef]
22. Gao, G.; Guo, H.; Gui, X.; Tan, Z.; Bai, B. Inverted multi-step bainitic austempering process routes: Enhanced strength and ductility. *Mater. Sci. Eng. A* **2018**, *739*, 298–305. [CrossRef]
23. Chu, C.; Qin, Y.; Li, X.; Yang, Z.; Zhang, F.; Guo, C.; Long, X.; You, L. Effect of two-step austempering process on transformation kinetics of nanostructured bainitic steel. *Materials* **2019**, *12*, 166. [CrossRef] [PubMed]
24. Gong, W.; Tomota, Y.; Harjo, S.; Su, Y.H.; Aizawa, K. Effect of prior martensite on bainite transformation in nanobainite steel. *Acta Mater.* **2015**, *85*, 243–249. [CrossRef]
25. Ravi, A.M.; Navarro-Lopez, A.; Sietsma, J.; Santofimia, M.J. Influence of martensite/austenite interfaces on bainite formation in low-alloy steels below Ms. *Acta Mater.* **2020**, *188*, 394–405. [CrossRef]
26. Avishan, B.; Khoshkebari, S.M.; Yazdani, S. Effect of pre-existing martensite within the microstructure of nano bainitic steel on its mechanical properties. *Mater. Chem. Phys.* **2021**, *260*, 124160. [CrossRef]
27. Lee, S.-J.; Park, J.-S.; Lee, Y.-K. Effect of austenite grain size on the transformation kinetics of upper and lower bainite in a low-alloy steel. *Scr. Mater.* **2008**, *59*, 87–90. [CrossRef]

28. Zhao, J.; Li, J.; Ji, H.; Wang, T. Effect of austenite temperature on mechanical properties of nanostructured bainitic steel. *Materials* **2017**, *10*, 874. [CrossRef] [PubMed]
29. Hu, F.; Hodgson, P.D.; Wu, K.M. Acceleration of the super bainite transformation through a coarse austenite grain size. *Mater. Lett.* **2014**, *122*, 240–243. [CrossRef]
30. Pashangeh, S.; Somani, M.; Banadkouki, S.S.G. Microstructural evolution in a high-silicon medium carbon steel following quenching and isothermal holding above and below the Ms temperature. *J. Mater. Res. Technol.* **2020**, *9*, 3438–3446. [CrossRef]
31. Tian, J.; Xu, G.; Hu, H.; Wang, X.; Zurob, H. Transformation kinetics of carbide-free bainitic steels during isothermal holding above and below Ms. *J. Mater. Res. Technol.* **2020**, *9*, 13594–13606. [CrossRef]
32. Kafadar, G.; Kalkanli, A.; Özdem, A.T.; Ögel, B. Effect of isothermal transformation treatment and tempering on the microstructure and hardness of a medium C and high Si steels. *ISIJ Int.* **2021**, *61*, 1679–1687. [CrossRef]
33. Hu, H.; Zurob, H.S.; Xu, G.; Embury, D.; Gary, R.; Purdy, G.R. New insights to the effects of ausforming on the bainitic transformation. *Mater. Sci. Eng. A* **2015**, *626*, 34–40. [CrossRef]
34. Hu, H.; Xu, G.; Wang, L.; Zhou, M.; Xue, Z. Effect of ausforming on the stability of retained austenite in a C-Mn-Si bainitic steel. *Met. Mater. Int.* **2015**, *21*, 929–935. [CrossRef]
35. Speer, J.G.; Edmonds, D.V.; Rizzo, F.C.; Matlock, D.K. Partitioning of carbon from supersaturated plates of ferrite, with application to steel processing and fundamentals of the bainite transformation. *Curr. Opin. Solid State Mater. Sci.* **2004**, *8*, 219–237. [CrossRef]
36. Caballero, F.G.; Garcia-Mateo, C.; Santofimia, M.J.; Miller, M.K.; de Andrés, C.G. New experimental evidence on the incomplete transformation phenomenon in steel. *Acta Mater.* **2009**, *57*, 8–17. [CrossRef]
37. Kim, B.; Celada, C.; Martín, D.S.; Sourmail, T.; Rivera-Díaz-del-Castillo, P.E.J. The effect of silicon on the nanoprecipitation of cementite. *Acta Mater.* **2013**, *61*, 6983–6992. [CrossRef]
38. Toji, Y.; Matsuda, H.; Raabe, D. Effect of Si on the acceleration of bainite transformation by pre-existing martensite. *Acta Mater.* **2016**, *116*, 250–262. [CrossRef]
39. Long, X.; Branco, R.; Zhang, F.; Berto, F.; Martins, R. Influence of Mn addition on cyclic deformation behaviour of bainitic rail steels. *Int. J. Fatigue* **2020**, *132*, 105362. [CrossRef]
40. Guo, H.; Zhou, P.; Zhao, A.; Zhi, C.; Ding, R.; Wang, J. Effect of Mn and Cr contents on microstructures and mechanical properties of low temperature bainitic steel. *J. Iron Steel Res. Int.* **2017**, *24*, 290–295. [CrossRef]
41. Garcia-Mateo, C.; Caballero, F.G.; Bhadeshia, H.K.D.H. Acceleration of low-temperature bainite. *ISIJ Int.* **2003**, *43*, 1821–1825. [CrossRef]
42. Qian, L.; Zhou, Q.; Zhang, F.; Meng, J.; Zhang, M.; Tian, Y. Microstructure and mechanical properties of a low carbon carbide-free bainitic steel co-alloyed with Al and Si. *Mater. Des.* **2012**, *39*, 264–268. [CrossRef]
43. Williamson, G.K.; Smallman, R.E. Dislocation densities in some annealed and cold-worked metals from measurements on the X-ray debye-scherrer spectrum. *Philos. Mag.* **1956**, *1*, 34–46. [CrossRef]
44. De, A.K.; Murdock, D.C.; Mataya, M.C.; Speer, J.G.; Matlock, D.K. Quantitative measurement of deformation-induced martensite in 304 stainless steels by X-ray diffraction. *Scr. Mater.* **2004**, *50*, 1445–1449. [CrossRef]
45. Dyson, D.J.; Holmes, B. Effect of alloying additions on the lattice parameter of austenite. *J. Iron Steel Inst.* **1970**, *208*, 469–474.
46. Xiong, X.C.; Chen, B.; Huang, M.X.; Wang, J.F.; Wang, L. The effect of morphology on the stability of retained austenite in a quenched and partitioned steel. *Scr. Mater.* **2013**, *68*, 321–324. [CrossRef]
47. Long, X.; Zhao, G.; Zhang, F.; Xu, S.; Yang, Z.; Du, G.; Branco, R. Evolution of tensile properties with transformation temperature in medium-carbon carbide-free bainitic steel. *Mater. Sci. Eng. A* **2020**, *775*, 138964. [CrossRef]
48. Sourmail, T.; Garcia-Mateo, C.; Caballero, F.G.; Morales-Rivas, L.; Rementeria, R.; Kuntz, M. Tensile ductility of Nanostructured bainitic steels: Influence of retained austenite stability. *Metals* **2017**, *7*, 31. [CrossRef]
49. Guo, H.; Feng, X.; Zhao, A.; Li, Q.; Ma, J. Influence of prior martensite on bainite transformation, microstructures, and mechanical properties in ultra-fine bainitic steel. *Materials* **2019**, *12*, 527. [CrossRef] [PubMed]
50. Zhao, F.; Morales-Rivas, L.; Yu, Q.; Wang, G.; Caballero, F.G.; San-Martin, D. Unforeseen influence of the prior austenite grain size on the mechanical properties of a carbide-free bainitic steel. *Mater. Sci. Eng. A* **2023**, *881*, 145388. [CrossRef]
51. Qian, L.; Li, Z.; Wang, T.; Li, D.; Zhang, F.; Meng, J. Roles of pre-formed martensite in below-Ms bainite formation, microstructure, strain partitioning and impact absorption energies of low-carbon bainitic steel. *J. Mater. Sci. Technol.* **2022**, *96*, 69–85. [CrossRef]
52. Zhou, Q.; Qian, L.; Tan, J.; Meng, J.; Zhang, F. Inconsistent effects of mechanical stability of retained austenite on ductility and toughness of transformation-induced plasticity steels. *Mater. Sci. Eng. A* **2013**, *578*, 370–376. [CrossRef]

Disclaimer/Publisher's Note: The statements, opinions and data contained in all publications are solely those of the individual author(s) and contributor(s) and not of MDPI and/or the editor(s). MDPI and/or the editor(s) disclaim responsibility for any injury to people or property resulting from any ideas, methods, instructions or products referred to in the content.

Article

Analysis of the Uniformity of Mechanical Properties along the Length of Wire Rod Designed for Further Cold Plastic Working Processes for Selected Parameters of Thermoplastic Processing

Konrad Błażej Laber

Department of Metallurgy and Metal Technology, Faculty of Production Engineering and Materials Technology, Czestochowa University of Technology, 19 Armii Krajowej Ave., 42-200 Czestochowa, Poland; konrad.laber@pcz.pl; Tel.: +48-34-325-07-97

Citation: Laber, K.B. Analysis of the Uniformity of Mechanical Properties along the Length of Wire Rod Designed for Further Cold Plastic Working Processes for Selected Parameters of Thermoplastic Processing. *Materials* **2024**, *17*, 905. https://doi.org/10.3390/ma17040905

Academic Editor: Joan-Josep Suñol

Received: 11 January 2024
Revised: 6 February 2024
Accepted: 8 February 2024
Published: 15 February 2024

Copyright: © 2024 by the author. Licensee MDPI, Basel, Switzerland. This article is an open access article distributed under the terms and conditions of the Creative Commons Attribution (CC BY) license (https://creativecommons.org/licenses/by/4.0/).

Abstract: This study presents the results of research, the aim of which was to analyze the uniformity of the distribution of selected mechanical properties along the length of a 5.5 mm diameter wire rod of 20MnB4 steel for specific thermoplastic processing parameters. The scope of the study included, inter alia, metallographic analyses, microhardness tests, thermovision investigations, and tests of the wire rod mechanical properties (yield strength, ultimate tensile strength, elongation, relative reduction in area at fracture), along with their statistical analysis, for three technological variants of the rolling process differing by rolling temperature in the final stage of the rolling process (Reducing Sizing Mill rolling block [RSM]) and by cooling rate using STELMOR® cooling process. The obtained results led to the conclusion that the analyzed rolling process is characterized by a certain disparity of the analyzed mechanical properties along the length of the wire rod, which, however, retains a certain stability. This disparateness is caused by a number of factors. One of them, which ultimately determines the properties of the finished wire rod, is the process of controlled cooling in the STELMOR® line. Despite technological advances concerning technical solutions (among them, increasing the roller track speed in particular sections), it is currently not possible to completely eliminate the temperature difference along the length of the wire rod caused by the contact of individual coils with each other. From this point of view, for the analyzed thermoplastic processing parameters, there is no significant impact by the production process parameters on the quality of the finished steel product. Whereas, while comparing the mechanical properties and microstructure of the wire rod produced in the different technological combinations, it was found that the wire rod rolled in an RSM block at 850 °C and cooled after the rolling process on a roller conveyor at 10 °C/s had the best set of mechanical properties and the smallest microstructure variations. The wire rod produced in this way had the required level of plasticity reserve, which enables further deformation of the given type of steel in compression tests with a relative plastic strain of 75%. The uniformity of mechanical properties along the length of wire rods designed for further cold plastic working processes is an important problem. This is an important issue, given that wire rods made from 20MnB4 steel are an input material for further cold plastic working processes, e.g., for the drawing processes or the production of nails.

Keywords: wire rod rolling; metallographic analysis; thermovision investigation; mechanical properties; steel for cold upsetting

1. Introduction

The quality of a wire rod is mainly determined by the temperature–strain rolling parameters in the finishing rolling blocks and the cooling rate of the wire rod on the roller conveyor using blown air. Rolling in finishing blocks is a complex process due to the interaction between metal and tools in the rolling mill stands having a common drive and the occurrence of inter-stand forces. High rolling speeds and small distances between stands in the rolling blocks determine the material's strengthening and softening processes. High rolling speeds also affect the heat transfer between the roll, the tools, and the surroundings, as well as the end-of-rolling temperature, which, together with subsequent cooling, influences the microstructure and properties of the finished product [1].

The uniformity of the mechanical properties along the length of the rolled band depends on a number of factors. Among the most important is the temperature distribution along the length of the rolled material throughout the whole rolling line, from the heating furnace to the cooling process of the finished product. The influence of the uniformity of the temperature distribution along the length of the feedstock on the properties of the finished product was discussed by the authors of the study, among others [2]. In their study, they proposed, among other things, ways to reduce differences in temperature by using special heat covers. According to the results presented in [2], it was determined, inter alia, that the temperature uniformity along the processed wire rod has a significant effect on the mechanical properties of the finished product. Uniformity of temperature along the length of the rolled band also influences the plasticity of the metal's flow and, thus, the dimensional accuracy of the finished product, as evidenced by the results of studies presented, among others, in [3]. Divergences in the temperature distribution along the length of the rolled metal also affect the energy and force parameters of the rolling process. With the view of reducing temperature inhomogeneity along the length of the rolled band, the authors of [3] proposed a solution consisting of heating in the furnace to a higher temperature during the final stretch of the feedstock.

An important stage determining the uniformity of the mechanical properties of the wire rod along its length is the process of controlled cooling in the STELMOR® line. This issue was researched by the authors of, inter alia, [4,5]. The results obtained indicated that, among other factors, the nonuniformity of the mechanical properties of the wire rod along its length is due to the characteristic arrangement of the coils (in the form of a "spiral") during the cooling process. As a result of the contact of the individual coils with each other, areas of varying (elevated) temperature, so-called 'hot spots', are created, which result in different microstructure development during cooling and consequently in heterogeneity of the mechanical properties of the finished product. The authors of [4,5] indicated ways to reduce the distinction in mechanical properties along the length of the wire rod by, among other things, changing the density of the wire coils' setting on the roller conveyor. In addition, [4] proposed a comprehensive model for predicting the properties of wire rods.

The authors of [6–12] also dealt with issues related to heat transfer during controlled cooling of wire rods on a roller conveyor. These works can be used in detailed studies of the conditions for cooling the wire rod, e.g., to improve the uniformity of the wire rod properties along its length. Study [8] presents a model for simulating the cooling process of wire rods on a roller conveyor. The study takes into account all types of heat transfer as well as coils' density. A system for testing and recording the temperature of wire rods during cooling on a roller conveyor is presented in [6]. It was used during industrial research aimed, inter alia, at determining the interdependence between the cooling rate and the mechanical properties of the finished product. Study [12] presents a system for controlling the wire rod attributes during cooling on a roller conveyor based on time–temperature-transformation (TTT) diagrams.

Despite technological advances in the applied technical solutions (such as increasing the roller track speed in particular sections, controlling the coil density of the wire rod, air distribution improving cooling uniformity, new nozzle sets bettering cooling efficiency, or adding fans of higher capability and equalizing chambers) presented in, inter alia, [13], complete elimination of the temperature difference along the length of the wire rod caused by the contact of individual coils with each other constitutes an unresolved problem.

In the technical literature on the rolling process, it is possible to find studies dealing with the influence of the production process parameters on the quality (mainly microstructure and mechanical properties) of the finished product [14–25]. Several groups of steels for wire rod production were described in [14], such as interstitial free (IF) ferritic steels, ferritic/martensitic steels, and pearlitic micro-alloyed steels. The possibilities of forming the microstructure of these steels are described, as well as the basic technological guidelines for obtaining a finished product with the desired attributes. Study [20] is mainly concerned with the possibility of improving the properties of low-carbon steel wire rods by introducing alloy additions into the steel, which enhance the mechanical properties of the finished product. The authors of this study have shown that an important factor in terms of the range of applications of the obtained wire rod is the significant increase in its capacity for further direct cold plastic working processes. Studies [15–19,21] deal with the rolling processes of wire rods made from high-carbon steels. In these studies, the effects of temperature and cooling conditions on the microstructure and properties of the wire rod were analyzed. Studies [26,27] present a model of the microstructure development during the rolling of wire rods made of high-carbon steel, type C70D, and the test results of using it.

However, few studies analyze the quality of the finished rolled product in terms of uniformity of microstructure and mechanical properties along its length, especially with regard to wire rods. A brief analysis of the quality of the finished product concerning microstructure and mechanical properties along its length can be found in [28], among others. However, this study deals with smooth round bars.

For this reason, the undertaken research topic, that is, the analysis of the uniformity of mechanical properties along the length of cold upsetting steel wire rods for selected thermoplastic processing parameters, is, in the author's opinion, valid. This is an important issue, given that wire rods made from 20MnB4 steel are an input material for further cold plastic processing, e.g., for the drawing industry or the production process of nails.

2. Materials and Methods

2.1. Materials

The tests presented in this study were carried out on a 5.5 mm diameter wire rod made of low-carbon, cold-upsetting steel of the 20MnB4 type with a chemical composition in accordance with the PN-EN 10263-4:2004 standard (Table 1) [29].

Table 1. Chemical composition of 20MnB4 steel grade [29].

Steel Grade	Steel Number	Melt Analysis, mass%						
		C	Si	Mn	P_{max}, S_{max}	Cr	Cu_{max}	B
20MnB4	1.5525	0.18 ÷ 0.23	≤0.30	0.90 ÷ 1.20	0.025	≤0.30	0.25	0.0008 ÷ 0.005

2.2. Characteristics of the Wire Rod Rolling Process—Main Process Parameters of Thermoplastic Processing

This study was carried out for an exemplary rolling mill of a combined type (combination of bar mill and wire rod mill). The input material for the rolling process was an ingot from the continuous casting process with a square cross-section, side of 160 mm, and length of 14.000 mm. The rolling process for the wire rod of 20MnB4 steel, with a final diameter of 5.5 mm, in the continuous rolling mill took place in 17 rolling passes, while the rolling in the wire rod mill took place in two rolling blocks: a 10-rolling stand NO-TWIST MILL (NTM) and 4-rolling stand REDUCING SIZING MILL (RSM).

The analyses presented in this study were carried out for three technological variants (V1–V3), differing both in the rolling temperature in the RSM rolling block for the wire rod and in the cooling rate of the finished product on the roller conveyor in STELMOR® technology. Tables 2 and 3 present the most important thermoplastic processing parameters for the whole process, studied, determined, and verified in studies [30–32], among others.

Table 2. Parameters of 5.5 mm diameter wire rod rolling process of 20MnB4 steel grade [1].

Pass No.	Temperature T [°C]	Strain ε [–]	Strain Rate $\dot{\varepsilon}$ [s^{-1}]	Break Time after Deformation t [s] [2]	Pass No.	Temperature T [°C]	Strain ε [–]	Strain Rate $\dot{\varepsilon}$ [s^{-1}]	Break Time after Deformation t [s] [2]
		Continuous rolling mill					NTM block of wire rod rolling mill		
1	1086	0.18	0.16	26.47	18	851	0.49	156.02	0.091
2	1057	0.39	0.35	19.89	19	860	0.51	171.25	0.074
3	1037	0.28	0.39	29.98	20	867	0.56	276.33	0.058
4	1023	0.59	0.96	11.33	21	883	0.54	303.93	0.048
5	1010	0.46	1.15	8.91	22	892	0.56	477.46	0.037
6	995	0.50	2.02	6.13	23	908	0.53	584.28	0.032
7	997	0.45	2.45	11.65	24	918	0.62	991.51	0.024
8	1005	0.48	4.71	3.35	25	941	0.57	1042.10	0.020
9	1009	0.44	5.57	2.63	26	956	0.62	1753.46	0.015
10	1022	0.54	10.39	1.85	27	982	0.56	1809.67	0.82
11	1030	0.48	12.07	3.09	V1	V2, V3		RSM block of wire rod rolling mill	
12	1049	0.50	20.53	2.28	28	796	845	0.53	2368.05	0.012
13	1052	0.51	24.74	3.18	29	831	873	0.48	2275.43	0.007
14	1069	0.50	46.34	1.35	30	850	894	0.13	1853.11	0.004
15	1072	0.41	47.13	1.11	31	853	895	0.10	1680.68	
16	1087	0.51	79.93	0.90					
17	1091	0.31	70.63	8.52					

[1] Table based on data published at works [30–32], [2] transport time of band between successive rolling stands.

Table 3. Parameters of the controlled cooling process after rolling of 5.5 mm diameter wire rod of 20MnB4 steel grade [1].

Temperature before RSM Rolling Block T [°C]		Cooling Method after Rolling Process [2]		
		Stage No. 1		
		Desired Temperature Value T [°C]	Cooling Time t, s	Cooling Rate C_r, °C/s
800	(V1)	575	475	0.4
850	(V2)	500	70	5
850	(V3)	500	35	10

[1] Table based on data published in [30], [2] In the second cooling stage, the studied steel was cooled to 200 °C at a rate of 1 °C/s.

In variant one (V1), the temperature of the wire rod at the entrance to the RSM rolling block (rolling pass no. 28) was approximately 800 °C. After the processed material exited the RSM rolling block, it was cooled with water under high-pressure to 750 °C for 0.3 s at a rate of 166.67 °C/s, using an accelerated cooling system. Further cooling in air took place on a roller conveyor at a rate of 0.4 °C/s, to a temperature of about 575 °C and then at a rate of 1.3 °C/s to a temperature of ca. 200 °C (heat insulating covers in the STELOR® line closed, fans switched off).

In variant two (V2), the temperature of the wire rod at the entrance to the RSM rolling block (rolling pass no. 28) was approximately 850 °C. After the processed material exited the RSM rolling block, it was cooled with blown air at a rate of 5.0 °C/s to a temperature of approximately 500 °C and then at a rate of 1.0 °C/s to a temperature of approximately 200 °C (thermal insulation covers in the STELOMR® line open, fans switched off).

In variant three (V3), the temperature of the wire rod at the entrance to the RSM rolling block (rolling pass no. 28) was approximately 850 °C. After the processed material exited the RSM rolling block, it was cooled with blown air at a rate of 10.0 °C/s to a temperature of approximately 500 °C and then at a rate of 1.0 °C/s to a temperature of approximately 200 °C (thermal insulation covers in the STELOMR® line open, fans switched on—rotational speed 75% of maximum).

2.3. Methods

So as to obtain a finished product with a uniform fine-grained ferritic–pearlitic microstructure without clear-marked banding, the final deformation stage (rolling in the RSM rolling block) should take place in the single-phase (austenitic) range, when its temperature is 30 ÷ 80 °C higher than the temperature at the beginning of the austenite transformation Ar_3 [1,33–36]. For this purpose, numerical modeling of the process under analysis was carried out using QTSteel® software(ITA Ltd., Ostrava, Czech Republic, Metaltech Services Ltd., Gateshead, UK, release 3.4.1) based on the finite element method [37]. This stage of the study aimed to determine the temperatures of phase transformations and the contribution of the individual phases of the microstructure to the thermoplastic processing parameters under analysis.

In addition (for verification), physical modeling of the analyzed wire rod rolling process using non-free torsion was carried out by using STD 812 torsion plastometer (manufactured by Bähr Thermoanalyse GmbH Hüllhorst, Germany, now TA Instruments New Castle, DE, USA), according to the methodology presented in [32]. The purpose of this stage of research was to determine (check) the microstructure of the steel under study immediately before the deformation process in the RSM rolling block. Thus, the tested material was hardened right before the deformation stage, simulating rolling the wire rod in the RSM rolling block. Then, the metallographic analysis and microhardness measurements of the resulting microstructure were carried out.

The subsequent stage of the research focussed on checking the uniformity of the temperature of the rolled band along its length in the rolling line. These tests were carried out using the ThermaCAM SC640 (FLIR Systems, Wilsonville, OR, USA) thermovision camera, which is equipped with an uncooled detector [38]. ThermaCAM Researcher Professional 2.10 (FLIR Systems AB) software was used to process these data.

This was followed by testing selected mechanical properties of the finished wire rod along its length (including their statistical analysis, by using Statistica ver. 13, TIBCO Software Inc., Palo Alto, CA, USA), produced according to technological variants V1–V3. These tests were carried out in a static tension test in accordance with the standard [39] using a Zwick Z/100 materials testing machine (produced by ZwickRoell, Wroclaw, Poland) [40]. In order to assess the capacity of the tested steel for further cold processing, upsetting tests were carried out according to the standard [41].

The scope of this study also included microstructure analysis (measurement of ferrite grain size and hardness using Vickers' method—loading force 9.81 N, loading time 5 s) of the wire rod along its length (in longitudinal section).

All metallographic analyses were conducted with light microscopy using a Nikon Eclipse MA 200 microscope (Nikon Metrology NV, Leuven, Belgium) with NIS-Elements software [40]. Hardness measurements were conducted using the Vickers' method with a FutureTech FM 700 microhardness tester (Kawasaki, Japan) [40].

2.4. Numerical Modelling—Mathematical Model of QTSteel® Software

In the QTSteel® program, when modeling the microstructure and mechanical properties of heat-treated or thermomechanical-treated steel, data from the cooling curves on the TTT diagram are used. Calculating the percentage content of the microstructure components is performed successively for the relevant sections of the cooling curve. To describe the kinetics of the transformation of individual components of the microstructure, the program uses the Avrami Equation (1) [42,43]:

$$X_i(T,t) = (1 - \exp(-k(T) \times t^{n(T)})) \times X_\gamma, \tag{1}$$

where: $X_i(T,t)$—volume fraction of individual components of the microstructure: ferrite, perlite, bainite, $k(T)$ and $n(T)$—parameters depending on the transformation mechanism and places of privileged nucleation and on the cooling rate, calculated based on TTT charts for a given temperature, T—temperature, t—time,—volume fraction of residual austenite.

The volume fraction of martensite during martensitic transformation is calculated by using the Koistinen–Marburger equation [37,43]:

$$X_m(T) = (1 - \exp(-b \times (T_{ms} - T)^n)) \times X_\gamma, \tag{2}$$

where: X_m—volume fraction of martensite, b, n—constant, T_{ms}—martensitic transformation start temperature, T—temperature, X_γ—volume fraction of residual austenite.

Vickers HV hardness is determined by means of a regression equation [42,43]:

$$HV = C_0 + X_f \times \sum(D_i \times c_i + X_p \times \sum E_i \times c_i + X_b \times \sum F_i \times c_i + X_m \times \sum G_i \times c_i, \quad (3)$$

where: HV—Vickers hardness, X_f, X_p, X_b, X_m—volume fractions: ferrite, perlite, bainite, martensite, C_0, D_i, E_i, F_i, G_i—constant, c_i—the percentage of alloying additions.

The tensile strength was determined based on Equation (4) [42]:

$$UTS = f(HV) = -a + b \times HV, \quad (4)$$

where: UTS—Ultimate tensile strength, HV—Vickers hardness, and a, b—constant.

Yield strength YS is determined by Equation (5) [42,43]:

$$YS = f(D_\alpha, C_r, X_f, \sum(X_p + X_b + X_m)), \quad (5)$$

where: D_α—ferrite grain size, C_r—cooling rate, X_f, X_p, X_b, X_m—volume fractions: ferrite, perlite, bainite, martensite.

Accurate results of research carried out using the DIL 805 A/D dilatometer, the aim of which was to develop TTT and DTTT diagrams and to determine the best cooling conditions for 20MnB4 steel, were published in [44]. According to published research, in the case of cooling wire rods on a roller conveyor, greater accuracy in predicting the microstructure and mechanical properties of the finished product is achieved based on DTTT charts, which take into account the deformation process preceding the cooling of the rolled band.

Taking into account the obtained results, the DTTT graph was used to determine the influence of the cooling conditions on the forming of the wire rod microstructure immediately after the deformation process. It was concluded that in order to obtain a ferritic–pearlitic microstructure in the finished product, the cooling rate should not exceed 15 °C/s. Faster cooling causes the formation of bainite, bainitic–martensitic, and martensitic structures in the material, which results in a decrease in the ability of the investigated steel for further cold plastic working processes or, in extreme cases, prevents it.

Numerical modeling using QTSteel software was carried out using input data shown in Tables 2 and 3 and published in [30].

3. Results and Discussion

3.1. Numerical Modelling Results—QTSteel® Software

Figure 1 shows part of the DTTT (Deformation Time Temperature Transformation) graph with phase transformation curves and cooling curves according to variants V1–V3, obtained as a result of numerical modeling of the thermoplastic processing of the analyzed rolling process for 5.5 mm diameter wire rods of 20MnB4 steel.

The percentage amounts of microstructure components, hardness, selected mechanical properties, and characteristic temperature values—obtained from numerical modeling of the thermoplastic processing of a 5.5 mm diameter wire rod made from 20MnB4 steel—are shown in Table 4.

According to the analysis of the test results obtained from numerical modeling of thermoplastic processing of a 5.5 mm diameter wire rod made from 20MnB4 steel according to variant V1, it was ascertained that the transformation of austenite to ferrite during cooling (Ar_3) began at the temperature of 744 °C. In contrast, the temperature of the onset of the transformation of austenite to pearlite during cooling (Ar_1) was 655 °C. The temperatures at the beginning and completion of ferrite to austenite transformation during heating were correspondingly: 719 °C and 827 °C. Bearing in mind the earlier stages of rolling (multi-sequence deformation over a wide temperature range, multi-stage accelerated cooling and reheating due to heat conduction from the center of the material towards its surface and as a result of deformation at high rates) [30] as well as the accelerated cooling after rolling in the RSM block, and upon analyzing the cooling curve (Figure 1) while taking into account the phase transition temperatures, it can be concluded that the rolling and cooling of steel in this variant took place in the single-phase range but relatively close to the temperature of the beginning of austenite to ferrite (Ar₃) transformation.

Table 4. Shares of microstructure components, hardness, selected mechanical properties, and characteristic temperature for the conditions of thermoplastic processing of a 5.5 mm diameter wire rod of 20MnB4 steel grade.

Temperature before RSM Rolling Block T [°C]		Contribution of Microstructure Components [%]		Vickers Hardness [HV]	Yield Strength YS [MPa]	Ultimate Tensile Strength UTS [MPa]	Plasticity Reserve YS/UTS	Characteristic Temperature			
		Ferrite	Pearlite					Ac_3	Ac_1	Ar_3	Ar_1
800	(V1)	94.5	5.5	173	319	516	0.618			744	655
850	(V2)	84.4	15.6	186	370	560	0.661	827	719	752	639
850	(V3)	83.7	16.3	194	392	589	0.666			742	631

where: Ac_1—temperature of the beginning of the transformation of ferrite into austenite during heating, Ac_3—temperature of the end of the transformation of ferrite into austenite during heating, Ar_3—temperature of the beginning of austenite transformation into ferrite during cooling, Ar_1—temperature of the beginning of the transformation of austenite into pearlite during cooling.

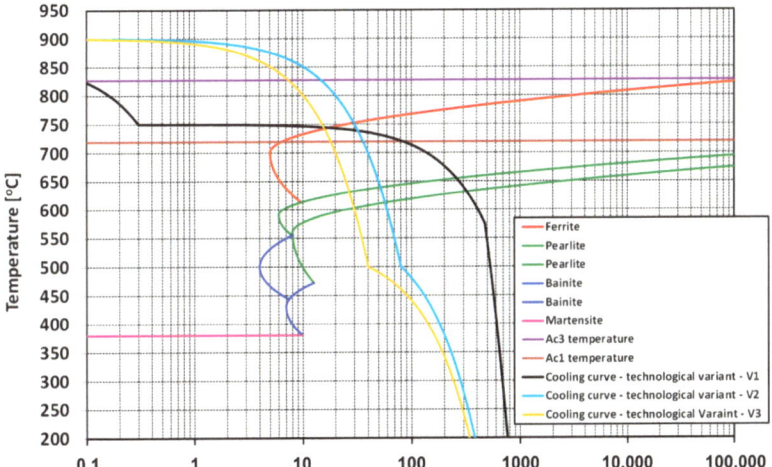

Figure 1. Part of the DTTT graph depicting investigated thermoplastic processing conditions for a 5.5 mm diameter wire rod of 20MnB4 steel.

Under these thermoplastic processing conditions, percentage amounts of ferrite and pearlite were 94.5% and 5.5%, respectively. The average hardness value was 173 HV. The mechanical properties of the material formed under these conditions were yield strength—319 MPa and ultimate tensile strength—516 MPa. The plasticity reserve amounted to 0.618.

Analysis of the results of numerical modeling according to the variant V2 led to the conclusion that the temperature at the onset of the transformation of austenite to ferrite during cooling (Ar₃) was 752 °C. In comparison, the temperature at the onset of austenite to pearlite transformation during cooling (Ar₁) was 639 °C. Taking into account the earlier rolling stages [30] and analyzing the cooling curve (Figure 1), it can be concluded that both the rolling and cooling of steel in this variant took place in a single-phase range. The proportions of ferrite and pearlite under these thermoplastic processing conditions were 84.4% and 15.6%, respectively. The average hardness value was 186 HV. The mechanical properties of the material formed under these conditions were 370 MPa yield strength and 560 MPa ultimate tensile strength, while the plasticity reserve amounted to 0.661. The increase in hardness and analyzed mechanical properties in this variant compared with variant V1 (despite the higher end-of-rolling temperature) may be due, among other factors, to the higher cooling rate.

Numerical modeling of thermoplastic processing of a 5.5 mm diameter wire rod made of 20MnB4 steel, according to variant V3, led to the following results: the temperature of the beginning of the transformation of austenite to ferrite during cooling (Ar_3) was 742 °C, while the temperature at the beginning of the transformation of austenite to pearlite during cooling (Ar_1) was 631 °C. Comparing the Ar_3 temperatures for variants V2 and V3, it can be noted that an increase in the cooling rate (from the same temperature) results in a decrease in the Ar_3 and Ar_1 transformations temperature, which is consistent with data published in, inter alia, [45]. Upon analyzing the earlier stages of the rolling process [30] and the cooling curve (Figure 1), it was determined that both rolling and final cooling of steel in this variant took place in the single-phase range. Under these conditions, the percentage ratios of ferrite and pearlite were 83.7% and 16.3%, respectively. The average hardness was 194 HV. The mechanical properties of the material formed under these conditions were yield strength—392 MPa and ultimate tensile strength—589 MPa. Plasticity reserve amounted to 0.666.

3.2. Physical Modelling Results—Multi-Sequence Non-Free Torsion—STD 812 Torsion Plastometer

As indicated, inter alia, in [1,33–36], in order to obtain a finished product with uniform fine-grained ferritic–pearlitic microstructure variations, the final stage of deformation (rolling in the RSM rolling block) should take place in the single-phase (austenitic) range, when its temperature is 30 ÷ 80 °C higher than the temperature at the beginning of the austenite transformation Ar_3.

In order to check (verify) the microstructural state of the steel under investigation, physical modeling of the analyzed rolling process was carried out immediately prior to rolling in the RSM rolling block. This modeling was conducted in multi-sequence torsion tests according to the methodology described in detail in [32], using an STD 812 torsion plastometer. During the physical modeling, the tested steel was hardened immediately prior to the deformation stage, simulating a rolling wire rod in the RSM rolling block. It is not possible to carry out this type of research directly in the rolling line of the analyzed wire rod. Metallographic analysis and microhardness measurements of the resulting microstructure were then carried out.

Physical modeling was carried out in accordance with the temperature and deformation parameters shown in Tables 2 and 3 and published, inter alia, in [30,32].

A schematic depiction of the physical modeling of wire rod rolling is shown in Figure 2.

In the beginning, the tested steel was heated to the temperature of 1165 °C, corresponding to the temperature in the leveling zone of a heating furnace (under industrial conditions). In order to obtain the same temperature over the entire volume of the sample working zone, the 20MnB4 steel was heated for 300 s. The next step was cooling for 30 s, to the temperature of 1086 °C, in simulation of the band cooling process during its transport from the furnace to the first stand of the rolling line. It was then deformed in 17 cycles, with deformation parameters shown in Table 2, replicating the rolling process in a continuous rolling mill. The subsequent stage consisted of accelerated cooling to the temperature of 851 °C, corresponding to the band temperature before the NTM rolling block of the continuous rolling mill. In the next physical modeling step, the rolling process in the NTM rolling block was simulated (Table 2), following which the accelerated cooling process between the NTM and RSM rolling blocks was modeled, according to the variant, down to 800 °C (variant V1) or to 850 °C (variants: V2 and V3) and then hardened at a cooling rate of approximately 130 °C/s in variant V1 or 146 °C/s in variants V2 and V3, using helium.

After physical modeling, samples of the material were subjected to metallographic testing. The samples were treated with ferric chloride for 90 s. In order to precisely identify the individual phases, additional microhardness measurements HV 0.01 (pressing force value 0.09807 N, time 10 s) were carried out.

Figure 3 shows examples of 20MnB4 steel microstructures after physical modeling.

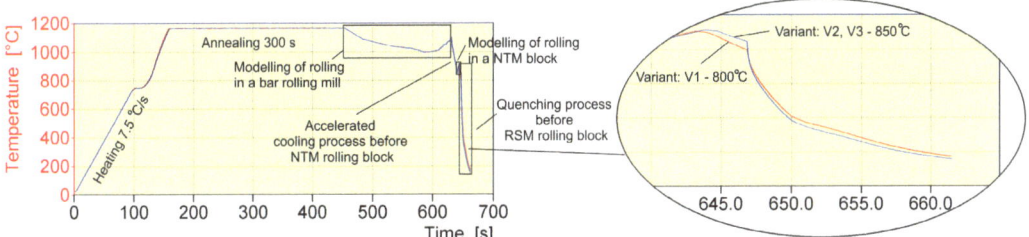

Figure 2. Scheme of physical modeling of rolling of 5.5 mm diameter wire rod of 20MnB4 steel grade.

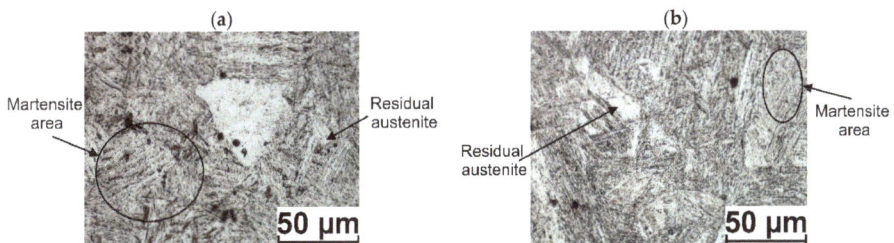

Figure 3. Example microstructure of 20MnB4 steel after physical modeling of wire rod rolling process: (**a**) technological variant—V1, (**b**) technological variant—V2, V3.

Microstructure analysis of 20MnB4 steel samples after physical modeling of the 5.5 mm in diameter wire rod rolling process shows that in all analyzed technological variants, the material after hardening had a martensitic structure over its entire cross-section with some residual austenite (not exceeding 4%). In the case of the V1 variant, the average microhardness of the tested samples was approximately 369.11 HV 0.01. The average microhardness of the tested samples for V2 and V3 variants was approximately 416.96 HV 0.01. Since no other phases were detected in the tested samples, it can be concluded that, under real conditions (in all analyzed variants), the investigated steel was deformed in the RSM rolling block of the wire rod mill in the single-phase (austenitic) state.

3.3. Thermovision Investigation Results—ThermaCAM SC640 Thermovision Camera

One of the factors affecting the uniformity of the microstructure and mechanical properties of the wire rod along its length is the temperature distribution along the length of the rolled feedstock. This section presents the results of measurements of the surface temperature distribution along the length of the rolled material in several places of the rolling line, including at the exit of the heating furnace, before rolling stands No. 1 and 15, before the NTM (No-Twist Mill) rolling block, before the wire rod coil former, and at the entrance to the rolling conveyor in the STELMOR® system. The testing was carried out using thermovision technology—a FLIR Systems ThermaCAM SC640 thermovision camera equipped with an uncooled detector [38]. Temperature measurements were recorded as measured video sequences, in which the areas of maximum recorded temperatures (marked by the presence of mill scale on the surface of the tested steel) were analyzed. ThermaCAM Researcher Professional software was used to process these data. Surface temperature investigations were preceded by the determination of the 20MnB4 steel emissivity within the temperature range under study (700 ÷ 1200 °C). A description of the methods for determining the emissivity can be found, inter alia, in [46,47]. Based on the results of these studies, it was determined that the emissivity of 20MnB4 steel in the analyzed temperature range varied from 0.80 to 0.85.

Examples of thermograms and a graph of the surface temperature distribution along the length of the rolled material in a number of places of the rolling line (for variants V2 and V3) are shown in Figures 4–6.

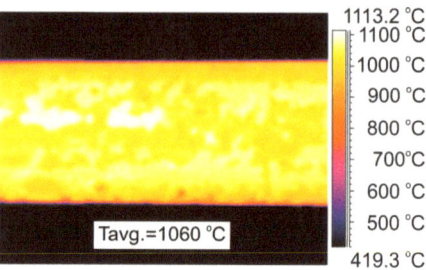

Figure 4. Example thermogram of temperature distribution on the surface of 20MnB4 steel grade band before rolling stand No. 1.

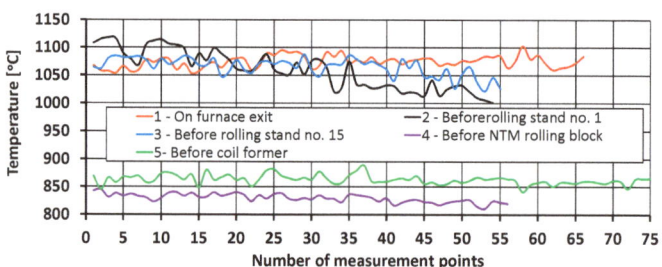

Figure 5. Diagram of surface temperature distribution along the length of the rolled material at various places in the rolling mill.

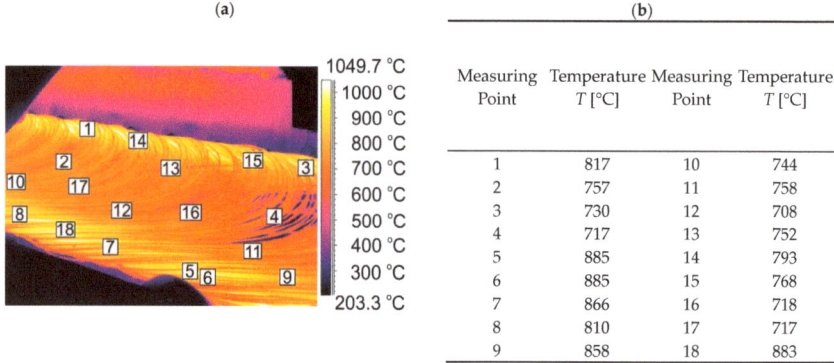

Measuring Point	Temperature T [°C]	Measuring Point	Temperature T [°C]
1	817	10	744
2	757	11	758
3	730	12	708
4	717	13	752
5	885	14	793
6	885	15	768
7	866	16	718
8	810	17	717
9	858	18	883

Figure 6. Example of temperature distribution on the surface of 20MnB4 steel wire rod at the entry to the STELMOR® roller conveyor: (**a**) thermogram, (**b**) temperature value at measuring points.

Based on the measurements of the surface temperature distribution along the length of the rolled material (Figure 5), it was determined that the feedstock heated in the furnace was characterized by a high uniformity of the temperature distribution along its length. The average surface temperature of the feedstock at the exit of the heating furnace was approximately 1074 °C, whereas the maximum difference in surface temperature along the length of the feedstock after the heating process was about 50 °C (Figure 5). According to industry data, the temperature in the last zone of the heating furnace (the so-called leveling zone) was approximately 1165 °C. The error between the temperature measured with the thermovision camera and industrial data was 7.8% and was mainly caused by the large amount of mill scale on the surface of the processed material.

As a result of air cooling during the transport of the feedstock to the first rolling stand, its average surface temperature decreased to approximately 1060 °C (Figure 4). At the same time, due to the long rolling time in the first rolling stand and the different air-cooling times of the beginning and end of the 14-m-long feedstock, the difference in surface temperature distribution along its length increased to about 116 °C (Figure 5).

In the subsequent stages of the rolling process, due to successive thermoplastic processing operations, a decrease in cross-sectional area, an increase in rolling speed, accelerated cooling, the generation and conduction of heat resulting from plastic strain, and shorter intervals between deformations, it was found that the surface temperature difference along the length of the rolled band decreased again. The average surface temperature of the rolled material in front of rolling stand No. 15 was approximately 1065 °C. In contrast, the temperature difference along the length of the processed material at this point of the rolling line was approximately 66 °C (Figure 5). The average surface temperature of the material before being processed in the NTM rolling block was approximately 830 °C, while the temperature difference along the band at this point of the rolling line was approximately 34 °C (Figure 5). The average surface temperature of the band in front of the wire coil former was about 863 °C, while the temperature difference along its length at this point in the rolling line was about 48 °C (Figure 5).

An analysis of the surface temperature measurements of the rolled steel grade, shown in Figure 6, led to the conclusion that the surface temperature difference again increased along the length of the wire rod at the entry to the STELMOR® roller conveyor. The increase in surface temperature along the length of the wire rod and its large variation at this point is due to the characteristic arrangement of the wire rod coils and the different heat transfer conditions in the individual areas of the roller conveyor. The temperature difference along the length of the wire rod at this point in the rolling line was as high as 177 °C (Figure 6).

3.4. Mechanical Properties Test Results and Upsetting Tests Results along Wire Rod Length—Zwick Z/100 Testing Machine

This section presents the results of tests of selected mechanical properties of the finished wire rod along its length (including their statistical analysis), produced according to technological variants V1–V3. The tests were carried out in a static tensile test according to the standard [39]. The gage length of the extensometer was 100 mm. Each time, a total of 20 samples were taken from the middle and end parts of the rolled band (Figure 7a,b). A general view of the 20MnB4 steel wire rod before and after the tensile test is shown in Figure 7c,d. Figure 8 shows examples of tensile curves of samples of 20MnB4 steel grade wire rod, produced according to technological variants V1–V3. Detailed results of the finished wire rod's mechanical properties tests along its length produced according to technological variants V1–V3 are presented in Table 5.

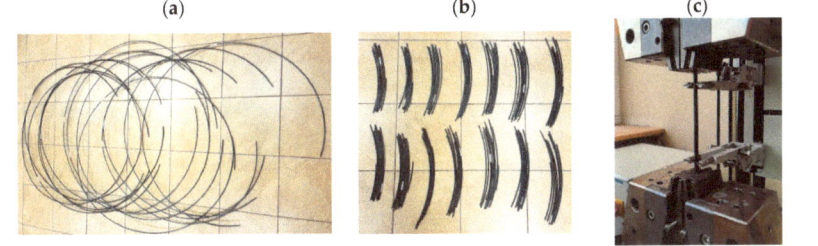

Figure 7. Mechanical properties tests along the length of 5.5 mm diameter wire rod of 20MnB4 steel grade: (**a**) wire rod coils, (**b**) samples for mechanical properties testing, (**c**) example sample before tensile test, (**d**) example sample after tensile test.

A basic statistical analysis of mechanical properties along the length of the wire rod is presented below. Basic statistical indicators (arithmetic average, standard deviation, variation coefficient, median) were calculated.

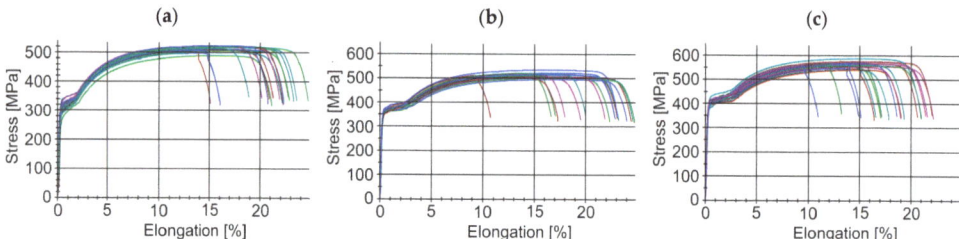

Figure 8. Examples of tensile curves of 20MnB4 steel grade wire rod samples: (**a**) technological variant—V1, (**b**) technological variant—V2, (**c**) technological variant—V3.

In addition, graphs were made of, inter alia, the changes in the analyzed mechanical properties along the length of the wire rod, as well as graphs of normal distribution.

Based on data presented in Tables 5 and 6, the average yield strength of the wire rod produced according to variant V1 was 302.88 MPa, while the average ultimate tensile strength was 513.17 MPa. In turn, the average value of unit elongation was approximately 22.30%. The average value of the relative reduction in area at fracture was ca. 68.87%. The average plasticity reserve of the wire rod was about 0.590.

An analysis of the mechanical properties of the wire rod produced according to variant V2 led to the conclusion that the average value of the yield strength of the wire rod was 359.30 MPa, which is approximately 19% higher than the average value of the yield strength of the wire rod produced according to variant V1. The average ultimate tensile strength of the wire rod produced according to variant V2 was similar to the average yield strength of the wire rod produced according to variant V1 and was 515.12 MPa. The average unit elongation of the wire rod produced according to variant V2 was approximately 23.02% and was ca. 3% higher than the average unit elongation of the wire rod produced according to variant V1. The average relative reduction in area at fracture of the wire rod produced according to variant V2 was 69.03% and was comparable with the corresponding value of the wire rod produced according to the V1 variant. The average plasticity reserve of the wire rod produced according to the V2 variant was approximately 0.698, about 18% higher than the average plasticity reserve of the wire rod produced according to variant V1.

Based on the analysis of results of the mechanical properties tests conducted on the wire rod produced according to variant V3, it was concluded that the average yield strength of the wire rod was 398.95 MPa, which is approximately 32% higher than the average yield strength of wire rod produced according to variant V1, and 11% higher than the average yield strength of wire rod produced according to variant V2. The mean value of the ultimate tensile strength of the wire rod produced according to variant V3 was 561.52 MPa, approximately 9% higher than the mean value of the ultimate tensile strength of the wire rod produced according to variants V1 and V2. The mean value of unit elongation of wire rods produced according to variant V3 was similar to the average value of unit elongation of wire rods produced according to variant V1, ca. 20.73%. This value was approximately 10% lower than the average unit elongation value of the wire rod produced according to the V2 variant. However, this did not adversely affect its capacity to be further subjected to the cold-forming process, which was confirmed using cold-upsetting tests with relative strain as high as 75%. The average value of the relative reduction in area at fracture of wire rod produced according to the V3 variant was approximately 72.75%, ca. 6% higher than the average relative reduction in area at fracture produced according to V1 variant and approximately 5% higher than the average value of the relative reduction in area at fracture in case of wire rod produced according to V2 variant. Whereas the average plasticity reserve of wire rod produced according to V3 variant was approximately 0.711, ca. 21% higher than the average plasticity reserve of wire rod produced according to V1 variant and ca. 2% higher than the average plasticity reserve of wire rod produced according to V2 variant.

Table 5. Mechanical properties along the length of 5.5 mm diameter wire rod of 20MnB4 steel grade.

Sample Number	Technological Variant—V1					Technological Variant—V2					Technological Variant—V3				
	Yield Strength YS [MPa]	Ultimate Tensile Strength UTS [MPa]	Unit Elongation A [%]	Relative Reduction in Area at Fracture Z [%]	Plasticity Reserve YS/UTS	Yield Strength YS [MPa]	Ultimate Tensile Strength UTS [MPa]	Unit Elongation A [%]	Relative Reduction in Area at Fracture Z [%]	Plasticity Reserve YS/UTS	Yield Strength YS [MPa]	Ultimate Tensile Strength UTS [MPa]	Unit Elongation A [%]	Relative Reduction in Area at Fracture Z [%]	Plasticity Reserve YS/UTS
1	2	3	4	5	6	7	8	9	10	11	12	13	14	15	16
1	314.93	502.74	19.10	70.46	0.626	354.12	507.40	24.43	70.32	0.698	389.76	547.35	20.02	71.71	0.712
2	267.18	489.56	21.10	68.10	0.546	351.74	505.57	22.80	67.94	0.696	401.69	570.55	20.78	72.00	0.704
3	281.48	498.20	21.04	68.72	0.565	364.18	522.59	23.42	69.60	0.697	394.54	548.26	20.66	72.11	0.720
4	335.63	518.86	24.24	69.97	0.647	352.72	501.56	24.89	70.51	0.703	380.49	541.51	19.44	70.84	0.703
5	333.23	522.58	22.73	69.95	0.638	359.33	505.05	21.06	67.02	0.711	399.55	556.18	23.40	73.16	0.718
6	296.57	510.49	22.42	67.20	0.581	361.14	514.56	24.61	68.58	0.702	402.74	567.44	20.59	73.72	0.710
7	317.25	522.13	22.74	69.17	0.608	353.16	502.45	24.74	69.78	0.703	400.64	555.63	18.16	72.31	0.721
8	297.08	511.19	22.28	67.72	0.581	355.45	514.39	24.00	69.69	0.716	390.24	569.73	20.83	72.88	0.685
9	280.09	517.29	23.27	67.51	0.541	358.10	496.54	24.16	71.32	0.696	400.97	560.57	16.53	72.69	0.715
10	313.50	522.80	21.47	68.35	0.600	358.80	507.77	22.00	68.76	0.707	405.96	566.54	22.75	73.07	0.717
11	286.24	507.12	21.68	67.78	0.564	362.95	510.95	21.69	67.77	0.710	392.96	541.29	19.00	70.96	0.726
12	295.42	506.49	22.85	67.42	0.583	355.48	510.26	22.00	68.29	0.697	403.71	558.66	21.37	73.44	0.723
13	326.01	507.31	19.54	69.84	0.643	368.36	535.31	24.74	68.76	0.688	390.20	567.80	21.55	73.08	0.687
14	291.13	500.29	23.24	69.05	0.582	355.06	520.59	20.61	67.99	0.682	381.31	562.00	19.93	73.62	0.678
15	287.95	522.09	23.16	69.52	0.552	360.88	517.94	22.94	69.20	0.697	398.01	553.20	20.25	74.29	0.719
16	312.53	522.65	22.88	69.15	0.598	347.54	503.24	21.29	68.34	0.691	404.93	575.94	23.80	74.73	0.703
17	307.24	520.99	25.14	70.27	0.590	356.16	521.38	22.84	68.36	0.683	406.82	566.57	20.78	69.84	0.718
18	309.41	523.65	21.74	68.97	0.591	362.79	517.90	22.75	68.89	0.701	404.21	560.22	21.75	72.10	0.722
19	291.87	514.10	21.86	68.26	0.568	376.00	546.00	24.70	71.80	0.689	418.14	587.66	21.64	74.92	0.712
20	312.91	522.95	23.47	69.93	0.598	372.00	541.00	21.00	67.70	0.688	412.18	573.28	21.43	73.52	0.719
Average value	302.88	513.17	22.30	68.87	0.590	359.30	515.12	23.02	69.03	0.698	398.95	561.52	20.73	72.75	0.711

Table 6. Basic statistical indicators of 5.5 mm diameter wire rod of 20MnB4 steel grade produced according to technological variants V1–V3.

Statistical Parameter	Technological Variant—V1					Technological Variant—V2					Technological Variant—V3				
	Yield Strength YS [MPa]	Ultimate Tensile Strength UTS [MPa]	Unit Elongation A [%]	Relative Reduction in Area at Fracture Z [%]		Yield Strength YS [MPa]	Ultimate Tensile Strength UTS [MPa]	Unit Elongation A [%]	Relative Reduction in Area at Fracture Z [%]		Yield Strength YS [MPa]	Ultimate Tensile Strength UTS [MPa]	Unit Elongation A [%]	Relative Reduction in Area at Fracture Z [%]	
1	2	3	4	5	6	7	8	9	10	11	12	13	
Arithmetic average	302.88	513.17	22.30	68.87		359.30	515.12	23.02	69.03		398.95	561.52	20.73	72.75	
Standard deviation	18.34	10.08	1.44	1.03		7.03	13.28	1.46	1.25		9.50	11.77	1.69	1.30	
Coefficient of variation	0.061	0.020	0.065	0.015		0.020	0.026	0.064	0.018		0.024	0.021	0.082	0.018	
Median	302.16	515.70	22.58	69.01		358.45	512.67	22.89	68.76		400.81	561.29	20.78	72.98	

Furthermore, upon analyzing data presented in Table 6, it was found that in the case of wire rods produced according to the V1 variant, the standard deviation values were 18.34 for yield strength, 10.08 for ultimate tensile strength, 1.44 for unit elongation and 1.03 for relative reduction in area at fracture. The variation coefficients of wire rods produced according to the V1 variant were 0.061 for yield strength, 0.02 for ultimate tensile strength, 0.065 for unit elongation, and 0.015 for relative reduction in area at fracture.

For wire rods produced according to the V2 variant, the standard deviation values were 7.03 for yield strength, 13.28 for ultimate tensile strength, 1.46 for unit elongation, and 1.25 for relative reduction in area at fracture. The variation coefficients in the case of wire rods produced according to the V2 variant were 0.02 for yield strength, 0.026 for ultimate tensile strength, 0.064 for unit elongation, and 0.018 for relative reduction in area at fracture.

The standard deviation values for wire rods produced according to the V3 variant were 9.50 for yield strength, 11.77 for ultimate tensile strength, 1.69 for unit elongation, and 1.3 for relative reduction in area at fracture. In contrast, the variation coefficients for wire rods produced according to the V3 variant were 0.024 for yield strength, 0.021 for ultimate tensile strength, 0.082 for unit elongation, and 0.018 for relative reduction in area at fracture.

Based on the analysis of data presented in Tables 5 and 6, it is difficult to unambiguously assess the uniformity of the analyzed mechanical properties along the length of wire rods subjected to different thermoplastic processing conditions. Each of the three technological variants is characterized by a certain dispersion of the analyzed mechanical properties along the wire rod length. The values of the standard deviation and the variation coefficient depend on both the end-of-rolling temperature and the cooling rate after the rolling process. Therefore, additional graphs of the variation in the analyzed mechanical properties along the wire rod length and of normal distribution were compiled.

Data presented in Figures 9–12 shows that each of the analyzed technological variants is characterized by some degree of nonuniformity of the mechanical properties along the length of 20MnB4 steel wire rod with a diameter of 5.5 mm, which, however, retains some stability. This nonuniformity is caused by a number of factors. One of the most important is the temperature distribution along the processed wire rod. The greatest nonuniformity of the temperature distribution along the wire rod was observed during the controlled cooling after the roller track process (Section 3.3). The nonuniformity of the temperature distribution causes, in turn, nonuniformity of the microstructure and, consequently, nonuniform properties of the finished wire rod.

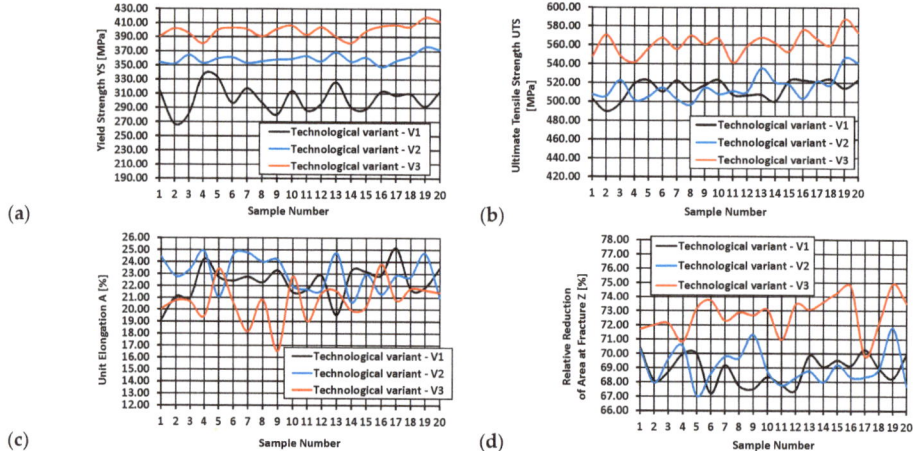

Figure 9. The course of changes in selected mechanical properties along the length of 5.5 mm diameter wire rod of 20MnB4 steel grade—technological variants V1–V3: (**a**) Yield Strength (YS), (**b**) Ultimate Tensile Strength (UTS), (**c**) Unit Elongation (A), (**d**) Relative Reduction in Area at Fracture (Z).

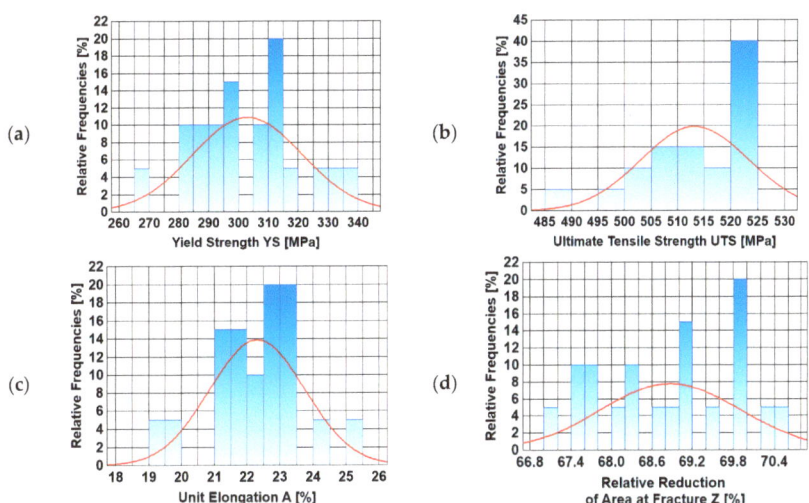

Figure 10. Normal distribution diagrams of selected mechanical properties along the length of a 5.5 mm diameter wire rod of 20MnB4 steel grade—technological variant V1: (**a**) Yield Strength (YS), (**b**) Ultimate Tensile Strength (UTS), (**c**) Unit Elongation (A), (**d**) Relative Reduction in Area at Fracture (Z).

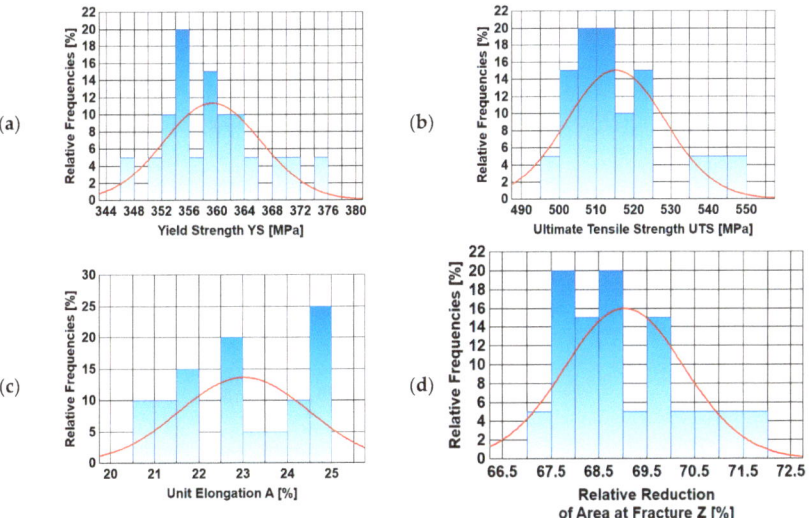

Figure 11. Normal distribution diagrams of selected mechanical properties along the length of a 5.5 mm diameter wire rod of 20MnB4 steel grade—technological variant V2: (**a**) Yield Strength (YS), (**b**) Ultimate Tensile Strength (UTS), (**c**) Unit Elongation (A), (**d**) Relative Reduction in Area at Fracture (Z).

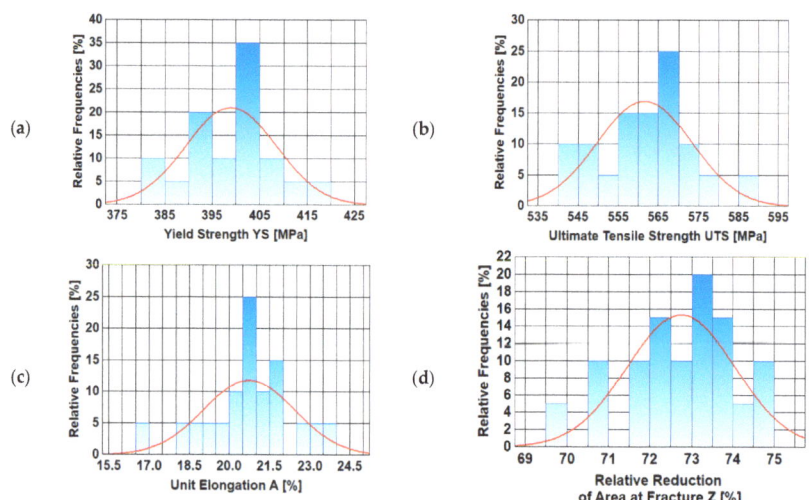

Figure 12. Normal distribution diagrams of selected mechanical properties along the length of a 5.5 mm diameter wire rod of 20MnB4 steel grade—technological variant V3: (**a**) Yield Strength (YS), (**b**) Ultimate Tensile Strength (UTS), (**c**) Unit Elongation (A), (**d**) Relative Reduction in Area at Fracture (Z).

An analysis of data shown in Figure 9a led to the conclusion that the highest yield strength value of wire rods produced according to the V1 variant was about 336 MPa, while the lowest was about 267 MPa. The highest yield strength value of wire rods produced according to the V2 variant was 376 MPa, and the lowest ca. 348 MPa. In contrast, the highest yield strength value of wire rods produced according to the V3 variant was about 418 MPa, and the lowest was around 381 MPa.

Based on data shown in Figure 9b, it was determined that the highest value of the ultimate tensile strength of the wire rod produced according to the V1 variant was about 524 MPa, and the lowest was around 490 MPa. The highest value of the ultimate tensile strength of the wire rod produced according to the V2 variant was 546 MPa, while the lowest was about 497 MPa. The highest value of ultimate tensile strength of the wire rod produced according to the V3 variant was about 588 MPa, and the lowest was around 541 MPa.

Based on data presented in Figure 9c, the highest unit elongation value of wire rods produced according to the V1 variant was 25.14%, and the lowest was 19.1%. The highest value of relative elongation of wire rods produced according to variant V2 was 24.89%, and the lowest was 20.61%. In turn, the highest unit elongation value of wire rods produced according to the V3 variant was 23.8%, and the lowest was 16.53%.

Based on data shown in Figure 9d, it was determined that the highest value of the relative reduction in area at fracture of wire rods produced according to the V1 variant was 70.46%, and the lowest was 67.2%. The relative reduction in area at fracture of wire rod produced according to the V2 variant was 71.8%, and the lowest was 67.02%. In turn, the highest value of the relative reduction in area at fracture of wire rod produced according to the V3 variant was 74.92%, and the lowest was 69.84%.

In order to determine whether the observed nonuniformity of mechanical properties along the wire rod is significant regarding the possibility of the tested steel undergoing further cold forming processing, additional cold upsetting tests were carried out for each technological variant according to the standard [41] as well as an assessment of the surface quality concerning the occurrence of any cracks. A minimum of three samples taken from the middle and end parts of the rolled material were tested each time. Figure 13 shows examples of samples after the cold upsetting process with a relative plastic strain of 50%, 67%, and 75%.

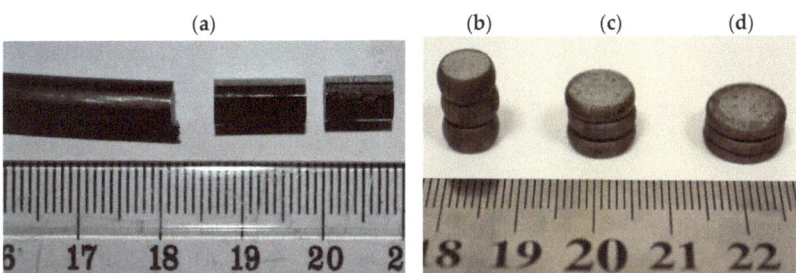

Figure 13. General view of wire rod of 20MnB4 steel grade samples for cold upsetting tests: (**a**) method of samples preparation from the finished product for cold upsetting tests, (**b**) view of samples after cold upsetting test (wire rod after the rolling process according to technological variant V2) with relative plastic strain value: (**b**) 50%, (**c**) 67%, (**d**) 75%.

Upon analyzing the test results, it was found that there were no cracks, scratches, or other surface defects on any surfaces of the technological variants analyzed, even after applying a relative plastic strain of 75% (the sample's hight indicator after cold upsetting 0.25).

In summary, the results of the tests regarding the mechanical properties of the finished wire rod along its length and of the upsetting test, it was found that for the studied thermoplastic processing parameters, there was no significant impact of the production process parameters on the quality of the finished steel product. This is partly due to the narrow rolling temperature range in the RSM block (800 °C and 850 °C). However, we have to keep in mind that the rolling temperature range in the RSM block covered by this study is typical for 20MnB4 steel. On the other hand, comparing the mechanical properties of the wire rod produced in the different technological variants, it was noted that the best complex of mechanical properties has the wire rod processed in the RSM rolling block at 850 °C and cooled after the rolling process on a roller conveyor at a rate of 10 °C/s (V3 technological variant). The wire rod produced in this way also has the required level of plasticity reserve, which enables further deformation of the studied steel grade in upsetting tests with a relative plastic strain of 75%.

3.5. Results of Metallographic and Hardness Tests of the Wire Rod along Its Length—Nikon Eclipse MA 200 Microscope, FM 700 Microhardness Tester

This section presents an analysis of the results of metallographic observations and measurements of the samples after the rolling process according to V1–V3 variants. A Nikon Eclipse MA-200 microscope with NIS-Elements software was used for the study [40]. In each case, metallographic observations and measurements of ferrite grain size (D_α) (perpendicular secant method [48]) were carried out on longitudinal sections of samples taken from the middle and end part of the rolled wire rod at a total of 30 points (5 points on the wire rod radius at several places along the wire rod length). In addition, Vickers hardness measurements were taken with FutureTech's FM 700 microhardness tester [40].

The aim of the testing was to determine the effect of the applied thermoplastic processing conditions on the microstructure and hardness of a 5.5 mm diameter wire rod made of 20MnB4 steel.

A diagram of the longitudinal section of the wire rod with the characteristic points plotted is shown in Figure 14. Detailed results of the ferrite grain size measurements (D_α) and hardness along the length of 20MnB4 steel wire rod with a diameter of 5.5 mm, produced according to technological variants V1–V3, are shown in Table 7. Photographs of the microstructure (in longitudinal section) of a 5.5 mm diameter wire rod made from 20MnB4 steel produced according to process variants V1–V3 are shown in Figures 15–17.

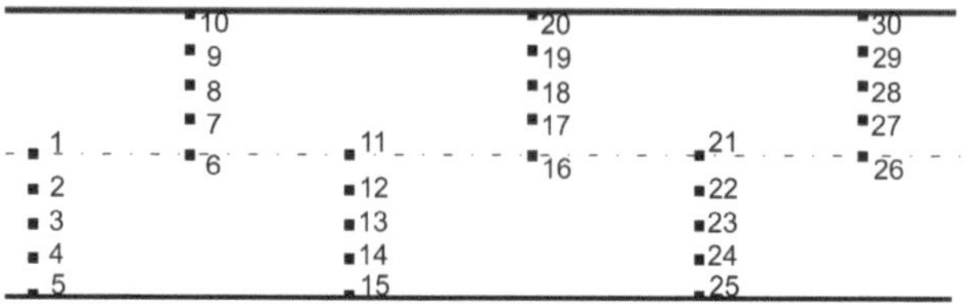

Figure 14. Scheme of a longitudinal section of wire rod with marked characteristic points.

Figure 15. Example of the microstructure of a 5.5 mm diameter wire rod of 20MnB4 steel grade—longitudinal section—technological variant V1: (**a**) magnification 500×, (**b**) magnification 200×.

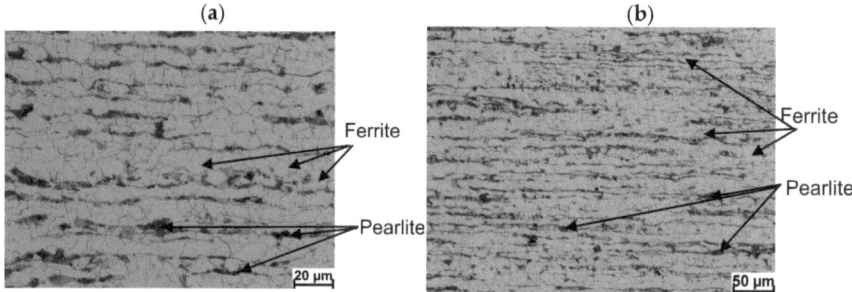

Figure 16. Example of the microstructure of a 5.5 mm diameter wire rod of 20MnB4 steel grade—longitudinal section—technological variant V2: (**a**) magnification 500×, (**b**) magnification 200×.

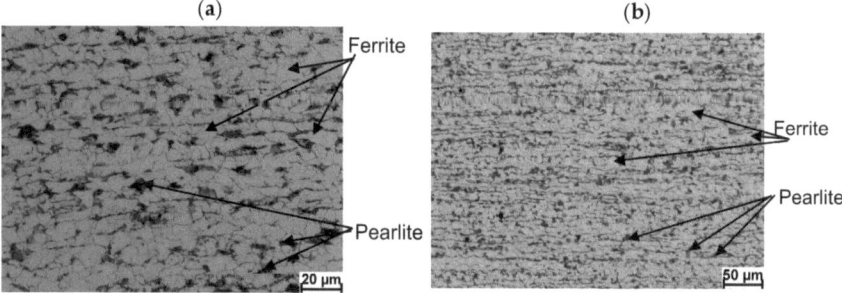

Figure 17. Example of the microstructure of a 5.5 mm diameter wire rod of 20MnB4 steel grade—longitudinal section—technological variant V3: (**a**) magnification 500×, (**b**) magnification 200×.

Based on the analysis of the test results shown in Table 7, the average D_α—ferrite grain size of the wire rod produced according to the V1 variant was 16.21 µm. The average ferrite grain size of the wire rod produced according to the V2 variant was approximately 39% smaller than that of the wire rod produced according to the V1 variant and was 9.86 µm. In turn, the average ferrite grain size of the wire rod produced according to the V3 variant was 8.07 µm and was approximately 50% smaller than the wire rod produced according to the V1 variant and around 18% smaller than the wire rod produced according to V2 variant. The greater fragmentation in the average ferrite grain observed in the case of the wire rod produced according to V2 and V3 variants was mainly due to the higher cooling rate on the roller track, which prevented grain growth. When analyzing the distribution of ferrite grain size along the radius of the wire rod, it was found that the largest ferrite grains were in the axis of the wire rod, while the smallest was observed at its surface. This was due, among other things, to the temperature distribution over the cross-section of the wire rod and the heat transfer conditions (higher temperature in the axis of the wire rod than at its surface).

Upon analyzing the results of the hardness measurements, it was determined that the average hardness of the wire rod produced according to the V1 variant was 177.20 HV. The average hardness of the wire rod produced according to the V2 variant was 182.18 HV, which is ca. 3% higher than the average hardness value of the wire rod produced according to the V1 variant. The average hardness of the wire rod produced according to the V3 variant was 198.17 HV, around 12% higher than the average hardness value of the wire rod produced according to variant V1 and ca. 9% higher than the average hardness value of the wire rod produced according to variant V2. An analysis of the hardness distribution along the radius of the wire rod led to the conclusion that it increases as the size of the ferrite grain becomes smaller.

Table 7. Measurements of ferrite grain size and hardness along the length of a 5.5 mm diameter wire rod of 20MnB4 steel grade were produced according to technological variants V1–V3.

Technological Variant—V1			Technological Variant—V2			Technological Variant—V3		
Measurement Location (According to the Figure 14)	Average Ferrite Grain Size D_α [µm]	Average Hardness Value [HV]	Measurement Location (According to the Figure 14)	Average Ferrite Grain Size D_α [µm]	Average Hardness Value [HV]	Measurement Location (According to the Figure 14)	Average Ferrite Grain Size D_α [µm]	Average Hardness Value [HV]
1	16.82	175.90	1	12.24	174.40	1	8.95	193.50
2	16.81	178.30	2	11.40	175.90	2	7.58	197.50
3	15.18	178.70	3	9.13	183.15	3	7.57	197.75
4	13.96	181.15	4	8.80	186.60	4	6.15	201.50
5	12.73	184.25	5	8.77	194.35	5	5.51	216.90
6	17.39	174.45	6	12.38	171.05	6	10.27	189.30
7	15.96	176.00	7	11.97	172.25	7	8.71	194.70
8	14.30	182.20	8	9.63	176.40	8	8.43	196.05
9	13.88	182.30	9	8.72	188.00	9	8.40	197.00
10	13.29	188.10	10	8.46	195.80	10	7.13	201.45
11	18.36	172.60	11	13.78	169.85	11	9.27	192.05
12	15.35	179.85	12	11.16	176.13	12	8.64	196.15
13	15.00	180.00	13	9.34	177.83	13	7.69	197.75
14	14.03	181.10	14	9.09	178.55	14	7.24	198.55
15	12.64	184.90	15	4.92	201.40	15	6.53	204.75
16	18.57	169.85	16	11.46	176.70	16	9.78	190.55
17	18.41	170.50	17	9.99	178.15	17	9.78	191.90
18	17.95	173.25	18	9.27	181.55	18	9.68	192.60
19	16.79	176.15	19	9.20	183.10	19	9.18	196.45
20	16.75	176.90	20	8.91	192.15	20	9.00	198.00
21	18.29	171.30	21	11.12	180.70	21	10.08	190.45
22	17.91	173.75	22	10.50	181.15	22	9.76	192.05
23	16.61	173.95	23	9.05	187.10	23	7.22	199.35
24	16.19	178.45	24	8.61	189.10	24	7.14	199.45
25	15.86	182.40	25	4.66	210.77	25	4.01	222.65
26	19.04	168.80	26	11.69	172.90	26	9.41	192.75
27	18.31	172.55	27	11.08	174.75	27	9.21	193.05
28	16.94	173.50	28	10.64	176.30	28	7.16	199.50
29	16.89	174.25	29	10.39	178.90	29	6.87	200.00
30	15.99	180.45	30	9.56	180.50	30	5.79	211.35
Average value	16.21	177.20	Average value	9.86	182.18	Average value	8.07	198.17

Metallographic analysis showed that the wire rod produced according to the V1 variant (Figure 15) had a nonuniform coarse-grained ferritic–pearlitic microstructure in the form of alternating bands of ferrite and pearlite. The observed banding was mainly due to the slow cooling of the wire rod after the rolling process. The average ferrite grain size ranged from 12.64 μm to 19.04 μm. Upon analyzing the microstructure of the wire rod produced according to the V2 variant (Figure 16), it was determined that increasing the cooling rate (despite the higher end-of-rolling temperature) resulted in favorable fragmentation and greater uniformity of the microstructure and significantly reduced its banding. The average ferrite grain size of the wire rod produced according to the V2 variant ranged from 4.66 μm to 13.78 μm. As depicted in Figure 17, increasing the cooling rate resulted in even greater fragmentation and uniformity of the ferrite grain size. The microstructure of the wire rod produced according to the V3 variant was also characterized by the lowest banding. The average ferrite grain size, in this case, ranged from 4.01 μm to 10.27 μm.

4. Conclusions

Based on the theoretical and experimental results presented in this study, the following conclusions were drawn:

1. Based on the analysis of the microstructure of 20MnB4 steel samples after physical modeling of the wire rod rolling process, it was determined that in all analyzed technological variants, the material after hardening had a martensitic structure throughout its mass with some residual autenite (not exceeding 4%). Since no other phases were identified in the test samples, it can be concluded that under real conditions (in all the analyzed variants), the tested steel was deformed in the RSM rolling block in a single-phase (austenitic) state.
2. Statistical processing of the obtained test results and analysis of the normal distributions of the technological variants of the wire rod rolling process analyzed in this study showed some nonuniformity of the analyzed mechanical properties along the length of the finished product.
3. One of the factors causing the nonuniformity of mechanical properties along the wire rod length is the different coil temperatures during the cooling of the wire rod, resulting from their characteristic arrangement on the roller conveyor—spiral.
4. For the analyzed thermoplastic processing parameters, increasing the rolling temperature in the RSM block from 800 °C (V1) to 850 °C (V2) and increasing the cooling rate to 10 °C/s (V3) resulted in an improved combination of the analyzed mechanical properties.
5. Under the analyzed thermoplastic processing conditions, wire rods produced according to technological variant V3 showed the lowest microstructure variation.
6. Increasing the cooling rate in the STELMOR® line in the V3 variant (despite the higher temperature of the rolled material in the RSM rolling block) resulted in greater fragmentation in the microstructure on the longitudinal section of the finished product by about 18% compared with the V2 variant and by about 50% compared with the V1 variant.
7. In the studied range of thermoplastic processing parameters, the observed nonuniformity of mechanical properties along the length of the 5.5 mm diameter wire rod of 20MnB4 steel does not adversely affect the capacity for further cold-forming, which has been confirmed by upsetting tests with a relative plastic strain of up to 75%.

Funding: This research study was financed from the resources of the National Research and Development Centre (Narodowe Centrum Badań i Rozwoju) in the years 201–2016 as Applied Research Project No. PBS2/A5/32/2013.

Institutional Review Board Statement: Not applicable.

Informed Consent Statement: Not applicable.

Data Availability Statement: Data are contained within the article.

Conflicts of Interest: The author declares no conflict of interest.

References

1. Gorbanev, A.A.; Zhuchkov, S.M.; Filippov, V.V.; Timoshpolskij, V.I.; Steblov, A.B.; Junakov, A.M.; Tishhenko, V.A. *Theoretical and Technological Basis of High Speed Wire Rod Production*; Izdatelstvo Vyshehjshaja Shkola (College Publishing House): Minsk, Belarus, 2003.
2. Zhang, H.; Feng, G.; Wang, B.; Liu, X.; Liu, X. Influence of Temperature Uniformity of Billet before Rolling on Microstructure and Properties of Hot Rolled Rebar. *J. Phys. Conf. Ser.* **2019**, *1213*, 052021. [CrossRef]
3. Dyja, H.; Mróz, S.; Sygut, P.; Sygut, M. *Technology and Modelling of the Rolling Process of Round Bars with Narrowed Dimensional Tolerance*; Publishing House of the Faculty of Process Engineering, Materials and Applied Physics of the Czestochowa University of Technology: Czestochowa, Poland, 2012.
4. Jain, I.; Lenka, S.; Ajmani, S.; Kundu, S. An Approach to Heat Transfer Analysis of Wire Loops Over the Stelmor Conveyor to Predict the Microstructural and Mechanical Attributes of Steel Rods. *J. Therm. Sci. Eng. Appl.* **2016**, *8*, 021019-1. [CrossRef]
5. Hwang, J.-K. Effect of Contact Point of Wire Ring on Cooling Behavior during Stelmor Cooling. *Materials* **2022**, *15*, 8262. [CrossRef] [PubMed]
6. Fang, C.J.; Lin, Y.Y. A novel temperature diagnostic system for Stelmor air-cooling of wire rods. *China Steel Tech. Rep.* **2022**, *25*, 66–72.
7. Lambert, N.; Wilmotte, S.; Economopoulos, M. *Improving the Properties of Wire by Suppressing the Recalescence on a Stelmor Conveyor*, Technical Steel Research, Mechanical Working; Contract No. 7210.EA/2/205; Commission of the European Communities: Brussels, Luxembourg, 1983.
8. Lindemann, A.; Schmidt, J. ACMOD-2D—A heat transfer model for the simulation of the cooling of wire rod. *J. Mater. Process. Technol.* **2005**, *169*, 466–475. [CrossRef]
9. Morales, R.D.; Lopéz, A.G.; Olivares, I.M. Mathematical simulation of Stelmor process. *Ironmak. Steelmak.* **1991**, *18*, 128–138.
10. Viéitez, I.; López-Cancelos, R.; Martín, E.B.; Varas, F. Predictive model of wire rod cooling. In Proceedings of the 28th ASM Heat Treating Society Conference, Detroit, MI, USA, 20–22 October 2015; pp. 518–524.
11. Vincent, J.C.; Kiefer, B.V. Modernization of Stelmor conveyor systems for expanded process capability and improved product quality. *Wire J. Int.* **2004**, *37*, 158–161.
12. Yu, W.H.; Chen, S.H.; Kuang, Y.H.; Cao, K.C. Development and application of online Stelmor controlled cooling system. *Appl. Therm. Eng.* **2009**, *29*, 2949–2953. [CrossRef]
13. *Morgan Stelmor Controlled Cooling Conveyor System—The Most Effective Cooling Conveyor on the Market*; Brochure No.: T06-0-N175-L2-P-V1-EN; Primetals Technologies USA LLC: Canonsburg, PA, USA, 2020.
14. Grosman, F.; Woźniak, D. Aspects of Rolling Wire Rod from Modern Steel. *Metall.-Metall. News* **2002**, *69*, 408–414.
15. Kazeminezhad, M.; Karimi Taheri, A. The effect of controlled cooling after hot rolling on the mechanical properties of a commercial high carbon steel wire rod. *Mater. Des.* **2003**, *24*, 415–421. [CrossRef]
16. Knapiński, M.; Koczurkiewicz, B.; Dyja, H.; Kawałek, A.; Laber, K. The analysis of influence production conditions on properties C70D 5,5 mm diameter wire rod. In Proceedings of the 23rd International Conference on Metallurgy and Materials–METAL 2014, Brno, Czech Republic, 21–23 May 2014.
17. Koczurkiewicz, B.; Dyja, H.; Niewielski, G. The influence of cooling conditions on perlite morphology of high carbon wire rods. In Proceedings of the 24th International Conference on Metallurgy and Materials–METAL 2015, Brno, Czech Republic, 3–5 June 2015; pp. 699–703.
18. Kuc, D.; Szala, J.; Bednarczyk, I. Influence of Rolling Temperature on the Properties and Microstructure of C70D Steel Intended for Wire Rod. *Metall.-Metall. News* **2016**, *83*, 348–350.
19. Laber, K.; Dyja, H.; Koczurkiewicz, B. Analysis of the industrial conditions of the multi-stage cooling process of the 5.5 mm in diameter wire rod of C70D high-carbon steel. *Mater. Test.* **2015**, *57*, 301–305. [CrossRef]
20. Majta, J.; Łuksza, J.; Bator, A.; Szynal, H. Microalloyed steel wire rod for subsequent cold forming. *Metall.-Metall. News* **2003**, *3*, 95–103.
21. Niewielski, G.; Kuc, D.; Hadasik, E.; Bednarczyk, I. Influence of hot working and cooling conditons on the microstructure and properties of C70D steel for wire rod. In Proceedings of the 26th International Conference on Metallurgy and Materials–METAL 2017, Brno, Czech Republic, 24–26 May 2017; pp. 490–495.
22. Kuc, D.; Niewielski, G.; Bednarczyk, I.; Schindler, I. Influence of rolling temperature and cooling conditions on 23MnB4 steel properties and microstructure designed for cold upsetting. *Metall.-Metall. News* **2016**, *9*, 413–416. [CrossRef]
23. Kuc, D.; Niewielski, G.; Schindler, I.; Bednarczyk, I. Influence of rolling temperature and cooling conditions on the properties and microstructure of steel 23MnB4 for cold-heading. In Proceedings of the 25th Anniversary International Conference on Metallurgy and Materials–METAL 2016, Brno, Czech Republic, 25–27 May 2016; pp. 452–457.
24. Kuc, D.; Szala, J.; Bednarczyk, I. Quantitative evaluation of the microstructure and mechanical properties of hot rolled 23MnB4 steel grade for cold upsetting. *Arch. Metall. Mater.* **2017**, *62*, 551–556. [CrossRef]
25. Woźniak, D.; Garbarz, B.; Żak, A.; Marcisz, J.; Adamczyk, M.; Walnik, B. Determination of The Values of Microstructural and Mechanical Parameters of a Pearlitic 0.7%C-0.60/0.70%Mn Steel Wire Rod Enabling to Achieve High Ability to Drawing with the Use of Large Cross Section Reduction. *J. Met. Mater.* **2019**, *71*, 16–24. [CrossRef]
26. Koczurkiewicz, B.; Dyja, H.; Kawałek, A.; Stefanik, A. Modeling of Microstructure Changes of C70D Wire Rod. Sovremennye Metody i Tekhnologii Sozdanija i Obrabotki Materialov: Minsk, Belarus, 2016; pp. 76–82.

27. Koczurkiewicz, B.; Stefanik, A.; Laber, K. The model of austenite microstructure of high-carbon steel. In Proceedings of the 25th Anniversary International Conference on Metallurgy and Materials–METAL 2016, Brno, Czech Republic, 25–27 May 2016.
28. Stradomski, G. *Modelling of Microstructure Development Processes during Thermoplastic Treatment of Round Bars Made of S355J0 Steel Grade*; Metallurgy, New technologies and Achievements, Collected monograph; Dyja, H., Ed.; Publishing House of the Faculty of Process Engineering, Materials and Applied Physics of the Czestochowa University of Technology: Czestochowa, Poland, 2009; pp. 53–77, ISBN 978-83-87745-13-4, ISSN 2080-2072.
29. *PN-EN 10263-4:2004*; Wire Rod, Rods and Wire for Upsetting and Cold Extrusion. Part 4: Technical Delivery Conditions for Steel for Heat Treatment. Polish Committee for Standardization: Warsaw, Poland, 2004.
30. Laber, K. *New Aspects of Wire Rod Production from Steel for Cold Heading*; Series: Monograph No. 79; Czestochowa University of Technology, Faculty of Production Engineering and Materials Technology Publishing House: Czestochowa, Poland, 2018, ISBN 978-83-63989-64-4, ISSN 2391-632X.
31. Laber, K.; Knapiński, M. Determining conditions for thermoplastic processing guaranteeing receipt of high-quality wire rod for cold upsetting using numerical and physical modelling methods. *Materials* **2020**, *13*, 711. [CrossRef] [PubMed]
32. Laber, K. Innovative Methodology for Physical Modelling of Multi-Pass Wire Rod Rolling with the Use of a Variable Strain Scheme. *Materials* **2023**, *16*, 578. [CrossRef] [PubMed]
33. Dobrzański, L.A. *Metal Science with the Basics of Materials Science*; Scientific and Technical Publishers: Warsaw, Poland, 1996.
34. Kajzer, S.; Kozik, R.; Wusatowski, R. *Rolling of Long Products. Rolling Technologies*; Silesian University of Technology Publishing House: Gliwice, Poland, 2004.
35. *PN-EN 10025*; Hot Rrolled Products of Non-Alloy Constructional Steels. Technical Conditions of Delivery. Polish Committee for Standardization: Warsaw, Poland, 2002.
36. Paduch, J.; Szulc, W. (Eds.) *Shaping New Qualities and Rationalizing of the Steel Products Production Costs Adapted to the Market's Competitive Requirements*; Part 3: Adaptation of Metallurgical Technologies to the Application and Quality Needs of the Market; Iron Metallurgy Institute: Gliwice, Poland, 2000; Volume 52, pp. 17,19,24–25.
37. QTSteel Software. In *Metallurgical Software for Simulating the Quench and Tempering of Steels*; User's Guide. Release 3.4.1–including the rolling option; ITA-Technology and Software: Ostrava, Czech Republic; MSL-Metaltech Services Ltd.: Gateshead, UK, 2015.
38. *User's Manual: FLIR 640 Series*; FLIR Systems, 30.06.2008, publication no. 1558550; Teledyne FLIR LLC: Wilsonville, OR, USA, 2008.
39. *PN-EN ISO 6892-1:2016-09*; Metals-Tensile Test-Part 1: Room Temperature Test Method. Polish Committee for Standardization: Warsaw, Poland, 2016.
40. Dyja, H.; Krakowiak, M. *The 60th Anniversary Chronicle-from the Faculty of Metallurgy to the Faculty of Process, Materials Engineering and Applied Physics*; Czestochowa University of Technology, Faculty of Process, Materials Engineering and Applied Physics Publishing House: Częstochowa, Poland, 2010.
41. *PN-83/H-04411*; Metal Upsetting Test. Polish Committee for Standardization, Measures and Quality, Alfa Standardization Publishing House: Warsaw, Poland, 1983.
42. Šimeček, P.; Hajduk, D. Prediction of mechanical properties of hot rolled steel products. *J. Achiev. Mater. Manuf. Eng.* **2007**, *20*, 395–398.
43. Šimeček, P.; Turoň, R.; Hajduk, D. Computer model for prediction of mechanical properties of long products after heat treatment. In Proceedings of the 68th ABM International Annual Congress, Belo Horizonte Minas Gerais, Brazil, 30 July–2 August 2013.
44. Laber, K.; Koczurkiewicz, B. Determination of optimum conditions for the process of controlled cooling of rolled products with diameter 16.5 mm made of 20MnB4 steel. In Proceedings of the 24th International Conference on Metallurgy and Materials-METAL 2015, Brno, Czech Republic, 3–5 June 2015; pp. 364–370.
45. Głowacka, M. *Metallography*; Gdansk University of Technology Publishing House: Gdansk, Poland, 1996; p. 219.
46. Minkina, W. *Thermovion Measurements—Instruments and Methods*; Częstochowa University of Technology Publishing House: Czestochowa, Poland, 2004.
47. Orlove, G.L. Practical thermal measurement techniques. In *Proceedings of the Thermosense V, Thermal Infrared Sensing Diagnostics, 21 March 1983*; SPIE: Paris, France, 1983; Volume 371, pp. 72–81.
48. *PN-84/H-04507/01*; Metals–Metallographic Grain Size Testing. Microscopic Methods for Determining Grain Size. Polish Committee for Standardization, Measures and Quality, Alfa Standardization Publishing House: Warsaw, Poland, 1985.

Disclaimer/Publisher's Note: The statements, opinions and data contained in all publications are solely those of the individual author(s) and contributor(s) and not of MDPI and/or the editor(s). MDPI and/or the editor(s) disclaim responsibility for any injury to people or property resulting from any ideas, methods, instructions or products referred to in the content.

Article

Effects of Zn, Mg, and Cu Content on the Properties and Microstructure of Extrusion-Welded Al–Zn–Mg–Cu Alloys

Krzysztof Remsak [1,*], Sonia Boczkal [1], Kamila Limanówka [1], Bartłomiej Płonka [1], Konrad Żyłka [1], Mateusz Węgrzyn [1] and Dariusz Leśniak [2]

[1] Lukasiewicz Research Network—Institute of Non-Ferrous Metals, 44-100 Gliwice, Poland
[2] Faculty of Non-Ferrous Metals, AGH University of Krakow, 30-059 Krakow, Poland
* Correspondence: krzysztof.remsak@imn.lukasiewicz.gov.pl

Abstract: The study presents the results of research on the influence of different contents of main alloying additions, such as Mg (2 ÷ 2.5 wt.%), Cu (1.2 ÷ 1.9 wt.%), and Zn (5.5 ÷ 8 wt.%), on the strength properties and plasticity of selected Al–Zn–Mg–Cu alloys extruded on a bridge die. The test material variants were based on the EN AW-7075 alloy. The research specimens, in the form of 100 mm extrusion billets obtained with the DC casting method, were homogenized and extrusion welded during direct extrusion on a 5 MN horizontal press. A 60 × 6 mm die cross-section was used, with one bridge arranged in a way to extrude a flat bar with a weld along its entire length. The obtained materials in the F and T6 tempers were characterized in terms of their strength properties, hardness, and microstructure, using EBSD and SEM. The extrusion welding process did not significantly affect the properties of the tested materials; the measured differences in the yield strength and tensile strength between the materials, with and without the welding seam, were up to ±5%, regardless of chemical composition. A decrease in plasticity was observed with an increase in the content of the alloying elements. The highest strength properties in the T6 temper were achieved for the alloy with the highest content of alloying elements (10.47 wt.%), both welded and solid. Significant differences in the microstructure between the welded and solid material in the T6 temper were observed.

Keywords: extrusion welding; Al–Mg–Zn–Cu; Al7075; alloying elements; EBSD; GOS

Citation: Remsak, K.; Boczkal, S.; Limanówka, K.; Płonka, B.; Żyłka, K.; Węgrzyn, M.; Leśniak, D. Effects of Zn, Mg, and Cu Content on the Properties and Microstructure of Extrusion-Welded Al–Zn–Mg–Cu Alloys. *Materials* **2023**, *16*, 6429. https://doi.org/10.3390/ma16196429

Academic Editors: Gábor Harsányi and Zongbin Li

Received: 31 July 2023
Revised: 15 September 2023
Accepted: 21 September 2023
Published: 27 September 2023

Copyright: © 2023 by the authors. Licensee MDPI, Basel, Switzerland. This article is an open access article distributed under the terms and conditions of the Creative Commons Attribution (CC BY) license (https://creativecommons.org/licenses/by/4.0/).

1. Introduction

Precipitation-hardened Al–Zn–Mg(Cu) alloys are characterized by their high strength, ductility, elastic modulus, corrosion resistance, as well as their fracture toughness. For this reason, they are widely used in the aerospace industry (airframe, fuselage), automotive industry (bumpers, body parts), among manufacturers of sports equipment, and in other industries where weight reduction while maintaining high-strength properties is important [1–3]. However, the 7XXX series aluminium alloys are also highly sensitive to the strain rate and deformation temperature, which directly translate into low processing efficiency, especially during the extrusion of welded profiles (on bridge or porthole dies) [4]. This is due to phenomena that produce a hot cracking effect at elevated working rates [5]. During the extrusion process, both in the press container and in the die itself, frictional forces arise that disturb the favorable state of hydrostatic stresses [6]. These forces cause tensile stresses to accumulate on the surface of the extruded band and cause additional heating of these areas. In extreme cases, local temperature increases in the material and the resulting stress release may result in the loss of continuity of the extruded strand [7]. This is of particular importance in the case of bridge–chamber dies, where the complicated internal geometry multiplies the adverse effects of frictional forces on the homogeneity of the material outflow [8]. The structural phenomena accompanying the extrusion process on porthole dies should also be considered. In the studies carried out so far, it was found that in the 6063 aluminium alloys extruded on porthole dies, the occurrence of dynamically recrystallized (DRX) grains was observed both in the welded area and beyond it. It was also noted that the amount of the recrystallized fraction may be related to the differentiation of the strain rate on the cross-section of the extruded profile [7]. Similar conclusions apply to Al–Zn–Mg alloys extruded on porthole dies. It has been shown that a higher strain temperature and a lower strain rate have a positive effect on the occurrence of dynamic recrystallization. Moreover, in the case of Al–Zn–Mg alloys, it was proven that in both the areas close to the walls of the profiles and in the welding zones, the recrystallized fraction is higher than in other zones [9]. Studies conducted on extruded Mg–Al–Zn alloys also indicate that complete dynamic recrystallization occurred in the welding areas [10]. Current studies on Al–Zn–Mg alloys extruded on porthole dies show a significant differentiation in the microstructure between the welded area and the outside of the weld [11].

The material for the extrusion of aluminium alloys are billets, cast using the DC (di-rect chill) method. It is a method of semicontinuous casting into a water-cooled crystal-lizer. The distinctive feature of that process is directional crystallization, taking place from the outer zones (intensively cooled zones near the crystallizer) to the center of the ingot. As a result, a dendritic structure is formed [12]. Dendrites from the Al-rich α phase grow along the temperature gradient. This directionality of solidification leads to microsegregation, and coarse intermetallic phases are formed that can significantly affect the properties and susceptibility to hot working, which directly affects the efficiency of technological processes. The above-mentioned segregation in Al–Zn–Mg(Cu) alloys during casting results in high concentrations of Cu, Mg, and Zn in interdendritic eutectic regions. This has a significant impact on reducing the corrosion resistance of the alloys, and can also be a place of crack initiation of the billets themselves [13]. In the case of Al–Zn–Mg(Cu) alloys, literature sources indicate that the main phases present in the material after casting are η-$MgZn_2$, T-$Al_2Mg_3Zn_3$, S-Al_2CuMg, and θ-Al_2Cu. The S, T, and η phases described in the literature are solid solutions with extended composition ranges, containing all four elements. It should be noted that the literature data clearly indicate that the mentioned phases are low-melting, especially η-$MgZn_2$, and their presence causes a significant decrease in the solidus temperature, even below 480 °C for the 7075 alloy [12,13]. It causes a large difference between the liquidus and solidus temperatures, which in turn leads to the need to lower the temperature of plastic-forming processes. As a result, it translates into a decrease in the speed of these processes (e.g., extrusion) and a decrease in the efficiency. Homogenization annealing is a way to increase the solidus temperature of Al–Zn–Mg(Cu) alloys. Data from the literature clearly indicate that the dendritic structure gradually dissolves during homogenization. Diffusion into the matrix in eutectic structures and gradual dissolution of the Mg$(Zn,Cu,Al)_2$ phase takes place. From the literature data, the conclusion drawn was that the driving force of phase transformations from Mg$(Zn,Cu,Al)_2$ to Al_2CuMg must be the supersaturation of copper in areas of the eutectic structures, which makes Al_2CuMg a stable phase [12,13].

In the case of precipitation-hardened 7XXX series alloys, the content, distribution, and size of the matrix precipitates have a key impact on their strength. It can be increased by increasing the content of the elements dissolved in the matrix during solution treatment, which can be separated from the solution in the aging stage [14,15]. During the supersaturation process, it is necessary to ensure sufficiently rapid cooling, which will ensure a uniform distribution of precipitates of the fine-grained n' phase during aging. Since the metastable n' phase should have the greatest impact on precipitation strengthening, the greatest strengthening effect will be obtained as a result of rapid cooling [16,17].

Thus far, the influence of the main alloy additions, that is, Zn, Mg, and Cu, on the strength properties of the 7XXX series alloys has been described in the literature [15,18–21]. For alloys with a constant Cu and (Zn+Mg) content, an increase in the Zn content will increase the strength of aged samples [15,18]. It was also found that with a constant Cu content, an increase in the share of (Zn+Mg) in the T6 temper will result in an increase in the number of precipitates, which will translate into an increase in the strength of Al–Zn–Mg–Cu alloys [17]. Of equal importance is that this can result in a large difference in plasticity between the matrix and PFZ areas (precipitation free zones), and consequently may lead to a reduction in fracture toughness also caused by coarse slip [17].

Al–Zn–Mg–Cu alloys, due to technological difficulties, are rarely used in extrusion processes on porthole dies. At the same time, the industry is systematically increasing its interest in high-strength Al alloys, also in the form of thin-walled closed profiles. For this reason, as well as due to the lack of relevant publications, there was a need to conduct research on the effects of different levels of the main alloying additives on extrusion-welded Al–Zn–Mg–Cu alloys.

This study investigated the influence of different contents of main alloying additions on the strength properties and plasticity of selected Al–Zn–Mg–Cu alloys extruded on bridge dies.

2. Materials and Methods

The tests were carried out on modified 7xxx series alloys, based on the 7075 alloy.

The initial stage of the research included casting variants of EN AW-7075 alloys in the form of 100 mm diameter billets on a semicontinuous production line consisting of 300 kg Monometer resistance furnace and a direct chill casting crystallizer (Monometer House, Rectory Grove, Leigh-on-Sea, Essex SS9 2HN, UK).

The chemical composition of the alloys was analyzed using optical emission spectrometry, and is presented in Table 1. Table 2 shows the contents of the main alloying elements and their proportions.

Table 1. Chemical compositions of 7075 alloy variants [wt.%].

Alloy No.	Si	Fe	Cu	Mn	Mg	Cr	Zn	Ti	Zr
1	0.08	0.15	1.16	0.00	1.99	0.18	4.97	0.02	0.161
2	0.08	0.15	1.91	0.00	2.33	0.18	5.67	0.02	0.166
3	0.10	0.21	1.53	0.00	2.3	0.18	5.78	0.02	0.151
4	0.10	0.22	1.50	0.00	2.21	0.17	7.76	0.02	0.159

Table 2. Main alloying elements contents and ratios.

Alloy No.	Cu/Mg	Mg+Zn	Mg+Zn+Cu	Zn/Mg
1	0.58	6.96	8.12	2.50
2	0.82	8	9.91	2.43
3	0.66	8.08	9.61	2.51
4	0.66	9.97	10.47	3.51

During the casting process, the metal was filtered through a 30 ppi ceramic filter. The alloys were cast in stages, two billets at a time, each approx. 2 m long (Figure 1).

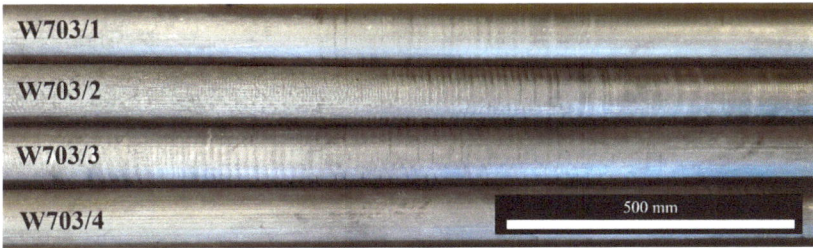

Figure 1. Cast billets of the investigated alloys. From the top to the bottom: alloy 1, 2, 3, 4. W703 is the casting batch number.

The billets were subjected to homogenizing annealing to maximize the solidus temperature level, which is of great importance in the context of maximizing the deformation rate of Al–Zn–Mg alloys [22,23]. This is important from the point of view of extrusion welding, as it significantly extends the permissible temperature range of the material in the deformation cavity. Homogenizing parameters were the subject of other research studies conducted [13]. The final annealing conditions are presented in Table 3.

Table 3. Homogenization annealing parameters for tested alloys.

Alloy No.	Heating 1 [°C-h]	Hold Time1 [h]	Heating 2 [°C-min]	Hold Time2 [h]
1	465-10	4	-	-
2	465-10	2	475-15	8
3	465-10	2	475-15	4
4	465-10	12	-	-

In the next stage of the research, extrusion of the flat bar on a die with a single bridge was carried out. The extrusion was performed on a 5 MN (500 t) horizontal hydraulic press (ZAMET BUDOWA MASZYN S.A., 83 Zagórska Str., 42-680 Tarnowskie Góry, Poland) in direct mode with extrusion force registration. For this purpose, a 60 × 6 mm die insert was designed and made for the existing tooling set, with a single bridge placed perpendicularly to the long edge of the extruded flat bar to allow the formation of a longitudinal weld in the middle of its width (Figure 2). The 60 × 6 mm flat bar was extruded from all variants of the 7075 alloy. The material was prepared in the form of Ø100 × 200 mm billets, then heated to 500 °C for extrusion, with a ram velocity of 1 mm/s. Table 4 presents the registered peak forces during the extrusion process.

Figure 2. A die insert with single bridge used for extrusion of 60 × 6 mm flat bar with weld.

Table 4. Forces registered during the extrusion.

Alloy No.	1	2	3	4
Registered peak extrusion load [MN]	4.11	4.78	4.40	5.06

The extruded sections were subjected to heat treatment of the T6 temper for all of the alloy variants. For this purpose, samples for solution treatment and artificial aging were taken from the extruded bars. The heat treatment parameters are listed in Table 5. To verify the effects of the heat treatment, a Brinell hardness test was performed. Based on the results obtained, the aging curves presented in Figure 3 were made for the individual variants of the alloys. They show that the tested variants of the alloys obtained more than 90% of their maximum hardness values after about 8 h. Further aging up to 24 h increases the hardness, but for industrial use it may be more economical to shorten the aging time. Therefore, for the purposes of this study, the aging time was set at 8 h.

Table 5. Heat treatment parameters.

Process	T6 Temper
Solutionizing	temperature 465 °C, hold time 2 h, water quenching
Artificial aging	temperature 120 °C, aging time 24 h

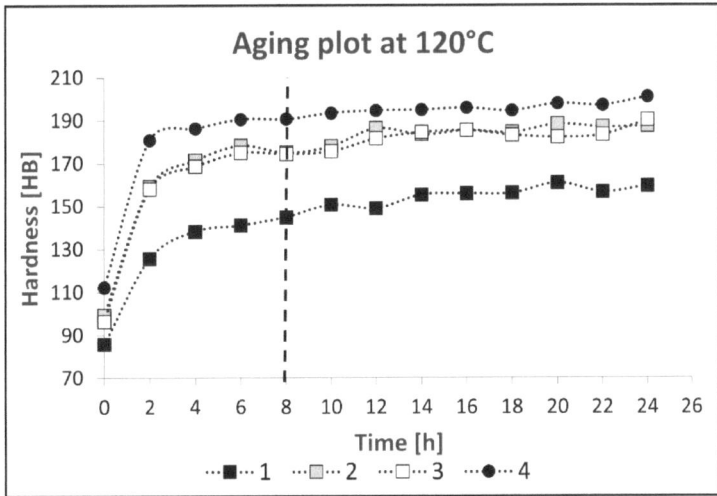

Figure 3. Aging plot for investigated alloys.

The heat-treated material and the reference material in the F temper were intended for further research.

A static tensile test was performed in accordance with the requirements of the standard PN-EN ISO 6892-1:2020-05 [24] on the Instron 5582—max load 100 kN. The strain deformation was measured with an extremely accurate video extensometer (Instron, Norwood, MA, USA), with crosshead speed 1 = 0.37 mm/min and crosshead speed 2 = 3 mm/min. The gauge length was 20 mm. A Brinell hardness test was conducted according to PN-EN-ISO-6506-1_2014 [25] on Duramin 2500E hardness tester (Struers, Ballerup, Hovedstaden, Denmark), with a ball diameter 2.5 mm, main load 31.25 kgF, and expanded uncertainty with the confidence of the result at the 95% level with the coefficient k = 2, determined indirectly by the M2 method in accordance with the PN-EN-ISO-6506-1_2014 standard.

The samples for the static tensile test were taken across the flat bar in such a way that the welded area was located in the middle of the length of the measurement base.

The reference material was taken from a flat bar that was extruded without a weld. The scheme of taking strength samples is presented below (Figure 4). The hardness test was carried out on the cross-section of the flat bars, in the middle of their thickness, with measurement points (1–13) placed every 5 mm, and in the central zone every 2.5 mm, as shown in Figure 5.

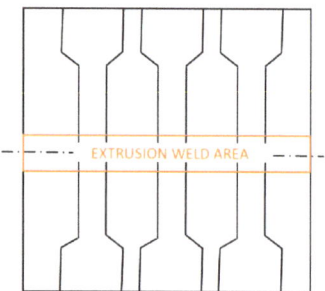

Figure 4. An example of sampling for static tensile test.

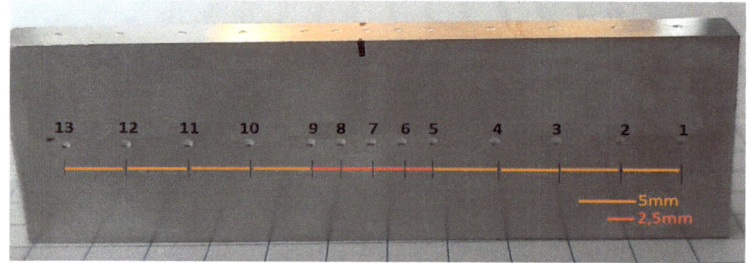

Figure 5. An example of the distribution of Brinell hardness test measurement points.

The microstructural characterization of the alloys was carried out with a high-resolution INSPECT F50 FEI scanning electron microscope with attachments for the chemical analysis via EDS and a Velocity plus EBSD camera (FEI Company, Hillsboro, OR, USA). The EBSD analysis was performed using EDAX Apex Advanced (ver. 2.5.1001.0001) and EDAX OIM Analysis 8 software (ver. 8.6.0101x64) and ICCD 2011 PDF database format. The samples for the tests were ground on SiC abrasive papers up to 4000 grit, and polished with diamond suspensions up to 1 µm. The final operation was polishing with a colloidal SiO_2 suspension with gradations of 0.01 µm to obtain a perfectly flat surface. The samples for crystallographic analysis were prepared on an RES101 ion milling instrument (Leica Microsystems, Wetzlar, Germany).

3. Results and Discussion

The extrusion load was recorded during the process, and the extrusion curves of the tested alloys were prepared (Figure 6). The highest yield resistance represented by an extrusion peak load up to 5.06 MN, was characteristic of alloy 4; it was slightly lower, 4.78 MN for alloy 2, then 4.40 MN for alloy 3 and 4.11 MN for alloy 1. It is clearly visible that the yield resistance depends on the level of the alloying components, which for alloy 4 is (Mg+Zn+Cu) 10.47%. In the case of alloys 2 and 3, for a similar level (Mg+Zn) of 8%, the decisive factor for the yield resistance is the Cu contents of 1.91% and 1.53%, respectively. The lowest yield resistance was recorded for alloy 1 with the lowest level of the main alloying elements (Mg+Zn+Cu), 8.12%.

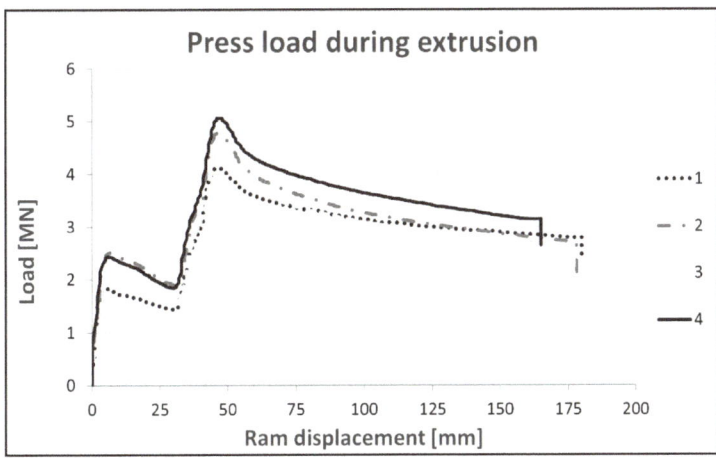

Figure 6. Press loads recorded during 60 × 6 mm section extrusion of tested aluminum alloys.

In the next stage of research, hardness measurements were made on the cross-section of the extruded flat bars, both in the F and T6 tempers. The hardness distribution curves are shown below in Figure 7. The analysis of the results shows a clear increase in the hardness of all of the alloy variants after heat treatment of the T6 temper. The lowest increase was recorded for alloy 1 (60 HB), and the highest for alloy 4. A slight (up to 5 HB) hardness variation was observed in the welded area (half the width of the flat bars), indicating the uniformity of the strength properties of the flat bar.

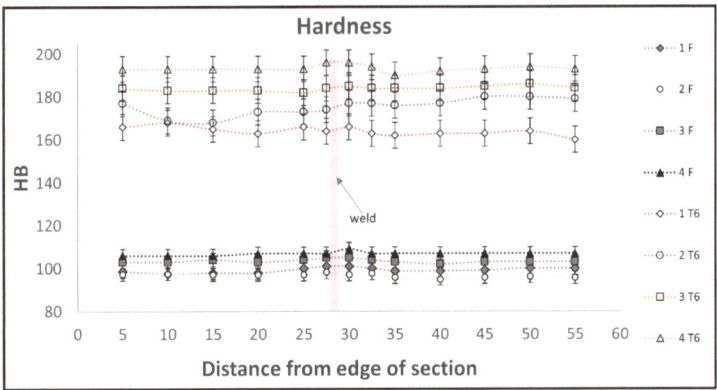

Figure 7. Hardness distributions [HB] on cross-sections of the flat bar 60 × 6 mm: at the top—after heat treatment of the T6 temper; at the bottom—in the F temper.

In the case of the tested variants of the Al–Zn–Mg–Cu alloy, heat treatment of the T6 temper resulted in an increase in tensile strength TS (Figure 8) and yield strength YS (Figure 9), both for the welded areas and the solid material. The largest increase in the tensile and yield strength TS and YS was recorded for alloy 4, and amounted to 244 MPa and 353 MPa, respectively, for the unwelded samples, and 305 MPa and 352 MPa, respectively, for the welded samples. For the other alloys tested, the increases in strength properties were smaller, and amounted to 150–220 MPa for the TS and 250–300 MPa for the YS. Differences in the increases of the described properties are directly caused by the amount of alloy additions in the individual alloys tested. Alloy 4 contains the highest addition level of 10.47 wt.%, with the content of Zn alone being 7.76 wt.%. According to the literature, the content of this alloying additive contributes most significantly to the precipitation strengthening of Al–Zn–Mg–Cu alloys, as it directly translates into an increase in the amount of the $MgZn_2$ phase, which is the main strengthening phase of these alloys. Heat treatment of the T6 temper did not significantly affect the differentiation of the strength properties in these areas of the individual alloy variants.

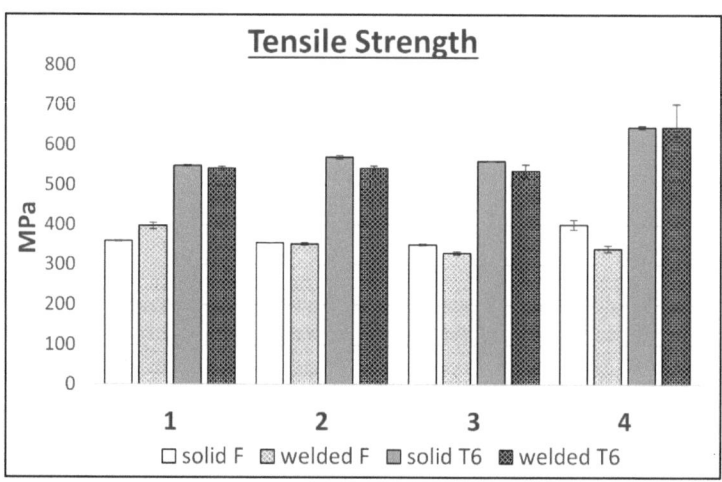

Figure 8. Comparison of the tensile strength results of flat bars with the weld and solid for 4 variants of alloys in the F and T6 tempers.

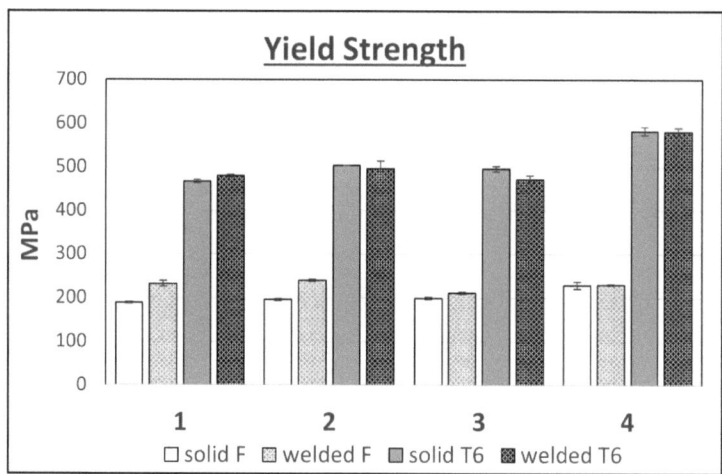

Figure 9. Comparison of the yield strength results of flat bars with the weld and solid for 4 variants of alloys in the F and T6 tempers.

A clear decrease in elongation in the welded areas is noticeable for each of the tested alloy variants (Figure 10). The unfavorable effect of increasing the content of alloy additions on the elongation of the tested alloys can also be observed. Variants 2 and 3 have similar amounts of main alloying elements, which are 9.91 wt.% and 9.61 wt.%, respectively, and are characterized by a similar elongation level for both the welded and solid samples in the F and T6 tempers. The highest elongation in all of the variants tested was recorded for alloy 1, and the lowest for alloy 4. The exception was the sample in the T6 state, which had an elongation greater by more than 2% than variants 2 and 3, whose elongations in each case were similar.

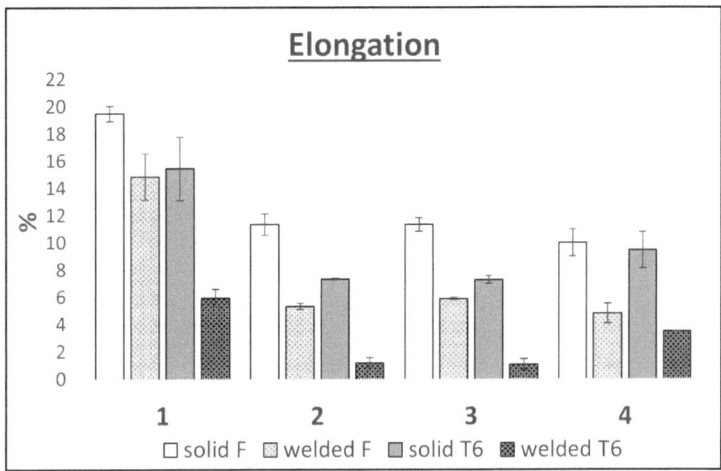

Figure 10. Comparison of the elongations of flat bars with the weld and solid for 4 variants of alloys in the F and T6 tempers.

Figure 11 shows SEM images of alloy 1 (a), alloy 2 (b), alloy 3 (c), and alloy 4 (d). The microstructure of alloys 3 and 4 have relatively smaller and more densely distributed fine precipitates in the Al matrix than those in alloys 1 and 2. All of the alloys reveal irregularly shaped and bright-coloured intermetallic particles.

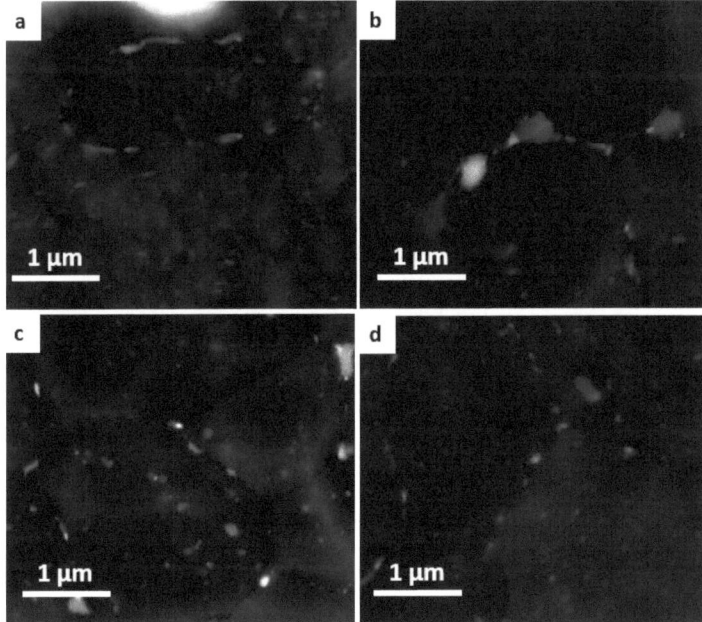

Figure 11. BSE images of investigated alloys: 1 (**a**), 2 (**b**), 3 (**c**), and 4 (**d**).

Figure 12 shows IPF images of the extruded alloys at the weld site from a cross-sectional view to the extrusion direction. The analyses were performed in the T6 temper. The EBSD analysis showed that the shape and size of the grain in the weld were different depending on the chemical composition of the alloys. In alloy 1 shown in Figure 12a, and in alloy 4 shown in Figure 12d, deformed and elongated grains with mainly the (001) and (111) orientations were analyzed. In the middle part of the microstructure, areas of equiaxed grains occurring within the boundaries of large, elongated grains were also observed. The grains occurred in a wide range of sizes, from 1 to over 200 mm. Meanwhile, the microstructures of alloys 2 (Figure 12b) and 3 (Figure 12c) were characterized by equiaxed grains with a random orientation, indicating the occurrence of dynamic recrystallization processes at the weld.

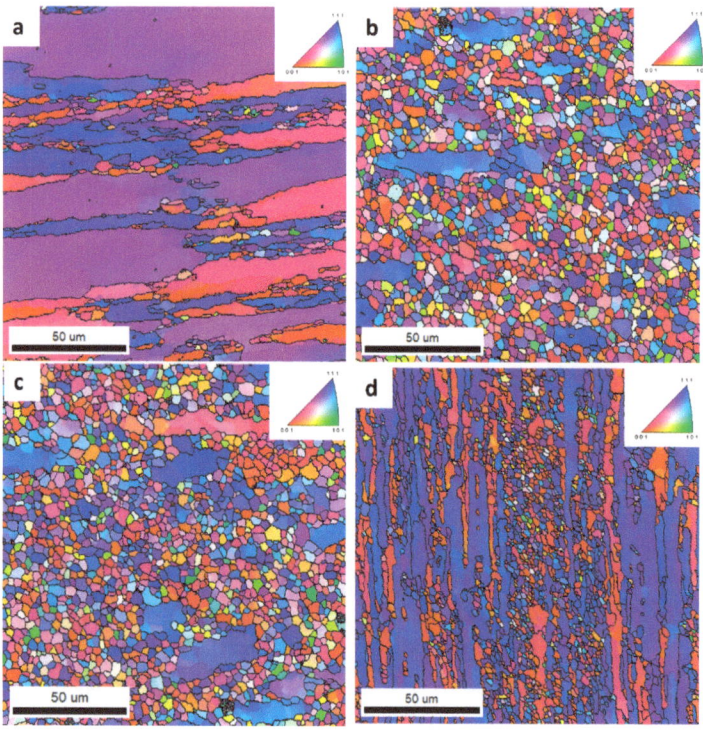

Figure 12. IPF images of investigated alloys: 1 (**a**), 2 (**b**), 3 (**c**), and 4 (**d**)—welded area.

The distributions of the average grain size, shown as histograms in Figure 13, confirm the significant differences in structure for the four alloys studied. The largest grains were found in alloy 1, with the finest grains occurring in a range of up to 25 μm in alloys 2 and 3. Alloy 4 had a wider grain size distribution compared to alloys 2 and 3, ranging up to 45 μm.

To compare the structures in the alloys analyzed, analogical scans were taken in the areas outside the welds (Figure 14). It was found that the structures in the four alloys analyzed showed no significant differences. Alloy 1 showed a slightly larger grain size compared to the other alloys analyzed. The grain size distribution shown in Figure 15 for Alloy 1 shows a wider grain size range up to 55 μm. In alloys 2 to 4, the range was approximately 32 μm.

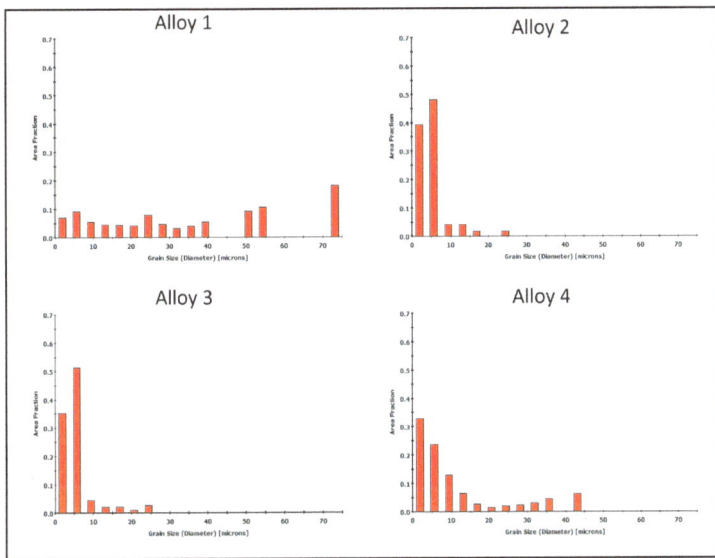

Figure 13. Grain size distributions for investigated alloys—welded area of extruded 60 × 6 mm profile. In red—area fraction of specific grain size.

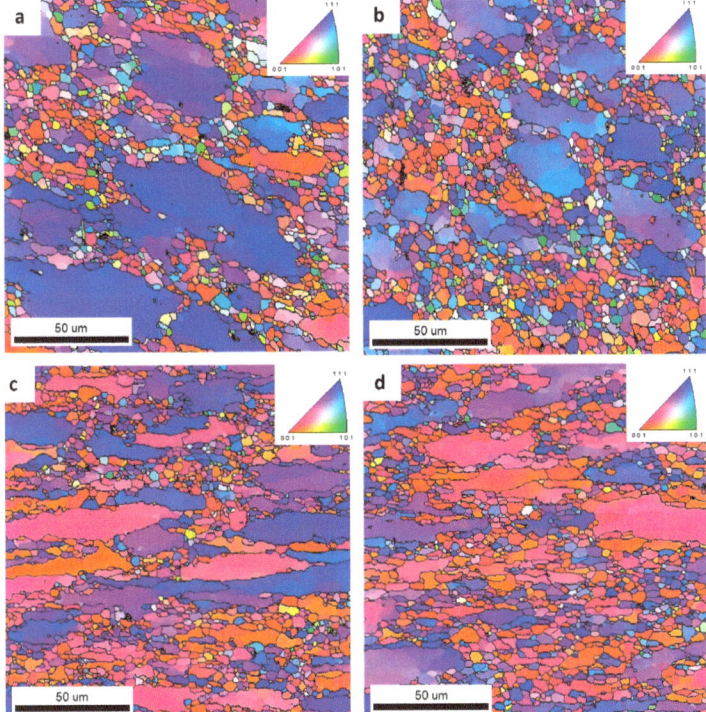

Figure 14. IPF images of investigated alloys: 1 (**a**), 2 (**b**), 3 (**c**), and 4 (**d**)—outside the welded area.

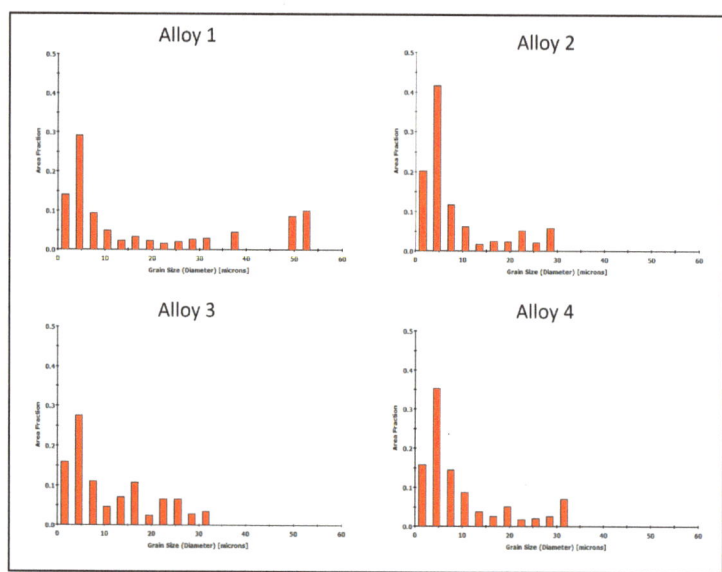

Figure 15. Grain size distributions for investigated alloys—outside the welded area of extruded 60 × 6 mm profile. In red—area fraction of specific grain size.

Figures 16 and 17 represent the GOS maps obtained from the IPF images shown in Figures 12 and 14. The GOS value represents the degree of average dispersion of the misorientation within the same grain in a welded area. Low GOS values can be obtained if some restoration of the microstructure, such as recovery or recrystallization, occurs in the chosen grain. Low GOS values (below 2° or 3°) have been proven to represent recrystallized areas [26]. As a result, as shown in Figures 16 and 17, the blue grains have the smallest GOS values and indicate dislocation-free grains with orientations spread between 0 and 2. In the experiment, the interval with the smallest GOS values with a threshold of 2° represents the recrystallized grains, while the GOS value above 2° represents the deformed grains [27]. The red grains, with orientation spreads between 5 and 35, indicate the heavily deformed grains.

As a result of these analyses, it was found (Figure 16) that welded alloys 2 and 3 had the highest proportions of grains with recrystallized fractions. Alloy 1 was characterized by large partially deformed grains. Alloy 4, on the other hand, showed fine bar-shaped grains that were partially deformed. Recrystallized grains were observed in the very center of the weld. The GOS values for the areas outside the weld (Figure 17) for the four alloys tested were very similar. The highest proportion of grains with a recrystallized fraction was found in alloy 2. However, the differences in the orientation distributions between the alloys were not significantly large.

The low content of the main alloying elements in alloy 1 translates into the lower plastic resistance of this material. This is reflected in the level of the recorded force during extrusion of this alloy on the press (4.11 MN). The lower plastic resistance mentioned above results in lower friction forces in the die deformation cavity and, consequently, in a smaller increase in the temperature of the extrusion. This, in turn, results in the formation of a deformed structure in alloy 1 with a low share of recrystallized grains, which was about 15%. In this case, no effect of the Zr addition on the fragmentation of the structure was observed, which is also the result of an insufficient increase in temperature during the extrusion. In sample 1, dynamic recovery occurs, and the shape of the grain is deformed.

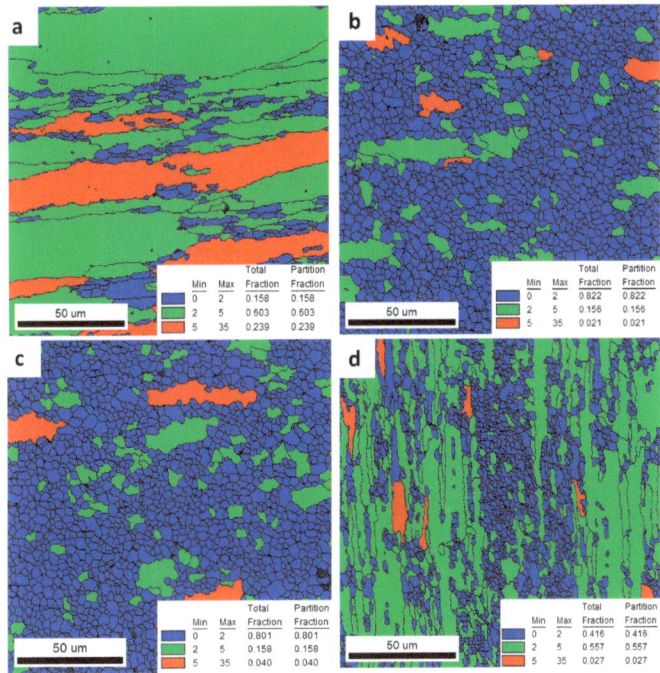

Figure 16. GOS maps of investigated alloys: 1 (**a**), 2 (**b**), 3 (**c**), and 4 (**d**)—welded area.

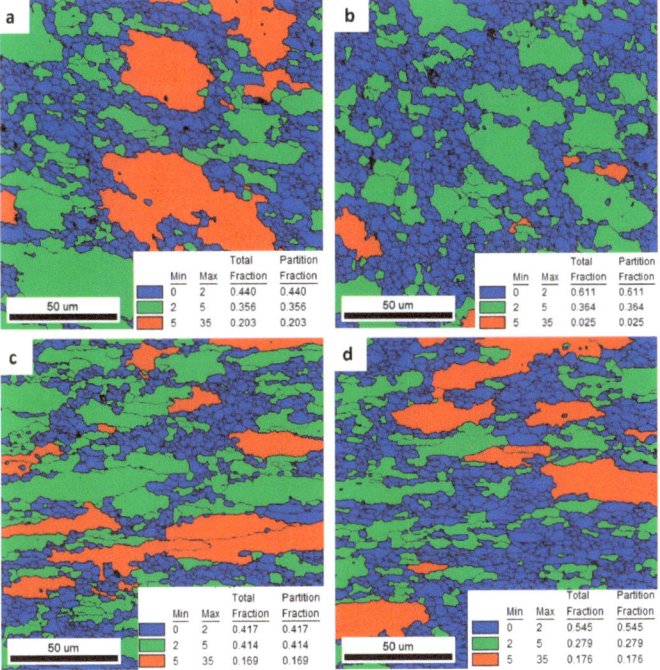

Figure 17. GOS maps of investigated alloys: 1 (**a**), 2 (**b**), 3 (**c**), and 4 (**d**)—outside the welded area.

Welded alloy 1 in the T6 temper had a tensile strength of about 540 MPa, and the elongation was up to 6%. The larger elongation may have been induced by the large grain size. In the case of alloy 1, the grain size distribution obtained from the EBSD test was disturbed. This is due to limitations of the scanned image analysis software, which does not take into account grains with unidentified boundaries. In this case, the largest surface fraction are for grains that are above 80 µm.

In the case of welded alloys 2 and 3, which nearly recrystallized, the T6 temper was also recorded with a tensile strength of about 540 MPa, but the elongation in that case was about 1%. This reduction in plasticity of alloys 2 and 3 relates to an inhibition of grain growth. The proportion of the recrystallized fraction in alloys 2 and 3 is approximately 81%.

For alloy 4, a slightly higher elongation of about 3% was found in the weld compared to alloys 2 and 3. This could be the result of a higher internal stress level due to the highest content of alloying elements in this alloy. From the observations, the grains in alloy 4 were also elongated along the weld line. The grain size along this line was greater than 100 µm, and less than 10 µm in the transverse direction. During the static tensile test, which was conducted in a direction transverse to the weld line, the grains underwent greater plastic deformation than the equiaxial fine grains in alloys 2 and 3.

All of the alloys contain alloying additions of Mg, Cu, and Zn, and additionally Zr. These elements can form $MgZn_2$ and Al_3Zr phases, while Cu remains in solid solution. Numerous studies [28–31] confirm that additives such as Zr are inhibitors of recrystallization. The nano-sized dispersoids formed by Zr can pin down and block the migration of the grain boundaries and, as a result, inhibit the recrystallization in Al–Zn–Mg–Cu alloys [30]. The $MgZn_2$ phase, on the other hand, has a strong precipitation strengthening effect for the Al–Zn–Mg–Cu alloys, and its degree is strongly dependent on the amount of Zn. The higher the Zn content in the alloy, the more the $MgZn_2$ precipitates in the material after the aging treatment. In conclusion, by increasing the Zn content in the Al–Zn–Mg–Cu alloys, the strength is increased [28].

In the case of welded alloy 4 in the T6 temper, the recrystallized grains fraction on the GOS image was about 41%. The introduction of a large amount of the alloying elements resulted in the formation of fine and densely distributed dispersoids that had a significant influence on the recrystallization behavior in alloy 4. Some of the grains remained fibrous, while others recrystallized. This phenomenon is related to the orientation of the grains, where the recrystallized grains are <111> oriented and the fibrous grains are <100> oriented; this was also confirmed by literature data [28]. The tensile strength of alloy 4 was almost 650 MPa, while the elongation was about 4%. The characteristic grain structure arrangement, perpendicular to the extrusion direction, was a result of the specific conditions in the welding chamber of the extrusion die. Consequently, there was a directional flow of the metal and vertical arrangement of the grains in the areas close to the weld.

4. Conclusions

I. Heat treatment of the T6 state of the tested materials resulted in a significant increase in the hardness of all of the alloy variants. The lowest increase was recorded for alloy variant 1 (60 HB), and the highest for alloy 4.

II. There was no significant differentiation in hardness in the area of the weld (half the width of the flat bars), which indicates the uniformity of the strength properties of the flat bar. Alloys containing Cu, both in the F and T6 states, are characterized by similar strength properties. The presence of a weld does not significantly affect their properties.

III. The highest strength properties (TS, YS, HB) characterize samples of alloy 4 in the T6 temper. The highest elongation (A) was found in alloy 1 (low level of alloying elements) in each tested variant.

IV. In the T6 temper, in alloys with a similar Cu content (alloys 3 and 4), an increase in the Zn content by 2% resulted in increases in the strength, yield strength, and elongation.

For alloys with a similar Zn content (alloys 2 and 3), the increase in properties (TS, YS, A) is determined by the Cu content.

V. A clear decrease in elongation in the samples examined across the weld was noted for each material tested.

VI. It was found in the T6 temper that the microstructure of extrusions without the weld in the four alloys analyzed showed no significant differences.

VII. Analysis of the microstructure of welded extrusions in the T6 temper confirmed the significant differences in the grain size distribution. The largest grains were found in alloy 1, with the finest grains occurring in a range of up to 25 μm in alloys 2 and 3. Alloy 4 had a wider grain size distribution compared to alloys 2 and 3, ranging up to 45 μm.

VIII. A high solute content of the main alloying elements in Al–Zn–Mg–Cu alloys, as well as the presence of Zr, can lead to a high resistance to recrystallization. A large amount of the alloying elements resulted in the formation of fine and densely distributed dispersoids that had a significant influence on the recrystallization behavior in alloy 4. Some of the grains remained fibrous while others became recrystallized. This phenomenon is related to the orientation of the grains, where the recrystallized grains are <111> oriented, and the fibrous grains are <100> oriented.

Author Contributions: Conceptualization, K.R., B.P. and D.L.; methodology, K.R., S.B. and B.P.; validation, K.R.; formal analysis, K.R. and S.B.; investigation, K.R., S.B., K.L., K.Ż. and M.W.; resources, B.P.; writing—original draft preparation, K.R., S.B., K.L. and M.W.; writing—review and editing, K.R., S.B. and B.P.; visualization K.L., K.Ż. and M.W.; supervision, K.R.; project administration, K.R. and B.P.; funding acquisition, B.P. All authors have read and agreed to the published version of the manuscript.

Funding: This study covers the research funded by the "Implementation Doctorate" program no. DWD/4/28/2020. The research material was created as a part of the project, THE NATIONAL CENTRE FOR RESEARCH AND DEVELOPMENT, grant number TECHMATSTRATEG2/406439/10/NCBR/2019 "Extrusion welding of high-strength shapes from aluminium alloys 7xxx series".

Institutional Review Board Statement: Not applicable.

Informed Consent Statement: Not applicable.

Data Availability Statement: Not applicable.

Conflicts of Interest: The authors declare no conflict of interest. The funders had no role in the design of the study; in the collection, analyses, or interpretation of data; in the writing of the manuscript; or in the decision to publish the results.

References

1. Staley, J.T.; Lege, D.J. Advances in aluminium alloy products for structural applications in transportation. *J. Phys. IV* **1993**, *3*, C7-179–C7-190. [CrossRef]
2. Becerra, G.; Ramos-Grez, J.; Montecinos, J. Texture Distribution and Plane Strain Mechanical Behavior of AA 7xxx Plates of Different Thicknesses. *J. Mater. Eng. Perform.* **2009**, *18*, 1144–1150. [CrossRef]
3. Reyes, A.; Hopperstad, O.S.; Lademo, O.G.; Langseth, M. Modeling of textured aluminum alloys used in a bumper system: Material tests and characterization. *Comput. Mater. Sci.* **2006**, *37*, 246–268. [CrossRef]
4. Dixon, B. Extrusion of 2xxx & 7xxx alloys. In Proceedings of the International Aluminum Extrusion Technology Seminar, Chicago, IL, USA, 16–19 May 2000; Volume 1, pp. 281–294.
5. Paulisch, M.C.; Lentz, M.; Wemme, H.; Andrich, A.; Driehorst, I.; Reimers, W. The different dependencies of the mechanical properties and microstructures on hot extrusion and artifcial aging processing in case of the alloys al 7108 and al 7175. *J. Mater. Process. Technol.* **2016**, *233*, 68–78. [CrossRef]
6. Flitta, I.; Sheppard, T. Nature of friction in extrusion process and its effect on material flow. *Mater. Sci. Technol.* **2003**, *19*, 837–846. [CrossRef]
7. Li, S.; Li, L.; Liu, Z.; Wang, G. Microstructure and Its Influence on the Welding Quality of 6063 Aluminum Alloy Port-hole Die Extrusion. *Materials* **2021**, *14*, 6584. [CrossRef]
8. Yu, J.; Zhao, G. Study on the welding quality in the porthole die extrusion process of aluminum alloy pro_les. *Procedia Eng.* **2017**, *207*, 401–406. [CrossRef]

9. Chen, L.; Zhang, J.; Zhao, G.; Wang, Z.; Zhang, C. Microstructure and mechanical properties of Mg-Al-Zn alloy ex-truded by porthole die with different initial billets. *Mater. Sci. Eng. A* **2018**, *718*, 390–397. [CrossRef]
10. Chen, G.; Chen, L.; Zhao, G.; Zhang, C.; Cui, W. Microstructure analysis of an Al-Zn-Mg alloy during porthole die extrusion based on modeling of constitutive equation and dynamic recrystallization. *J. Alloys Compd.* **2017**, *710*, 80–91. [CrossRef]
11. Leśniak, D.; Libura, W.; Leszczyńska-Madej, B.; Bogusz, M.; Madura, J.; Płonka, B.; Boczkal, S.; Jurczak, H. FEM Numerical and Experimental Work on Extrusion Welding of 7021 Aluminum Alloy. *Materials* **2023**, *16*, 5817. [CrossRef]
12. Woźnicki, A.; Leszczyńska-Madej, B.; Włoch, G.; Grzyb, J.; Madura, J.; Leśniak, D. Homogenization of 7075 and 7049 Aluminium Alloys Intended for Extrusion Welding. *Metals* **2021**, *11*, 338. [CrossRef]
13. Ghosh, A.; Ghosh, M. Microstructure and texture development of 7075 alloy during homogenisation. *Philos. Mag.* **2018**, *98*, 1470–1490. [CrossRef]
14. Starke, E.A.; Staley, J.T. Application of modern aluminum alloys to aircraft. *Prog. Aerosp. Sci.* **1996**, *32*, 131–172. [CrossRef]
15. Clinch, M.R.; Harris, S.J.; Hepples, W.; Holroyd, N.J.H.; Lawday, M.J.; Noble, B. Other applications: Optimizing the mechanical properties of 7xxx series alloys by controlling composition and precipitation kinetics. In Proceedings of the 11th International Conference on Aluminium Alloys, Aachen, Germany, 22–26 September 2008; Hirsch, J., Skrotzki, B., Gottstein, G., Eds.; pp. 148–160.
16. Sha, G.; Cerezo, A. Early-Stage Precipitation in Al-Zn-Mg-Cu Alloy (7050). *Acta Mater.* **2004**, *52*, 4503–4516. [CrossRef]
17. Chen, J.; Zhen, L.; Yang, S.; Shao, W.; Dai, S. Investigation of precipitation behavior and related hardening in AA 7055 aluminum alloy. *Mater. Sci. Eng. A* **2009**, *500*, 34–42. [CrossRef]
18. Ludtka, G.M.; Laughlin, D.E. The Influence of Microstructure and Strength on the Fracture Mode and Toughness of 7XXX Series Aluminum Alloys. *Metall. Trans. A* **1982**, *13A*, 411–425. [CrossRef]
19. Clinch, M.R.; Harris, S.J.; Hepples, W.; Holroyd, N.J.H.; Lawday, M.J.; Noble, B. Influence of Zinc to Magnesium Ratio and Total Solute Content on the Strength and Toughness of 7xxx Series Alloys. *Mater. Sci. Forum* **2006**, *519–521*, 339–344. [CrossRef]
20. Shu, W.X.; Hou, L.; Zhang, C.; Zhang, F.; Liu, J.; Liu, J.T.; Zhuang, L.; Zhang, J.S. Tailored Mg and Cu contents affecting the microstructures and mechanical properties of high-strength Al-Zn-Mg-Cu alloys. *Mater. Sci. Eng. A* **2016**, *657*, 269–283. [CrossRef]
21. Wagner, J.A.; Shenoy, R.N. The effect of copper, chromium, and zirconium on the microstructure and mechanical properties of Al-Zn-Mg-Cu alloys. *Metall. Trans. A* **1991**, *22*, 2809–2818. [CrossRef]
22. Chen, K.; Liu, H.; Zhang, Z.; Li, S.; Todd, R.I. The improvement of constituent dissolution and mechanical properties of 7055 aluminum alloy by stepped heat treatments. *J. Mater. Process. Technol.* **2003**, *142*, 190–196. [CrossRef]
23. Wang, T.; Yin, Z.; Sun, Q. Effect of homogenization treatment on microstructure and hot workability of high strength 7b04 aluminium alloy. *Trans. Nonferrous Met. Soc. China* **2007**, *17*, 335–339. [CrossRef]
24. *PN-EN ISO 6892-1:2020-05*; Metals—Tensile Test—Part 1: Room Temperature Test Method. Polish Committee for Standardization: Warsaw, Polish, 2020.
25. *PN-EN-ISO-6506-1_2014*; Metallic Materials—Brinell Hardness Test—Part 1: Test Method. ISO: Geneva, Switzerland, 2014.
26. Ding, S.; Taylor, T.; Khan, S.A.; Sato, Y.; Yanagimoto, J. Further understanding of metadynamic recrystallization through thermomechanical tests and EBSD characterization. *J. Mater. Process. Technol.* **2022**, *299*, 117359. [CrossRef]
27. Dalai, B.; Moretti, M.A.; Åkerstrom, P.; Arvieu, C.; Jacquin, D.; Lindgren, L.E. Mechanical behavior and microstructure evolution during deformation of AA7075-T651. *Mater. Sci. Eng. A* **2021**, *822*, 141615. [CrossRef]
28. Sun, Y.; Bai, X.; Klenosky, D.; Trumble, K.; Johnson, D. A Study on Peripheral Grain Structure Evolution of an AA7050 Aluminum Alloy with a Laboratory-Scale Extrusion Setup. *J. Mater. Eng. Perform.* **2019**, *28*, 5156–5164. [CrossRef]
29. Humphreys, F.J.; Hatherly, M. Chapter 9: Recrystallization of Two-Phase Alloys. In *Recrystallization and Related Annealing Phenomena*; Elsevier: Amsterdam, The Netherlands, 2012; pp. 311–314.
30. Sun, Y.; Johnson, D.R.; Trumble, K.P. Effect of Zr on Recrystallization in a Directionally Solidified AA7050. *Mater. Sci. Eng. A* **2017**, *700*, 358–365. [CrossRef]
31. Humphreys, F.J.; Hatherly, M. Chapter 7: Recrystallization of Single-Phase Alloys. In *Recrystallization and Related Annealing Phenomena*; Elsevier: Amsterdam, The Netherlands, 2012; pp. 228–229.

Disclaimer/Publisher's Note: The statements, opinions and data contained in all publications are solely those of the individual author(s) and contributor(s) and not of MDPI and/or the editor(s). MDPI and/or the editor(s) disclaim responsibility for any injury to people or property resulting from any ideas, methods, instructions or products referred to in the content.

Article

Experimental Investigation on Microstructure Alteration and Surface Morphology While Grinding 20Cr2Ni4A Gears with Different Grinding Allowance Allocation

Rong Wang [1], Size Peng [1], Bowen Zhou [1], Xiaoyang Jiang [2], Maojun Li [2,*] and Pan Gong [3,*]

1. Hunan Xingtu Aerospace and Spacecraft Manufacturing Co., Ltd., Zhuzhou 412000, China
2. State Key Laboratory of Advanced Design and Manufacture for Vehicle Body, Hunan University, Changsha 410082, China
3. State Key Laboratory of Materials Processing and Die & Mold Technology, School of Materials Science and Engineering, Huazhong University of Science and Technology, Wuhan 430074, China
* Correspondence: maojunli@hnu.edu.cn (M.L.); pangong@hust.edu.cn (P.G.)

Abstract: Transmission gear is a key component of vehicles and its surface integrity affects the safety of the transmission system as well as the entire mechanical system. The design and optimization of allowances in form grinding are important for improving dimensional accuracy and machining efficiency during the manufacturing of heavy-duty gears. This work aims to investigate the effects of grinding allowance allocation on surface morphology, grinding temperature, microstructure, surface roughness, and microhardness fluctuation during the form grinding of 20Cr2Ni4A gears. Results indicated that grinding temperature was primarily influenced by rough grinding involving significant grinding depths exceeding 0.02 mm. The ground surface exhibited slight work hardening, while thermal softening led to a reduction in microhardness of around 40 HV. Ground surface roughness Ra varied from 0.930 μm to 1.636 μm, with an allowance allocation of the last two passes exerting the most significant influence. Analysis of surface and subsurface microstructures indicated that a removal thickness of 0.02 mm during fine grinding was insufficient to eliminate the roughness obtained from rough grinding. Evident ridges, gullies, and surface defects such as material extraction, adhesion, and plastic deformation were also observed. The proposed grinding strategy was validated in practical manufacturing with good surface quality and geometrical accuracy.

Keywords: allowance allocation; microstructure; surface morphology; form grinding

1. Introduction

Heavy-duty gear components are extensively utilized in vehicles, and the demand for efficient and precise machining of these ones is continuously growing alongside industrial advancements. Form grinding plays a pivotal role in the manufacturing and processing of gears. Prior to the grinding process, samples are processed with forging, rough machining, and heat treatments. Consequently, grinding is the final process to ensure the ultimate accuracy and surface quality of gear products. To improve manufacturing efficiency, the total grinding allowance is designed to be minimized due to the lower material removal rate and the higher specific energy associated with grinding compared to other machining methods such as turning and milling [1]. Both the grinding forces and residual stress from previous heat treatments will affect machining quality and geometrical accuracy. Previous studies have shown that allowance allocation in grinding is highly related with stress generation, highlighting the importance of considering allowance allocation in different grinding stages to enhance part accuracy [2].

Research works investigated approaches to improve machining accuracy by optimizing machining allowances. For instance, Batueva et al. [3] proposed a machining model stabilizing the cutting force component by adjusting the machining allowance of complex areas with varying shapes on the machined surface, thereby enhancing machining accuracy. Ruzzi et al. [4] analyzed the effects of grinding parameters on surface integrity during the surface grinding of Inconel alloys, where the design of the experiment method was applied using grinding speed, work speed, grinding depth, and up/down grinding as variations. Results showed that grinding speed was the most significant factor affecting work hardening. Luu et al. [5] devised a novel method for designing a conical skiving cutter that allowed pre-defined grinding allowance for skived gears by modifying the normal rack, resulting in a uniform grinding allowance on active involute sections. Lv et al. [6] presented a novel path route planning method that generated the grinding route based on residual height error compensation, a proposed geometric algorithm, and the machining allowance threshold. Optimizing the machining allowances can enhance efficiency and reduce energy consumption, thereby supporting sustainable development in the manufacturing industry [7]. Guo et al. [1] introduced a systematic energy-efficient method based on minimizing energy consumption while ensuring acceptable surface roughness to quantify grinding stock allowances in the turning–grinding process, which was validated by reducing energy consumption up to 16.6%. Hood et al. [8] used electroplated diamond superabrasive wheels for grinding an intermetallic alloy in order to improve both the grinding efficiency and surface quality. Guerrini et al. [9] conducted a gear machining experiment by varying the skiving–grinding allowance allocation ratios and optimizing the process parameters, demonstrating the feasibility of dry grinding for gear machining while ensuring accuracy and avoiding grinding burn. Li et al. [10] aimed to achieve high-efficiency and high-quality glass–ceramic grinding by optimizing the appropriate grinding parameters based on the study of parameter influence on processing behavior during rough grinding, semi-fine grinding, and fine grinding processes.

However, the research on machining allowances have not comprehensively addressed the dual goals of improving machining efficiency and accuracy. For heavy-duty gears, the integrity of ground surface directly affects contact, wear, and the generation of cracks, ultimately influencing the fatigue life of the whole component [11]. Guerrini et al. [9,12] proposed a defined threshold energy level to prevent grinding burns on gear surfaces via a dry grinding experiment, analyzing the effects of different processing parameters on the white layer and hardness gradient in the subsurface. Riebel et al. [13] studied the influence of wheel groove depth/width on the grinding performance during creep grinding and found that coolant-induced force was significant for improving the grinding performance. Murtagian et al. [14] proposed a grinding model to analyze the effects of abrasive particle size and shape, grinding feed, and the depth of the cut on subsurface plastic deformation depth. Yang et al. [15] investigated the effect of grinding parameters on surface roughness and elucidated the formation mechanism of plastic deformation and defects on ground surface. Tao et al. [16] proposed a transient analog model to simulate the transient material removal behavior in continuous forming grinding, predicting the grinding morphology of the tooth surface. They found that the top tooth surface was relatively smoother than the root one, and the surface roughness decreased along the feed direction.

Based on the aforementioned literature review, prior research focusing on ground surface integrity have not adequately examined the influence of grinding allowance allocation on surface quality. To fill this gap, the effects of grinding allowance allocation on grinding temperature, surface roughness, and microhardness fluctuation were analyzed. The research work aimed to study the influence characteristics/mechanisms of different grinding strategies on surface morphology and microstructure and to further understand the corresponding material removal mechanisms during the form grinding process. Several principles for grinding allowance allocation in the process are summarized, providing guidance for optimizing the manufacturing process of gear products and improving production efficiency.

2. Experimental Works

2.1. Workpiece Material and Equipment

A raw vehicle transmission gear made of 20Cr2Ni4A was selected for the experiment, and it was cut into 12 small pieces via the wire electrical discharge machining technique, which generated very limited heat and defects for the grinding samples and would not further affect the grinding performance. The chemical composition of 20Cr2Ni4A is presented in Table 1. The material exhibits good toughness due to its low carbon content. The gear has undergone carburizing and quenching processes, resulting in the formation of a carburized layer on the surface. This carburized layer enhances the material's abrasion resistance and imparts high hardness. Microhardness measurements indicated that the maximum hardness of the surface layer for the gear sample used in this study was ~650 HV. The effective thickness of the carburized layer, characterized by the microhardness higher than 550 HV, was 1100 μm.

Table 1. Chemical composition of 20Cr2Ni4A steel (wt.%).

C	Si	Mn	Cr	Ni	Fe	S
0.17–0.23	0.17–0.37	0.30–0.60	1.25–1.65	3.25–3.65	≥95.00	≤0.03

The gear grinding trials were performed using a three-axis automatic hydraulic high-precision grinding machine (DY-510ASM, China). The machine had a maximum rotation speed of 4000 r/min and a rated power of 7.5 kW. The gear tooth profile was obtained using an alumina forming grinding wheel with #80 abrasive mesh. The maximum diameter of the grinding wheel was 350 mm, and its profile coincided with the gear involute. Before each trial, the grinding wheel was dressed to ensure consistent working conditions throughout the grinding process. During the grinding process, temperature signals were captured using a GG-K-30-1000-CZ type thermocouple and a multiple-channel temperature recorder (NAPUI-HE130T-16, China), which has the acquisition frequency of 0–10,000 Hz and the resolution of 0.01 °C. As the grinding time in each pass was very short, the acquisition frequency was set at 8000 Hz in order to obtain real-time temperature variations. To position the thermocouple, a narrow groove with a width of 1.0 mm was cut in the middle of the tooth surface. The thermocouple with a diameter of 1.0 mm was placed in the groove and sealed/fixed using heat-resistant waterproof adhesive, as shown in Figure 1. The sensor end of the thermocouple was positioned ~0.5 mm away from the tooth surface. Temperature measurements were recorded at a sampling frequency of 1 Hz. A white light interference was employed to observe ground surface morphology. A roughness meter was used to measure surface roughness (Ra), with each sample being measured three times, followed by taking the average value. The teeth of each grinding set were cut into samples and embedded in resin. The cross-section of the ground surface was ground and polished. Subsequently, it was corroded using a 4% nitric acid–alcohol solution for 15 s. To analyze the material microstructure, a scanning electron microscope (COXEM EM-30N, China) was utilized. The microscope was operated with magnifications of ~2000–5000 times. Additionally, a hardness tester was used to measure the microhardness, with 1000 g load for 10 s.

2.2. Grinding Parameters

In the gear manufacturing process, it is crucial to eliminate any deformations caused by heat treatment and ensure the dimensional accuracy of the final product. In this study, the total material removed from the raw sample to the finished one was set as 0.400 mm. The grinding parameters were selected according to on-site machining parameters. The grinding parameters commonly used on the grinding machine range from 0.19 m/s to 0.35 m/s for the feed speed, 0.01 mm to 0.04 mm for the grinding depth, and the spindle speed is 900 r/min to 4000 r/min. Three different combinations of allowances were selected to design the grinding allowance allocation strategy. These included rough grinding–semi-fine grinding–fine grinding (referred to as P1, P3, P5, and P8), rough grinding–semi-fine grinding 1—semi-fine grinding 2—fine grinding (referred to as P2, P6, P7, and P9), and rough grinding–fine grinding (referred to as P4), as shown in Figure 2. Nine grinding allowance allocation plans were designed based on these combinations, with details listed in Table 2. To investigate the influence of improving machining efficiency in the rough grinding and semi-fine grinding stages on grinding temperature and surface integrity, the grinding depth for the rough grinding stage was set at 0.025 mm, 0.03 mm, and 0.035 mm, respectively. The grinding depth for each stage decreased relating to the previous stage.

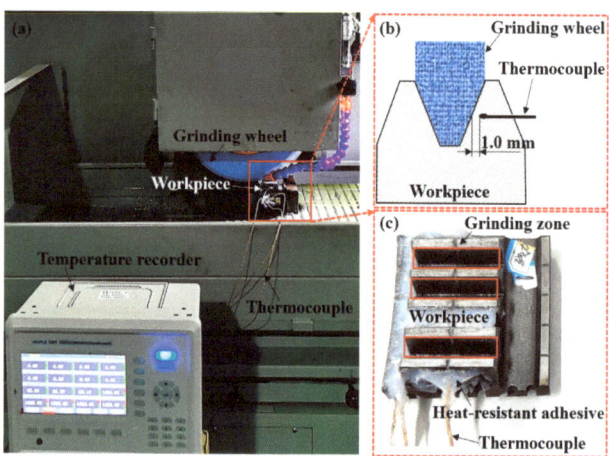

Figure 1. (**a**) Grinding equipment, (**b**) schematic diagram of temperature measurement using thermocouple and (**c**) workpiece sample for grinding trials.

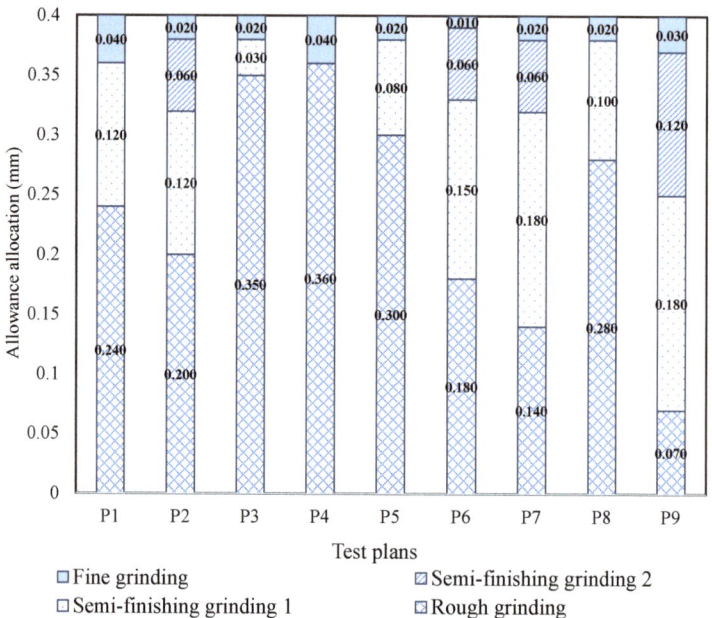

Figure 2. Allowance allocation in different grinding stages.

Table 2. Grinding depth and number of passes of different test plans.

Test Plans	Rough Grinding		Semi-Finishing Grinding 1		Semi-Finishing Grinding 2		Fine Grinding	
	a_p (mm)	Passes	a_p (mm)	Passes	a_p (mm)	Passes	a_p (mm)	Passes
P1	0.020	12	0.015	8	-	-	0.010	4
P2	0.025	8	0.020	6	0.015	4	0.010	2
P3	0.025	14	0.015	2	-	-	0.010	2
P4	0.030	12	-	-	-	-	0.010	4
P5	0.030	10	0.020	4	-	-	0.010	2
P6	0.030	6	0.025	6	0.015	4	0.005	2
P7	0.035	4	0.030	4	0.015	4	0.010	2
P8	0.035	8	0.025	4	-	-	0.010	2
P9	0.035	2	0.030	6	0.020	6	0.015	2

3. Results and Discussion

3.1. Grinding Temperature

According to the recorded grinding temperature and the details of allowance allocation for different grinding trials, it was found that the grinding depth and the number of consecutive grinding times in different grinding stages showed a significant influence on grinding temperature. Due to the negative rake angle of the abrasive particles, a large quantity of energy was generated in the contact area between the grinding wheel and the workpiece during grinding with a large cutting depth, and most of the energy was converted to a high temperature in the grinding zone [17]. Additionally, continuous grinding on the tooth grooves multiple times blocked the fluid entering the contact area and promoted grinding temperature [18]. According to the grinding temperature model established by Yang et al. [19], when a transient point heat source Q_d occurred at $O(0, 0, 0)$ in an infinite homogeneous space, as shown in Figure 3, the temperature change at any point $M(x, y, z)$ in the space over time τ could be calculated as:

$$T = \frac{Q_d}{c\rho(4\pi\alpha\tau)^{3/2}} e^{-\frac{x^2+y^2+z^2}{4\alpha\tau}} \quad (1)$$

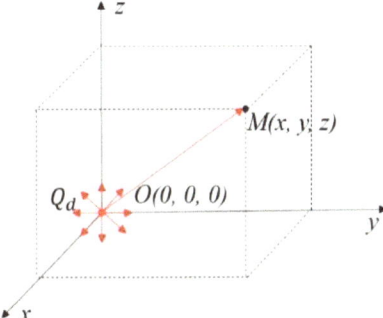

Figure 3. Schematic diagram of instantaneous point heat source temperature field.

In which T was the temperature change, c was the specific heat capacity of workpiece material, ρ was the density of workpiece material, and α was the thermal conductivity.

As shown in Figure 4, when the heat source was replaced by an instantaneous infinitely long linear heat source along the z-axis, and the uniform heat output was Q_s, the temperature change to any point $M(x,y,z)$ in space over time τ was the integral of the point heat source along an infinitely long straight line, as shown below:

$$T = \frac{Q_s}{c\rho(4\pi a\tau)^{3/2}} e^{-\frac{x^2+y^2}{4a\tau}} \int_{-\infty}^{+\infty} e^{-\frac{(z-z_i)^2}{4a\tau}} dz_i \quad (2)$$

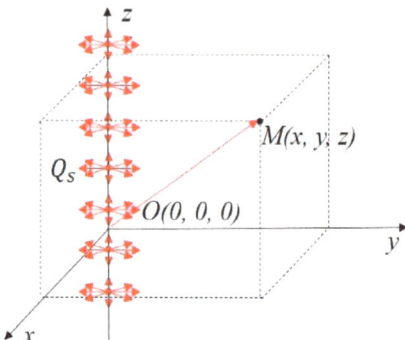

Figure 4. Schematic diagram of temperature field of instantaneous infinite linear heat source.

Equation (2) can be simplified as

$$T = \frac{Q_s}{c\rho(4\pi a\tau)} e^{-\frac{x^2+y^2}{4a\tau}} \quad (3)$$

From Equation (3), it can be seen that the temperature rise at any point was only related to the distance from that point to the linear heat source. In order to facilitate calculation, the gear involute in this experiment was simplified to a straight line. Since the width of the tooth surface was much smaller than the diameter of the grinding wheel, the linear speed of each part of the grinding wheel was considered to be the same. Thus, as shown in Figure 5, the grinding area was simplified as a constant heating linear segment where the heat source moves uniformly on the tooth surface with a heating power of Q_s.

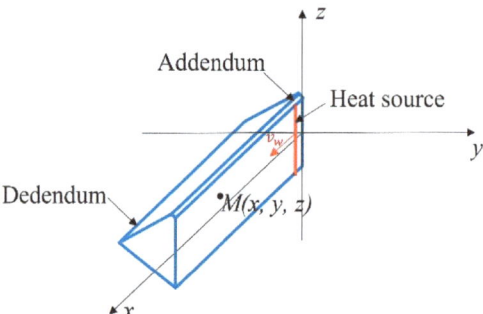

Figure 5. Schematic diagram of simplified tooth surface moving line heat source.

According to Equation (3), in a flash of $d\tau_i$ the temperature change dT caused by the heat source differential section $q_s d\tau_i$ at point M was

$$dT = \frac{q_s d\tau_i}{c\rho(4\pi a\tau)} e^{-\frac{(x-v_w \tau_i)^2+y^2}{4a\tau}} \quad (4)$$

The heat source started moving from the moment τ_i and continued moving for a period τ to reach another moment t. In the model of Figure 6, the distance from the heat source to the point $M(x, y, z)$ on the x-axis was defined as $X = x - v_w t$, where v_w was the feed speed. Equation (4) could be transformed to Equation (5):

$$dT = \frac{q_s d\tau_i}{c\rho(4\pi\alpha\tau)} e^{-\frac{(X+v_w\tau)^2+y^2}{4\alpha\tau}} \tag{5}$$

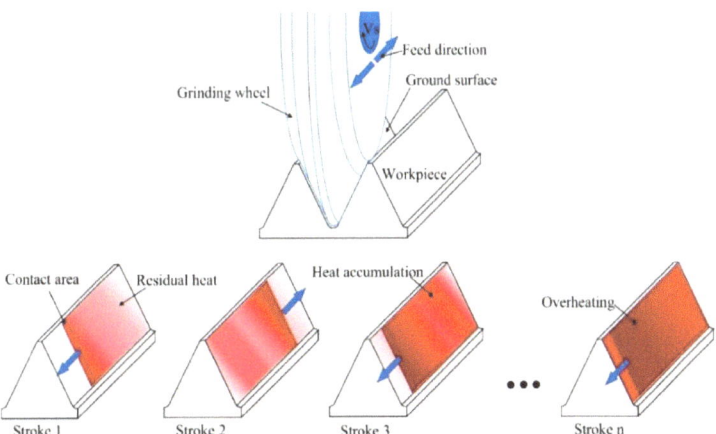

Figure 6. Schematic diagram of heat accumulation mechanism in continuous grinding.

According to Equation (5), during the movement of the heat source from the moment to move to the moment t, the temperature rise of $M(x, y, z)$ caused by the moving constant linear heat source could be calculated as:

$$T = \frac{q_s}{4\pi\lambda} \int_0^t \frac{1}{\tau} e^{-\frac{(X+v_w\tau)^2+y^2}{4\alpha\tau}} d\tau_i \tag{6}$$

Equation (6) indicates that the temperature at the tooth surface and any internal point is continuously influenced by the heat source throughout the grinding process, from the beginning of a grinding stroke to the end of it. Therefore, even after the grinding area has passed through a measurement point, the heat source still affects the temperature at that point. Furthermore, due to the inability to dissipate heat instantaneously, the grinding heat conducted into the workpiece persists for a certain period of time after grinding. The contact area between the grinding wheel and the workpiece is considered a continuous heat source, resulting from the heat generated by friction and cutting. As depicted in Figure 6, during a grinding stroke, the heat source moves from one end of the tooth to the other, with most of the heat being transferred to the workpiece, leading to an increase in grinding temperature. When a grinding stroke is completed and immediately followed by the next one, a new heat source is generated and moves uniformly across the tooth surface. In this work, the workbench's movement track length is 200 mm, and the feed speed is 0.19 m/s, resulting in an interval of approximately one second between two consecutive grinding strokes. As a result, at the start of a new grinding stroke, the temperature at the previously ground surface has not yet decreased to the ambient temperature, and the residual heat in the workpiece has not dissipated completely. Consequently, in the new stroke, more heat is generated and transferred into the workpiece, accumulating with the residual heat. With continuous grinding, the accumulated heat in the workpiece leads to high grinding temperatures. This phenomenon is a result of the heat accumulation and insufficient dissipation between consecutive grinding strokes, contributing to the generation of elevated grinding temperatures.

Figure 7 presents the maximum temperatures recorded during all grinding trials. As the temperature changes during grinding passes were relatively low and the grinding time in each pass was very short, the temperature variation/duration over time was not presented in this work. The maximum temperatures ranged from 104 °C to 229 °C, indicating that the different grinding allowance allocation strategy had a significant impact on it. Specifically, trial P3 exhibited a considerably higher temperature compared to the other ones. The allowance in P3 was divided into three stages, with a grinding depth of 0.025 mm in the rough grinding stage (14 passes), 0.015 mm in the semi-fine grinding stage (2 passes), and 0.01 mm in the fine grinding stage (2 passes). As the grinding depth increased, more abrasive particles were involved in the cutting process, generating and transferring more heat into the workpiece [20–22]. Additionally, the limited thermal conductivity of the workpiece material and the limited time interval between each grinding stroke led to heat accumulation. The unique characteristic of the allowance allocation strategy in P3 was the removal of a substantial amount of material via continuous grinding with a large grinding depth in the rough grinding stage, accounting for approximately 90% of the total allowance, which resulted in a sharp increase in temperature due to the significant accumulation of heat. For trial P7, the allowance was divided into four stages including 0.035 mm in the rough grinding stage (4 passes), 0.03 mm in semi-fine grinding stage 1 (6 passes), 0.015 mm in semi-fine grinding stage 2 (4 passes), and 0.01 mm in the fine grinding stage (2 passes). Similar to the strategy in trial P3, a significant amount of material was removed via continuous grinding with a large grinding depth in the rough grinding and semi-fine grinding stage 1, with a total of 10 grinding strokes having a grinding depth of not less than 0.03 mm, accounting for 80% of the total allowance, which subsequently produced a relatively high grinding temperature.

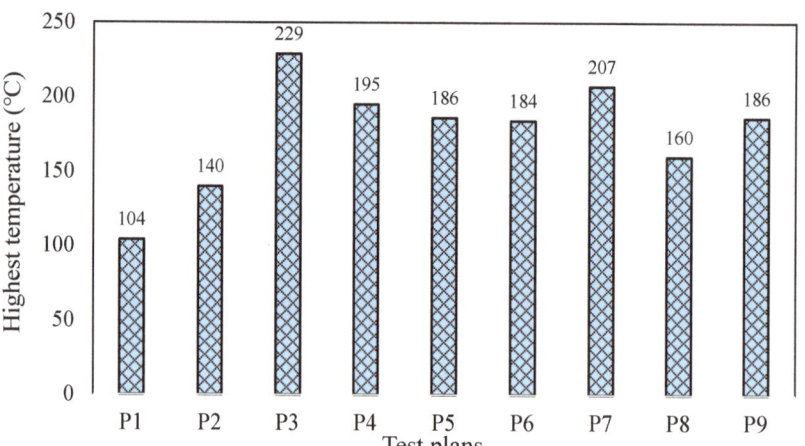

Figure 7. Maximum temperature recorded in all trials with different test plan/allowance.

Trial P1 exhibited the lowest temperature among all trials, as the allowance allocation ratio for each stage was rough grinding: semi-fine grinding 1: fine grinding = 6:3:1, and the grinding depth for each stage did not exceed 0.02 mm. The grinding depth in the rough grinding stage was only 0.02 mm, with 12 grinding strokes. Compared to trial P3, the generation and accumulation of grinding heat were significantly reduced, leading to a lower grinding temperature. Trial P2 showed a slightly higher temperature when the allowance allocation ratio for each stage was rough grinding: semi-fine grinding 1: semi-fine grinding 2: fine grinding = 10:6:3:1. Except for rough grinding, the grinding depth for each stage did not exceed 0.02 mm. In the rough grinding stage, the grinding depth was 0.025 mm, with 8 grinding strokes resulting in a maximum temperature of 140 °C.

3.2. Ground Surface Topography

The morphology of the ground gear surface plays a crucial role in determining the surface quality, contact fatigue strength, crack resistance, and transmission stability of the final product. The precision grinding parameters used in the final grinding pass have a significant impact on surface integrity and gear performance. Figure 8 presents the three-dimensional morphology of the ground surfaces obtained from four trials with different grinding allowance allocation. Although no macroscopic cracks were observed in the profiles, it is difficult to determine whether the microcracks were presented due to limited magnification. Similarly, it is challenging to assess grinding burns based solely on the color or burning traces on the surface. Serious grinding burns are typically caused by high temperatures, resulting in residual tensile stresses within the surface layer, which can lead to the formation of microcracks [13]. The ground surface profiles mainly consist of flat surfaces, gullies, and ridges. The formation of gullies and ridges can be attributed to material plastic deformation induced by grinding heat, where the material is pushed to both sides by abrasive particles. Another possible reason for the formation of gullies and ridges is the presence of surface fluctuations caused by a large grinding depth in the preceding passes, which cannot be fully eliminated in the final finish grinding stage. For instance, in the case of trial P5, a substantial amount of material (0.3 mm) was removed in the rough grinding stage over 10 passes. In the subsequent semi-fine grinding stage, 0.08 mm of material was removed over four passes, and only two passes were used for surface repair in the fine grinding stage, with a total grinding depth of 0.01 mm. It is evident that the surface fluctuations cannot be completely mitigated, resulting in visible gullies and ridges on the ground surface of the samples from trial P5.

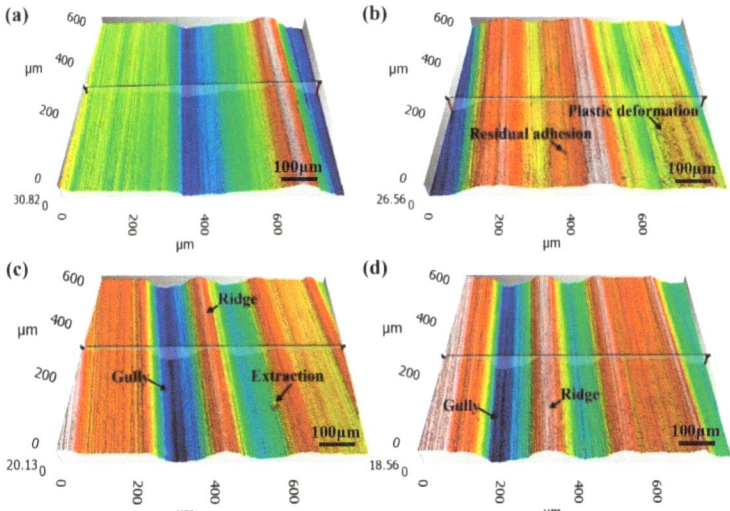

Figure 8. Typical ground surface topography recorded from samples in trials (**a**) P2, (**b**) P3, (**c**) P5, and (**d**) P7.

Various defects can be observed on the ground surface, including surface residual adhesion and plastic deformation, as shown in Figure 8b, as well as the material extraction depicted in Figure 8c. The extraction of granular material and the presence of residue adhesion are likely the result of material adhesion to the grinding wheel due to the influence of grinding heat. These defects and surface plastic deformation are closely related to high temperatures. Overall, the surface morphology and defects observed on the ground surface provide valuable insights into the quality of the grinding process and the potential influence of parameters on the resulting surface characteristics.

The surface roughness (Ra) values of all trials with different grinding allowance allocations are presented in Figure 9, with the lowest one recorded in trial P1. It was observed that from rough grinding to fine grinding, the allowance allocated for each pass and the grinding depth of each stage gradually decreased in a gradient manner in trial P1. This allocation feature for each grinding pass contributed to improving the surface quality achieved in the preceding pass. Comparing the allocation characteristics in trials P5 and P7, it can be observed that the allocation in the last two passes largely determined the final surface roughness. It is evident that the surface roughness obtained by P7 (0.015 mm × 4 grinding strokes + 0.010 mm × 2 grinding strokes) is smaller than that of P5 (0.020 mm × 4 grinding strokes + 0.010 mm × 2 grinding strokes). Furthermore, comparing the Ra value of trials P6 and P7 reveals that at a grinding depth of 0.005 mm in the finish grinding stage, the rough surface resulting from the previous grinding pass cannot be completely polished in just two grinding strokes without increasing the number of ones. The low roughness value in trial P1 indicates that a smooth surface can be achieved with a grinding depth of 0.01 mm in the fine grinding stage. The roughness values recorded in trials P2 and P7 suggest that adopting a grinding depth of 0.015 mm in the final semi-finish grinding stage is feasible. This implies that the ground surface roughness can be improved without extending the machining time. On the premise that fine grinding is capable of repairing the surface quality obtained from previous grinding passes, processing time can be saved in the rough grinding stage. This approach is similar to the method reported by Li et al. [23], where fine grinding was employed to remove the burn layer caused by rough grinding and enhance the overall grinding efficiency.

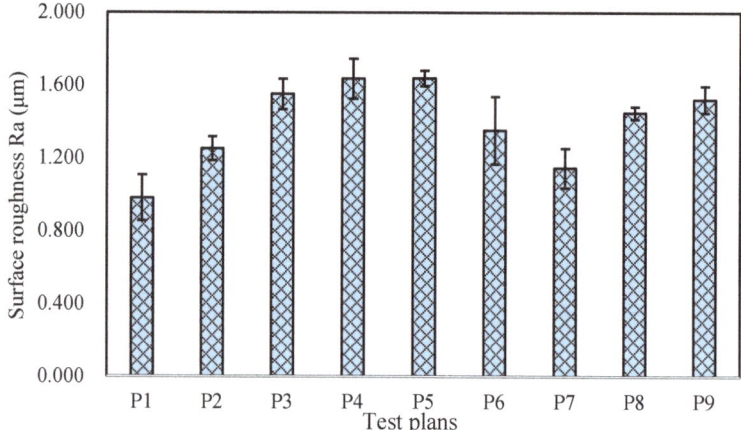

Figure 9. Surface roughness Ra for all trials with different grinding allowance allocation/test plans.

3.3. Microstructure Morphology

Typical microstructure morphology within the subsurface layer is shown in Figure 10. The profiles of ground subsurface in trials P1 and P2 exhibit relatively smooth contours with no obvious pits or bumps. This can be attributed to the polishing effect brought about by the last two grinding passes, which improved the surface roughness. In contrast, the surface profiles in trials P3 and P4 are generally rough, and they consist mainly of short smooth lines, bumps, and pits. In the case of the samples from trial P3, significant sharp cusps and pits are observed in the subsurface, primarily resulting from plastic deformation at high temperature. A darker-colored thermally softened layer is visible below the contour, consistent with the measurement of the high grinding temperature. This indicates that a large amount of grinding heat penetrated into the interior of the workpiece during the rough grinding stage, affecting the deeper base material. The contour of the sample from trial P7 is smoother compared to the one in trial P3, mainly due to the polishing effect in the second semi-finishing and finishing stages. However, a dark layer caused by thermal effects can still be observed, suggesting that a significant amount of grinding heat infiltrated the workpiece and affected the deep materials during the rough grinding stage. In the case of trial P9, the sample contour is distributed with numerous small bumps and pits. These are primarily caused by the large grinding depth in the rough grinding stage. In the fine grinding stage, the polishing effect on the rough ground surface is not sufficient with a grinding depth of 0.015 mm. Defects such as pits, cusps, dark layers, and plastic deformation are present, which can be attributed to the thermal effects and polishing effect determined via the allowance allocation. Overall, the subsurface profiles provide visual evidence of the effects of allowance allocation on surface quality and the subsurface characteristics of ground samples. Smooth contours and minimal defects are observed in trials with appropriate allowance allocation, while rough surfaces and various defects are apparent in those with inadequate allowance allocation or excessive grinding depths.

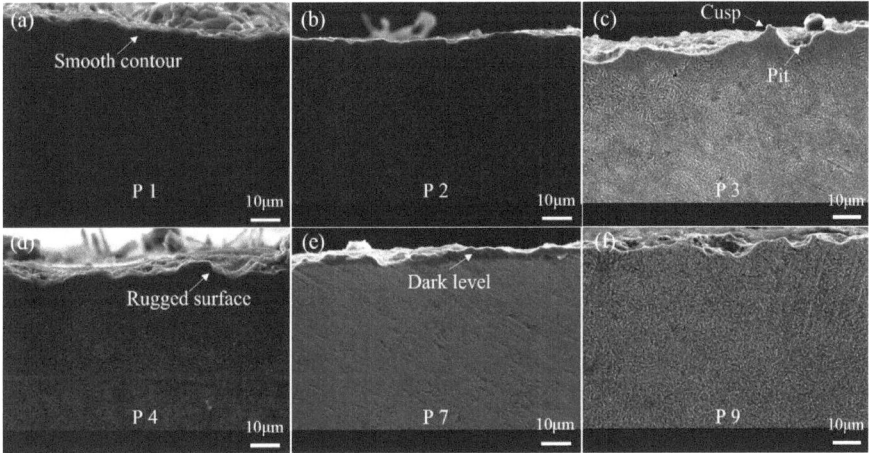

Figure 10. Microstructure morphology of subsurface from samples with different grinding allowance allocations in trials (**a**) P1, (**b**) P2, (**c**) P3, (**d**) P4, (**e**) P7, and (**f**) P9.

Figure 11 presents a detailed morphology of the ground subsurface from trials P3 and P9 with higher magnifications. In Figure 11a,b, sharp tooth-shaped plastic deformation is evident in the surface profile, contributing to surface irregularities. The grains beneath the ground surface are relatively fine, which may be the result of grain refinement caused by the extrusion of the grinding wheel. In Figure 11d, a layer with a thickness of several microns can be observed below the ground surface, and the grain morphology within this layer is unclear. This layer is typically known as the grinding white layer, which results from dislocation and grain refinement caused by grinding heat and mechanical extrusion. The white layer can improve surface hardness, but it can also increase brittleness, and a thermally softened layer tends to form beneath it. Although phase transformation can also contribute to the generation of this special layer, the measurement results of grinding temperature suggest that it is challenging for the general grinding process to reach temperatures necessary for phase transformation. Figure 11e illustrates the mechanism of plastic deformation and grain refinement. During abrasive cutting and plowing, materials in front of the abrasive's movement are compressed, forming chips that flow to both sides. It generates a significant amount of grinding heat, which further intensifies plastic deformation. With the occurrence of dislocations and slippage, the original crystal stacking structure of the workpiece material is disrupted, causing large grains to break into smaller ones and form new stacking structures. Grain refinement occurs within a limited depth below the surface due to the small grinding depth in the finish grinding stage. Grain refinement reduces the volume of grains, thereby improving surface microhardness and strength. However, compared to the mechanical effects, the depth of grinding heat conduction is relatively deeper, which may lead to the formation of a heat-softened layer with coarse grains beneath the hardened layer.

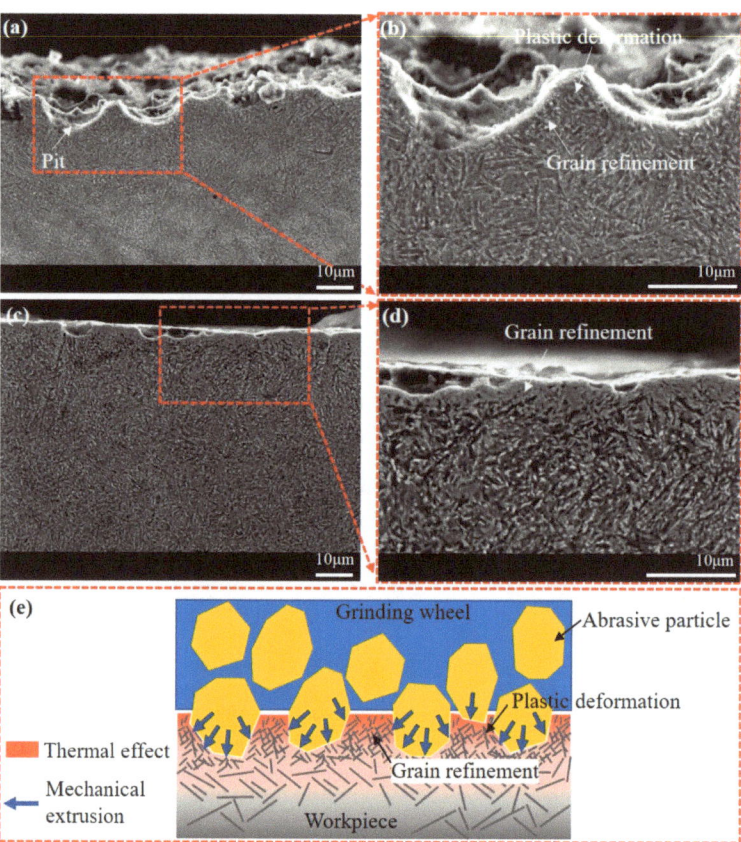

Figure 11. Subsurface morphology of samples from trials (**a**,**b**) P3 and (**c**,**d**) P9, and (**e**) schematic diagram showing mechanism of plastic deformation and grain refinement.

As shown in Figure 12, the dark layer observed on the ground surface from trial P7 indicates the presence of thermal effects. The high grinding temperature resulting from a grinding depth of 0.035 mm in the rough grinding stage may have caused grinding burn. Although the burn layer was partially removed during the second semi-finishing grinding and fine grinding stages, a shallow dark layer remained. Figure 12c,d illustrate the subsurface morphology of the samples from trial P4. The shallowest layer represents a grain refining layer generated via mechanical extrusion, followed by a thermally softened layer with coarse grains. Deeper into the subsurface, the grains become finer. This change suggests that the subsurface from trial P4 can be divided into three layers including a mechanically affected layer, a thermally affected layer, and an unaffected layer. Figure 12e depicts the mechanism relating to this effect, indicating that thermal effects occur in a deeper layer compared to the mechanical ones. This finding supports the understanding that the depth of grinding heat conduction is probably greater than the depth of mechanical extrusion and grain refinement.

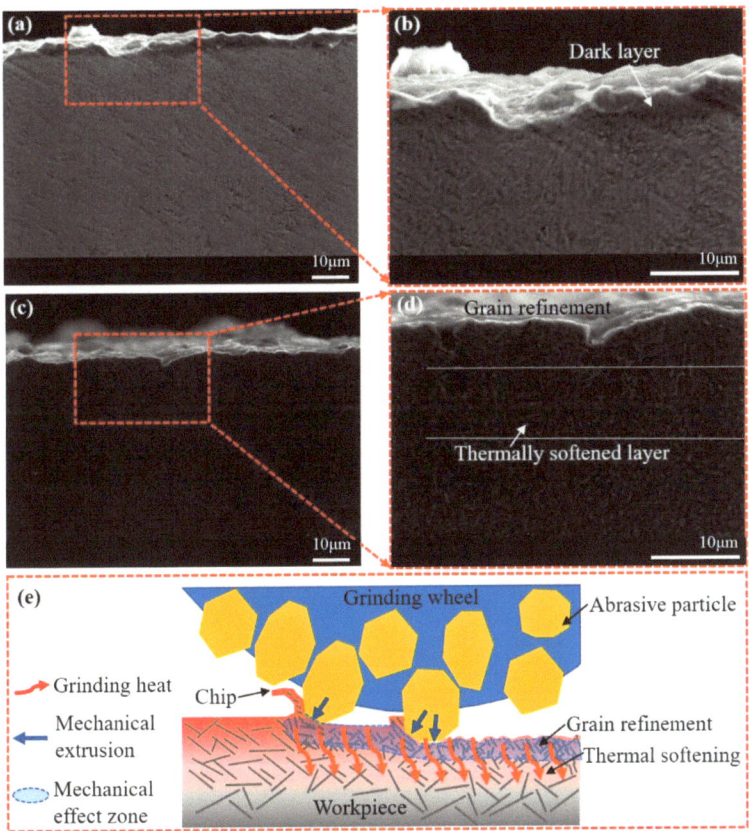

Figure 12. Subsurface morphology of samples from trials (**a**,**b**) P7 and (**c**,**d**) P4, and (**e**) schematic diagram showing generation mechanism of thermal effect.

Figure 13 depicts the presence of pores and pits observed on the subsurface and the corresponding mechanisms. Pits on the grinding surface can be attributed to abrasive particles embedding into the material, as indicated by previous research [15]. Additionally, thermal stress caused by grinding heat can accelerate the wear of the grinding wheel. Another possible reason for the presence of pores and pits is the transformation of deep gaps and ridges formed during the rough grinding stage into holes and pits via mechanical extrusion during the fine grinding stage. For trials P5 and P8, they had a total grinding allowance of 0.02 mm accounting for only 5% of the total material thickness, where it becomes challenging to achieve the desired surface quality if the damaged surfaces cannot be effectively repaired during the fine grinding stage. Figure 13e illustrates a possible mechanism for the generation of pores and pits. In the fine grinding stage, with a small grinding depth, the proportion of scratching and plowing increases. The material in the ridges flows to both sides under the pressure of the abrasive particles but is not completely removed. Meanwhile, the grooves adjacent to the ridges are not sufficiently squeezed. As a result of plastic deformation, gullies remain on the ground surface, and the cross-section surface appears as pits and pores.

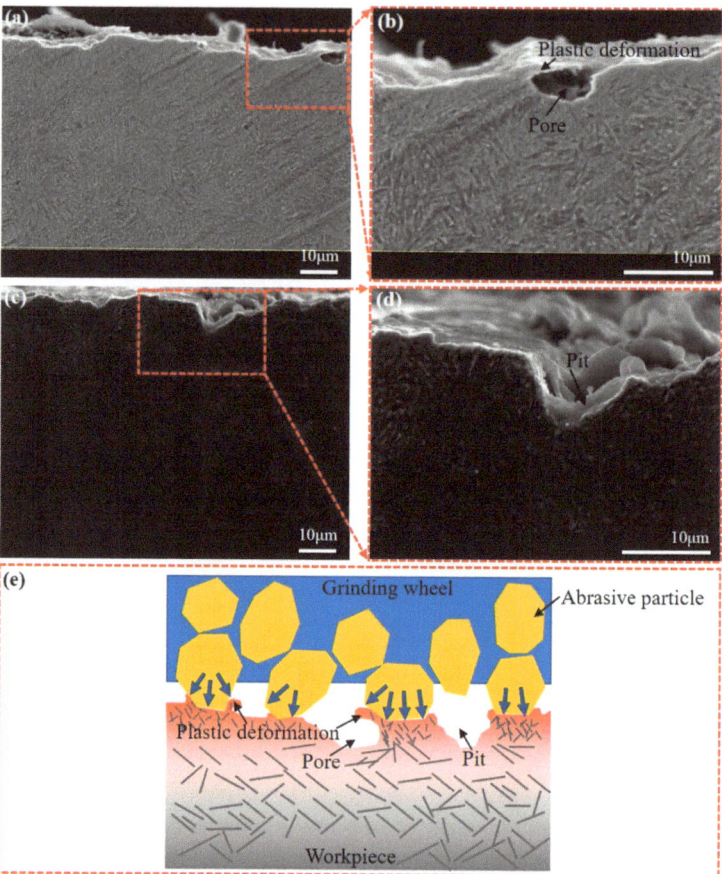

Figure 13. Subsurface morphology of samples from trials (**a**,**b**) P8 and (**c**,**d**) P5, and (**e**) schematic diagram showing generation mechanism of pores and pits.

3.4. Microhardness Variations

Figure 14 presents the microhardness gradient curve of the ground subsurface from all trials with different grinding allowance allocations. The workpiece surface, which had undergone carburization and quenching treatment, exhibited microhardness beyond 600 HV, gradually decreasing with the increasing measurement depth. Up to a measurement depth of 1100 μm, the hardness value fluctuated around 450 HV. The curve shows that the peak microhardness value of each sample was higher than 550 HV, within a depth range of 100 μm to 700 μm. Comparing the microhardness curve of the workpiece surface before grinding (as shown in Figure 14a), the initial effective carburized layer thickness was approximately 1100 μm. Since the material removed by grinding was 400 μm, the change in the residual carburized layer thickness due to the grinding process was not significant.

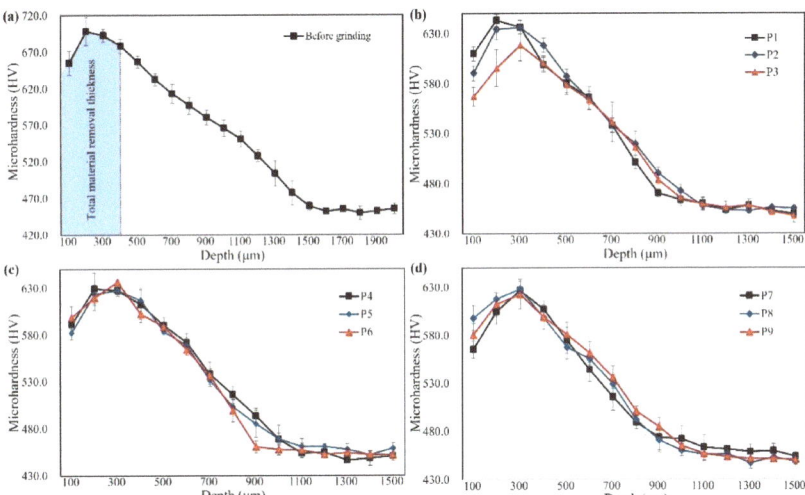

Figure 14. Microhardness variation within the subsurface layer when grinding quenched gear samples with different grinding allowance allocation, including samples (**a**) before grinding and from trial (**b**) P1 to P3, (**c**) P4 to P6, and (**d**) P7 to P9.

The sample from trial P1 exhibited the highest peak microhardness on the ground subsurface, reaching close to 650 HV at a depth of 200 μm, which was 15 HV higher than the value at the same depth before grinding (at a depth of 600 μm). The increased hardness in trial P1 suggests the occurrence of work hardening during grinding. It is worth noting that in all the curves, the hardness value at a depth of 100 μm was lower than that at 200 μm. This may be because the measured indentation size was relatively large when located at the edge of the gear material and the resin, resulting in a smaller measured value than the actual one. For the other samples except the one from trial P1, another possible reason was the occurrence of thermal softening within a depth of 200 μm below the surface layer. The change in microhardness on the ground subsurface was the combined result of grinding heat effects and mechanical effects during grinding. When the thermal effect was dominant, the material's hardness would be reduced, while the mechanical effect on hardness was opposite to the thermal effect. Since the grinding temperature in trial P1 was relatively low, work hardening occurred due to the mechanical effect contributed by the grinding wheel. In contrast, the surface hardness curve obtained in trial P3 showed thermal softening during grinding, with a hardness value at a depth of 200 μm approximately 40 HV lower than the hardness value at a depth of 600 μm before grinding, consistent with the measurement results of the grinding temperature. The hardness gradient curves of the subsurface obtained from the other trials also exhibited slight thermal softening. The thermal softening effect was most pronounced in trial P7, with a reduction of about 30 HV at a depth of 200 μm. The difference in the hardness gradient among different trials indicates that the allocation mode of the grinding allowance covertly influences the microhardness gradient of the ground subsurface, based on the combined effects of the mechanical and thermal factors.

3.5. Grinding Strategy Validation

In order to verify the results of the previous experiment, three 20Cr2Ni4A gear ring blanks marked A, B, and C were randomly selected at the manufacturing site for machining. The original allowance distribution strategy was selected for grinding gear A, and the optimized distribution strategy was used for gear B and gear C. The specific process parameters were shown in Table 3. During the grinding of gear B and gear C, the number of strokes in the rough grinding stage was reduced by increasing the grinding depth, and the spindle speed of the grinding wheel was increased appropriately.

Table 3. Grinding parameters of gears in on-site machining.

Gear	Strategy	Grinding Depth (mm)	Strokes	Spindle Speed (r/min)	Total Strokes
A	Rough grinding	0.015	12	2700	17
	Semi-finish grinding	0.013	4	2400	
	Fine grinding	0.010	1	2100	
B	Rough grinding	0.025	2	2700	14
	Semi-finish grinding	0.020	6	2700	
	Semi-finish grinding	0.014	4	2400	
	Fine grinding	0.01	2	2100	
C	Rough grinding	0.018	8	2700	14
	Semi-finish grinding	0.015	4	2700	
	Fine grinding	0.01	2	2700	

After grinding, the dimensions of the gears were recorded including cumulative tooth pitch deviation, tooth thickness deviation, profile deviation, and lead profile deviation. Improving the geometrical accuracy of gear dimensions can reduce transmission errors and heat production during operating, and it can also reduce mechanical shock and noise. The results are presented in Table 4 and Figure 15. The gear rings processed via the optimized strategy presented good dimension accuracy, verifying the feasibility of achieving higher dimensional accuracy while improving grinding efficiency.

Table 4. Results of dimensional accuracy of machined gears.

Gear	Cumulative Pitch Deviation (mm)	Tooth Thickness Deviation (mm)	Profile Deviation (mm)	Lead Profile Deviation (mm)
A	0.0297	0.0309	0.0083	0.0099
B	0.0137	0.0131	0.0069	0.0092
C	0.0186	0.0164	0.0056	0.0088

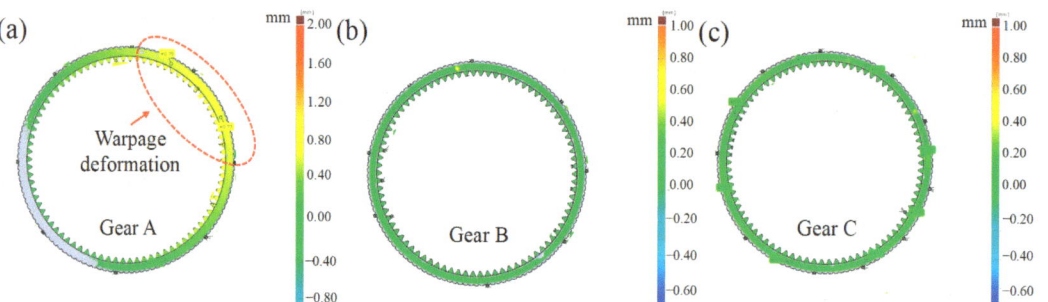

Figure 15. Deformation distribution of gears manufactured with different grinding strategies for (a) gear A, (b) gear B, and (c) gear C.

4. Conclusions

(1) The allocation mode of the grinding allowance has a significant influence on the maximum grinding temperature. Multiple grinding passes with large cutting depths in the rough grinding and first semi-fine grinding stages result in a relatively higher temperature, and the corresponding grinding temperature model under different grinding allowance allocations was proposed.

(2) The allocation mode of the grinding allowance in the finish grinding stage affects the ground surface morphology, limiting the removal of material thickness (within 0.02 mm) in the fine grinding stage and making it difficult to polish the rough surface formed by rough grinding, with the surface roughness Ra exceeding 1.60 μm. Grain refinement and defects such as plastic deformation, dark layer, material pull-out, and adhesion, caused by thermal effects, are observed on the ground surface.

(3) Different grinding temperature levels result in variations in surface hardness. The trial with the lowest grinding temperature exhibited a work hardening effect, with an increase in microhardness of 15 HV in the ground subsurface. Conversely, the trial with the highest grinding temperature showed a thermal softening effect, with a decrease in microhardness of 40 HV within the subsurface layer.

(4) The feasibility of achieving higher dimensional accuracy while improving grinding efficiency was validated when using the optimized grinding strategy, with the distribution of allowance designed via gradual reduction according to the grinding pass/stroke sequence.

Author Contributions: Conceptualization, M.L.; Methodology, R.W., S.P. and X.J.; Formal analysis, B.Z. and X.J.; Data curation, R.W. and S.P.; Writing—original draft, R.W.; Writing—review & editing, M.L. and P.G.; Funding acquisition, M.L. and P.G. All authors have read and agreed to the published version of the manuscript.

Funding: This research received no external funding.

Institutional Review Board Statement: Not applicable.

Informed Consent Statement: Not applicable.

Data Availability Statement: The data that support the findings of this study are available from the corresponding author upon reasonable request.

Conflicts of Interest: The authors declare no financial or commercial conflict of interest.

References

1. Guo, Y.; Duflou, J.R.; Lauwers, B. Energy-based optimization of the material stock allowance for turning-grinding process sequence. *Int. J. Adv. Manuf. Technol.* **2014**, *75*, 503–513. [CrossRef]
2. de Mello, A.V.; de Silva, R.B.; Machado, Á.R.; Gelamo, R.V.; Diniz, A.E.; de Oliveira, R.F.M. Surface grinding of Ti-6Al-4V alloy with SiC abrasive wheel at various cutting conditions. *Procedia Manuf.* **2017**, *10*, 590–600. [CrossRef]
3. Batueva, V.V.; Guzeev, V.I.; Batuev, V.A. Procedure of engineering design of fine milling of crooked spatial surfaces with a stepped allowance. *Procedia Eng.* **2017**, *206*, 1099–1104. [CrossRef]
4. de Souza Ruzzi, R.; da Silva, R.B.; da Silva, L.R.R.; Machado, Á.R.; Jackson, M.J.; Hassui, A. Influence of grinding parameters on Inconel 625 surface grinding. *J. Manuf. Process.* **2020**, *55*, 174–185. [CrossRef]
5. Luu, T.; Wu, Y. A novel correction method to attain even grinding allowance in CNC gear skiving process. *Mech. Mach. Theory* **2022**, *171*, 104771. [CrossRef]
6. Lv, C.; Zou, L.; Huang, Y.; Liu, X.; Li, Z.; Gong, M.; Li, H. A trajectory planning method on error compensation of residual height for aero-engine blades of robotic belt grinding. *Chin. J. Aeronaut.* **2022**, *35*, 508–520. [CrossRef]
7. Fathallah, B.B.; Dakhli, C.E.; Terres, M.A. The effect of grinding parameters and gas nitriding depth on the grindability and surface integrity of AISI D2 tool steel. *Int. J. Adv. Manuf. Technol.* **2019**, *104*, 1449–1459. [CrossRef]
8. Hood, R.; Cooper, P.; Aspinwall, D.K.; Soo, S.L.; Lee, D.S. Creep feed grinding of γ-TiAl using single layer electroplated diamond superabrasive wheels. *CIRP J. Manuf. Sci. Technol.* **2015**, *11*, 36–44. [CrossRef]
9. Guerrini, G.; Landi, E.; Peiffer, K.; Fortunato, A. Dry grinding of gears for sustainable automotive transmission production. *J. Clean. Prod.* **2018**, *176*, 76–88. [CrossRef]
10. Li, P.; Chen, S.; Jin, T.; Yi, J.; Liu, W.; Wu, Q.; Peng, W.; Dai, H. Machining behaviors of glass-ceramics in multi-step high-speed grinding: Grinding parameter effects and optimization. *Ceram. Int.* **2021**, *47*, 4659–4673. [CrossRef]
11. Lerra, F.; Grippo, F.; Landi, E.; Fortunato, A. Surface integrity evaluation within dry grinding process on automotive gears. *Clean. Eng. Technol.* **2022**, *9*, 100522. [CrossRef]
12. Guerrini, G.; Lerra, F.; Fortunato, A. The effect of radial infeed on surface integrity in dry generating gear grinding for industrial production of automotive transmission gears. *J. Manuf. Process.* **2019**, *45*, 234–241. [CrossRef]
13. Riebel, A.; Bauer, R.; Warkentin, A. Investigation into the effect of wheel groove depth and width on grinding performance in creep-feed grinding. *Int. J. Adv. Manuf. Technol.* **2020**, *106*, 4401–4409. [CrossRef]

14. Murtagian, G.R.; Hecker, R.L.; Liang, S.Y.; Danyluk, S. Plastic deformation depth modeling on grinding of gamma Titanium Aluminides. *Int. J. Adv. Manuf. Technol.* **2010**, *49*, 89–95. [CrossRef]
15. Yang, M.; Chen, J.; Wu, C.; Xie, G.; Jiang, X.; Li, M.; He, H. Residual stress evolution and surface morphology in grinding of hardened steel following carburizing and quenching processes. *Mater. Sci. Eng. Technol.* **2022**, *53*, 1421–1436. [CrossRef]
16. Tao, Y.; Li, G.; Cao, B.; Jiang, L. Simulation of tooth surface topography in continuous generating grinding based on the transient analogy model. *J. Mater. Process. Technol.* **2023**, *312*, 117833. [CrossRef]
17. Ibrahim, A.M.M.; Omer, M.A.E.; Das, S.R.; Li, W.; Alsoufi, M.S.; Elsheikh, A. Evaluating the effect of minimum quantity lubrication during hard turning of AISI D3 steel using vegetable oil enriched with nano-additives. *Alex. Eng. J.* **2022**, *61*, 10925–10938. [CrossRef]
18. Kizaki, T.; Katsuma, T.; Ochi, M.; Fukui, R. Direct observation and analysis of heat generation at the grit-workpiece interaction zone in a continuous generating gear grinding. *CIRP Ann.* **2019**, *68*, 417–422. [CrossRef]
19. Yang, S.; Chen, W.; Nong, S.; Dong, L.; Yu, H. Temperature field modelling in the form grinding of involute gear based on high-order function moving heat source. *J. Manuf. Process.* **2022**, *81*, 1028–1039. [CrossRef]
20. Jiang, X.; Liu, K.; Si, M.; Li, M.; Gong, P. Grinding Force and Surface Formation Mechanisms of 17CrNi2MoVNb Alloy When Grinding with CBN and Alumina Wheels. *Materials* **2023**, *16*, 1720. [CrossRef]
21. González-Santander, J.L.; Espinós-Morató, H. Depth of thermal penetration in straight grinding. *Int. J. Adv. Manuf. Technol.* **2018**, *96*, 3175–3190. [CrossRef]
22. Jiang, X.; Liu, K.; Yan, Y.; Li, M.; Gong, P.; He, H. Grinding temperature and surface integrity of quenched automotive transmission gear during the form grinding process. *Materials* **2022**, *15*, 7723. [CrossRef] [PubMed]
23. Li, G.; Wang, L.; Yang, L. Multi-parameter optimization and control of the cylindrical grinding process. *J. Mater. Process. Technol.* **2002**, *129*, 232–236. [CrossRef]

Disclaimer/Publisher's Note: The statements, opinions and data contained in all publications are solely those of the individual author(s) and contributor(s) and not of MDPI and/or the editor(s). MDPI and/or the editor(s) disclaim responsibility for any injury to people or property resulting from any ideas, methods, instructions or products referred to in the content.

Article

Modeling of Closure of Metallurgical Discontinuities in the Process of Forging Zirconium Alloy

Grzegorz Banaszek [1], Kirill Ozhmegov [1], Anna Kawałek [1], Sylwester Sawicki [1,*], Alexandr Arbuz [2,3] and Abdrakhman Naizabekov [4]

1. Metal Forming Department, Częstochowa University of Technology, ul. J.H. Dąbrowskiego 69, 42-201 Częstochowa, Poland; grzegorz.banaszek@pcz.pl (G.B.); kvozhmegov@wp.pl (K.O.); kawalek.anna@pcz.pl (A.K.)
2. Mechanical Engineering Department, AbylkasSaginov Karaganda Technical University, 56 Nursultan Nazarbayev Ave., Karaganda 100027, Kazakhstan; mr.medet@outlook.com
3. Core Facilities Department, Nazarbayev University, 53 KabanbayBatyr Ave, Astana 010000, Kazakhstan
4. Rudny Industrial Institute, 50Let Oktyabrya Street 38, Rudny 111500, Kazakhstan; naizabekov57@mail.ru
* Correspondence: sylwester.sawicki@pcz.pl

Abstract: This article presents the results of testing the conditions of closing foundry voids during the hot forging operation of an ingot made of zirconium with 1% Nb alloy and use of physical and numerical modeling, continuing research presented in a previous thematically related article published in the journal *Materials*. The study of the impact of forging operation parameters on the rheology of zirconium with 1% Nb alloy was carried out on a Gleeble 3800 device. Using the commercial FORGE®NxT 2.1 program, a numerical analysis was performed of the influence of thermo-mechanical parameters of the hot elongation operation in trapezoidal flat and rhombic trapezoidal anvils on the closure of foundry voids. The analysis of the obtained test results was used to formulate recommendations on the technology of hot forging and the anvilgeometry, ensuring closure of foundry voids. Based on their research, the authors conclude that the shape of the deformation basin and the value and hydrostatic pressure have the greatest influences on the closure of foundry voids.

Keywords: closurefoundry voids; zirconium ally; Forge; numerical modeling; forging process

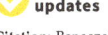

Citation: Banaszek, G.; Ozhmegov, K.; Kawałek, A.; Sawicki, S.; Arbuz, A.; Naizabekov, A. Modeling of Closure of Metallurgical Discontinuities in the Process of Forging Zirconium Alloy. *Materials* 2023, *16*, 5431. https://doi.org/10.3390/ma16155431

Academic Editors: Wenming Jiang and Guozheng Quan

Received: 21 June 2023
Revised: 24 July 2023
Accepted: 28 July 2023
Published: 2 August 2023

Copyright: © 2023 by the authors. Licensee MDPI, Basel, Switzerland. This article is an open access article distributed under the terms and conditions of the Creative Commons Attribution (CC BY) license (https://creativecommons.org/licenses/by/4.0/).

1. Introduction

Zirconium alloy products, due to their physical and mechanical properties, are used in industries in which high requirements are placed on mechanical properties and corrosion resistance in aggressive environments. One of the main areas of use for zirconium alloy products is as construction elements for the production of fuel assemblies (nuclear reactor cores). The use of alloys on baize zirconia for the manufacture of the construction elements of poly rods is related to the small active cross-section for the absorption of thermal neutrons. They are characterized by the following additive properties: impermeability of fission products (compared to steel, they have a 15-fold smaller active cross-section for neutron capture) and resistance to radiation damage; low activity under neutron irradiation; sufficient mechanical strength;a low creep rate at high temperatures; sufficient thermal conductivity;andvery slow reactions with fuel and coolant (water), i.e., significant resistance to corrosion [1–3]. In addition to meeting the requirements for ensuring high mechanical and anti-corrosion properties, the products must have a uniform structure without metallurgical discontinuities [2]. The heterogeneity of the structure in the finished products may lead to stress concentrations, which may cause failure of fuel assemblies, which in turn may eventually lead to accidents [4].

There are many zirconium-based industrial alloys used in the nuclear industry. In general, the alloys are differentiated by the presence and content of elements such as Nb, Sn, Fe, Cr, Ni, and O. The choice of an alloy is related to the purpose of a given constructional element for a specific type of reactor. PWR zirconium with 1% Nb(M5) is used for making the plugs for cladding tubes of reactor fuel assemblies [5].

The technology for manufacturing tubes and rods from Zr–1%Nb alloy includes the following operations (Figure 1) [2,6]. During the hot forging process, the cast structure is fragmented, and the cross-section of the forging is reduced. The next stage is hot extrusion, during which there is a significant change in the cross-section. The next stage is cold rolling with intermediate and final annealing, during which a specific structure and the properties of the finished products are formed.

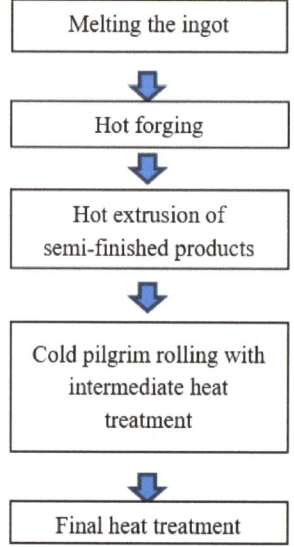

Figure 1. The technology operations in manufacturing tubes and rods from the Zr–1%Nb alloy.

Authors [6,7] have stated that, in the process of ingot melting, the formation of a piping cavity in the axis of the ingot and of central porosity is inevitable. The formation of metallurgical discontinuities is related to the gradient of the crystallization rate in relation to the volume of the ingot. After melting, the ingots are subjected to ultrasonic testing to find metallurgical discontinuities. However, it is known from the authors' experience that there may be central porosities with diameters of up to Ø10 mm in the axial part, as well as metallurgical voids with diameters of Ø1.0 mm at a certain distance from the axis of the ingot.

An analysis of the literature shows that methods for closing foundry voids during hot elongation of steel bars and aluminum and magnesium alloys are known [8–15], while no workswere found on closing foundry voids in the elongation operation of zirconium bars with the exception of the authors' work [16].

Banaszek et al. [15] carried out numerical modeling of the closure of foundry voids during the forging operation of Mg alloys bars and physical verification of the obtained results.

In the work [15], the conditions were determined for closing foundry voids in the forging process of M5 alloy rods with the use of rhombic and flat anvils. To do so, the investigators used a physical simulator of metallurgical processes, the Gleeble 3800, and FORGE software, with which computer simulations of this process were carried out. The use of a combination of computer and physical modeling to determine the parameters of the hot elongation process is effective in terms of materials and time costs [14,17,18].

In the work, the authors showed the results of studies of the influence of the geometry of flat trapezoidal and rhombic trapezoidal anvils on the closure of foundry voids in the process of hot forging of M5 alloy rods. Based on the testing results, recommendations were made for the technology of hot forging processes, ensuring the closing of foundry voids occurring in ingots made of M5 alloy.

2. Purpose and Scope of Work

The objective of the study was to formulate recommendations for the technology of forging elongation that would ensure the closing of foundry voids occurring in zirconium alloy, observed after the vacuum process of double remelting in an electric arc furnace. The authors proposed that the operations of hot forging elongation should be performed with the use of flat trapezoidal and rhombic–trapezoidal tools.

To achieve the goals of the study, tests investigatingthe forging of ingots in two anvil assemblies were carried out. Modeling of the elongation investigation was performed using the Forge program based on FEM, with the distributions of the hydrostatic pressure and effective strain being determined on a cross-section of the deformed zirconium alloy after each elongation process.

To test the forging process, the datafrom plastometric tests of the zirconium alloy were obtained using the GLEBLLE 3800, based on which diagrams showing the relation between stress and real deformation were developed, and the coefficients of the plasticizing stress function were determined.

Based on the results of numerical studies, guidelines for the technology of forging shaped anvils for closing foundry voids were developed.

The test results should help to improve the structural and mechanical properties of zirconium alloy during the appropriate selection of the tool shape and technological parameters of the forging process.

3. Methodology of Experimental Research

The zirconium alloy applied for the investigation was an M5 alloy with the following composition and chemical configuration: zirconium with 1.1% Nb, 0.05% Fe, and 0.6% O.

The impact of thermomechanical parameters of the hot elongation of the M5 alloy rod on the plasticizing stress σ_p was determined using a Gleeble 3800 plastometer, 323 NY-355, Poestenkill, NY 12140, USA (Figure 2). Specimens of the M5 alloy in a crystallized state (Figure 3), each of a working portion with a diameter of 10 mm and a height of 12 mm, were used for the tests.

Figure 2. The Gleeble 3800 plastometer testing equipment.

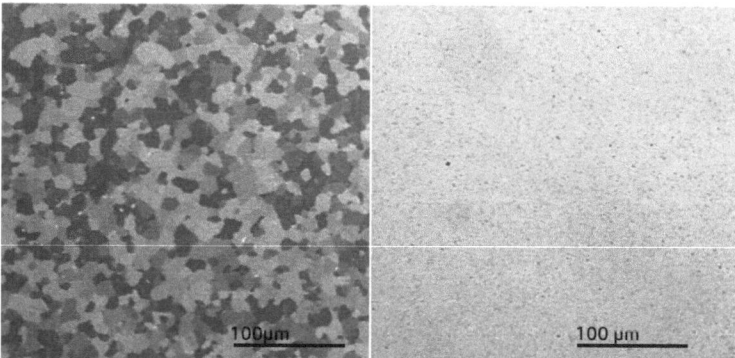

Figure 3. The microstructure of starting specimens of the M5 alloy.

The tests were carried out in the temperature range of 770–950 °C [2,16]. On the basis of the Zr–Nb phase equilibrium scheme at the 870–950 °C temperature range, the structure of the metal is characterized by grains in the shape of the A2 unit cell. At forging temperatures of 750–870 °C, the structure of Zr is in the zone of α + β [2]. During the deformation of the zirconium with 1% Nb alloy sample in the specified temperatures, the limit value of the deformation decreases by half [19–22]. In this article, the authors show that the forging process of these alloys at temperatures less than 750 °C results in the occurrence of cracks.

Tests were carried out for the deformation speed range from $0.5\ \mathrm{s}^{-1}$ to $5.0\ \mathrm{s}^{-1}$, even though the average deformation speed for the hot elongation of the Zr alloys rods on the pressure is $0.5\ \mathrm{s}^{-1}$ [16]. However, in local places of the deformed rods in the shaped tools, the deformation speed is slightly higher, which is why it was decided to expand the research area.

The functionality of this process is the ability to obtain relatively large strain values (up to $\varepsilon = 1.2$ [19]); moreover, it is the most favorable strain state for describing the properties of the alloys during plastic deformation [17].

Origin software was used to process the final test results. This software allowed for a comprehensive analysis of the experimental data obtained, reducing the number of samples tested to two for a single point with an error limit less than 2–3%.

To use of the real results from plastometric tests of the M5 Zr alloys, an approximation of flow curves $\sigma_P - \varepsilon$ was carried out with the use of the generalizing connection–functions of Henzel A. and SpittelT. [19]:

$$\sigma_P = A \cdot e^{m1 \cdot T} \cdot T^{m9} \cdot \varepsilon^{m2} \cdot e^{m4/\varepsilon} \cdot (1+\varepsilon)^{m5 \cdot T} \cdot e^{m7 \cdot \varepsilon} \cdot \dot{\varepsilon}^{m3} \cdot \dot{\varepsilon}^{m8 \cdot T} \qquad (1)$$

where σ_p is stress, T is temperature, ε is strain, $\dot{\varepsilon}$ is strain rate, A, and m1–m9 are function coefficients.

The approximation of the testing effects was carried out according to the method [14] in FORGE@2023 software, Transvalor S.A., E-Golf Park, 950 avenue Roumanille, CS 40237 Biot, 06904 Sophia Antipolis cedex, France.

4. Testing Results

Figures 4–6 illustrate the $\sigma_P - \varepsilon$ curves of the zirconium alloys deformed at T = 770–950 °C for the deformation speed $\dot{\varepsilon}$ from $0.5\ \mathrm{s}^{-1}$ to $5.0\ \mathrm{s}^{-1}$. These figures show increases in the temperature of test T from 770 °C to 950 °C, during which the value of the plasticizing stress σ_p decreases about 2.5 times. An increase in the deformation speed from $0.5\ \mathrm{s}^{-1}$ to $5.0\ \mathrm{s}^{-1}$ led to an increase in the plasticizing stress σ_p, and at T = 770 °C, this increase amounted to ~29%. At T = 850 °C, this increase was ~36%, and at T = 950 °C, it was 43%. The conducted research showed that the increase in the deformation speed affected the plasticizing stress σ_p, which increased, and in addition, the magnitude of this increase was correlated with the temperature.

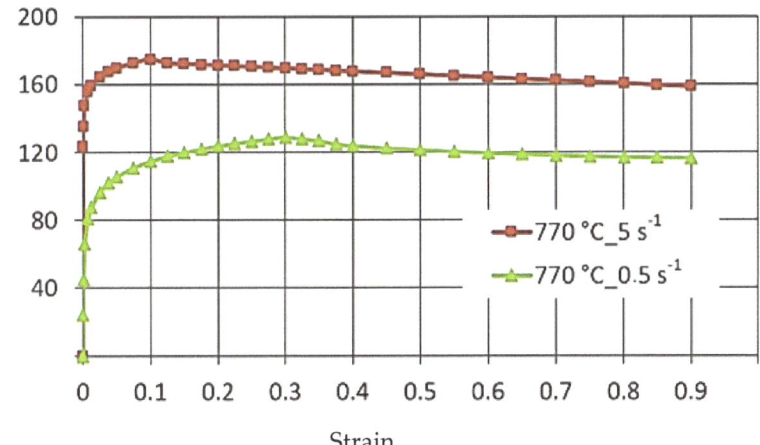

Figure 4. The $\sigma_p - \varepsilon$ flow curves of the M5 alloy, obtained for T = 770 °C in the deformation speed range $\dot{\varepsilon}$ from 0.5 to 5.0 s^{-1} using the Gleeble 3800.

The $\sigma_p - \varepsilon$ plasticity curves for the tested temperature range varied. For the deformation speed $\dot{\varepsilon} = 0.5$ s^{-1} in the tested temperature range, the $\sigma_p - \varepsilon$ curves reached the max. stress value for a programmed deformation value, and as the temperature increased, the max. plasticizing stress value σ_p moved toward actual smaller values.

The data shown in Figure 5 demonstrate that the curve obtained at deformation speeds of $\dot{\varepsilon} = 0.5$ s^{-1} and $\dot{\varepsilon} = 5.0$ s^{-1} are of a different character. The max. value of yield stress during deformation speed $\dot{\varepsilon} = 0.5$ s^{-1} was observed for the value of real deformation $\varepsilon = 0.12$, and with a continued increase in deformation, a decrease in the value of yield stress was obtained. For the deformation speed $\dot{\varepsilon} = 5$ s^{-1}, the value of yield stress increases to the value of real deformation $\varepsilon = 0.3$, where the curve flattening effect takes place.

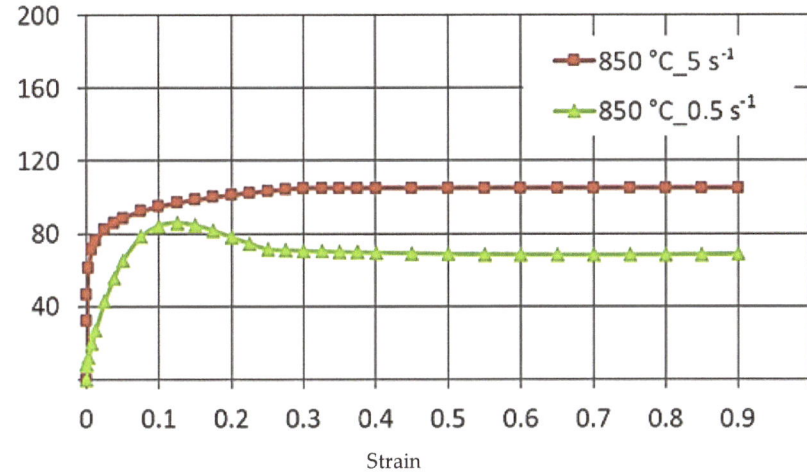

Figure 5. The $\sigma_p - \varepsilon$ flow curves of the M5 alloy, obtained for the temperature T = 850 °C in the deformation speed range $\dot{\varepsilon}$ from 0.5 to 5.0 s^{-1} using the Gleeble 3800 metallurgical simulator.

This difference may be due to the dynamic softening process being delayed with the increase in strain rate. Dynamic recrystallization is inhibited, and the softening of the material proceeds in accordance with the mechanism of dynamic polygonization. At the temperature T = 862 °C, a polymorphic transformation takes place in zircon, and the crystal lattice changes from the hexagonal A3 lattice to the spatially centered cubic lattice A2 [2]. The non-monotonicity of the curve may be caused by the beginning of the polymorphic transformation [11].

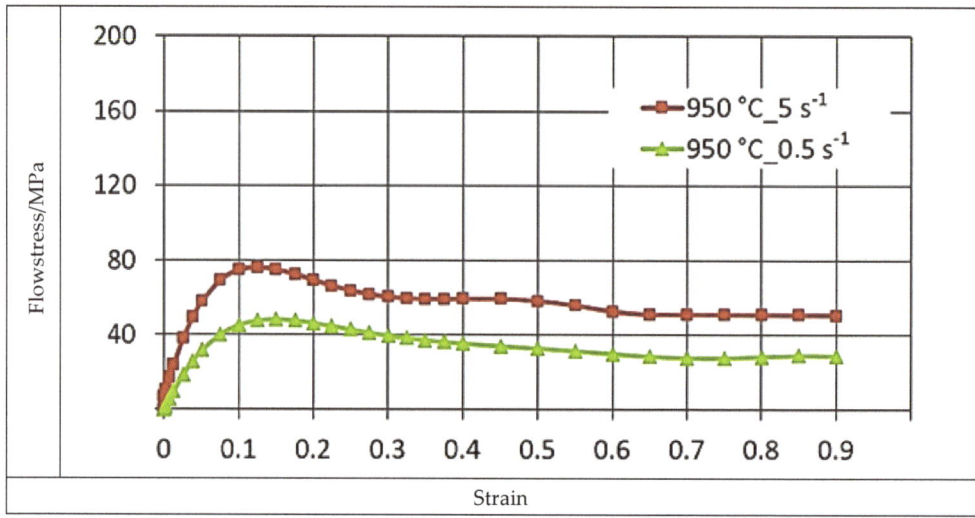

Figure 6. The $\sigma_p - \varepsilon$ flow curves of the M5 alloy, obtained for T = 950 °C in the deformation speed range $\dot{\varepsilon}$ at 0.5 to 5.0 s^{-1} using the Gleeble 3800.

As an effect of approximation of the investigational zirconium alloy curves, Hensel–Spittel coefficients were determined for the temperature and speed of the following plastic processing conditions: T = 770–950 °C, and $\dot{\varepsilon}$ = 0.5–5.0 s^{-1} (Figure 7).

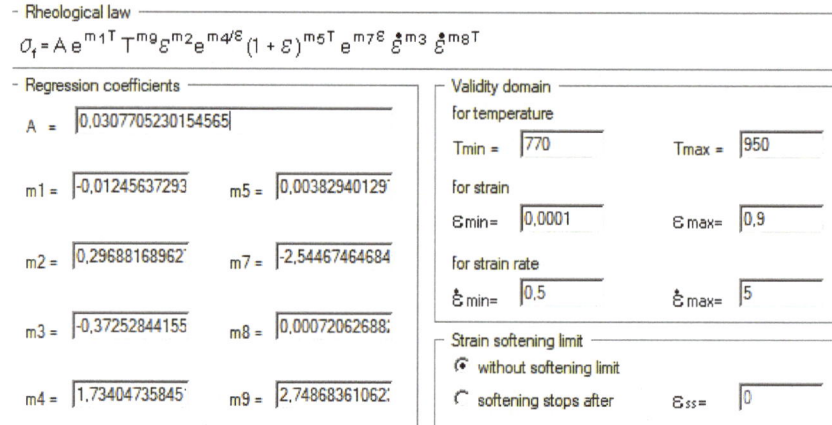

Figure 7. Values of the equation coefficients for the zirconium alloy.

The multifactorial Equation (1), using the coefficients shown in Figure 8 in the process of computer simulation, describes the changes in the value of the plasticizing stress of the M5 alloy depending on the thermomechanical parameters of the tested technological processes. The approximation results for the M5 alloy in the temperature range of T = 770–950 °C and deformation rates from 0.5 s^{-1} to 5.0 s^{-1} are shown in Figure 8 in the form of three-dimensional graphs.

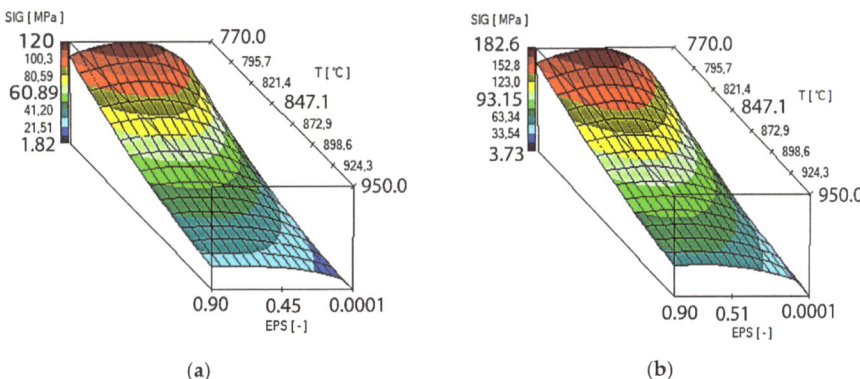

Figure 8. Results of the approximation of the M5 alloy stress–strain curves in the temperature range T = 770–950 °C in the form of three-dimensional graphs: (**a**) $\dot{\varepsilon} = 0.5$ s^{-1}; (**b**) $\dot{\varepsilon} = 5.0$ s^{-1}.

It can be assumed that the coefficients of the approximating Function (1) are wellchosen if the average error does not exceed 15% [18]. For the range of parameters examined in the article, the average error of approximation is ~10%.

5. Methodology of Conducted Numerical Research

This article provides a theoretical analysis of the process of elongating Zr alloy bars for two different anvils to determine the possibility of closing foundry voids in Zr alloys. Tools were characterized by various geometrical surfaces of the deformation valley, which resulted in obtaining various directions and orientations of the pressure and friction forces in the deformed Zr alloy in the deformation place. The various directions and orientations of forces influenced the mechanism of welding of the foundry voids inside the deformed bars. Constant parameters of the forging operation were assumed, such as a forging initiation temperature of 950 °C, a value of relative compressionof 35%, and a value of the feed angleof 8 mm/s [23,24].

The parameters for the lengthening operation were adopted based on the authors' experience in this field [15,16].

The computer program FORGE was applied to model the forging process in shaped anvils [25]. Forges enable the runningofthermomechanical simulations of, among other processes, plastic forming processes [26]. A description of the temperature, force, stress, strain, and thermomechanical and frictional laws used in the investigation appeared in [21].

For thermal calculations, the Galerkin equation was used, while the strengthening curves were approximated by the Henzel–Spitel equation (Figure 8).In this paper, to simulate the elongation operation, a thermo–visco–plastic model of the deformed body, which is based on the theory of large plastic deformations, was used. To generate the grid of finite elements, tetrahedral elements with the bases of triangles were used. In the generated model input, a number of nodes equal to 8958 was used for the simulation, while the number of tetrahedral elements adopted for the simulation was 40,760. The coefficient of friction between the surface of the tools and the bars, adoptedaccording to Coulomb's law, was $\mu = 0.3$. The value of the friction coefficient used for numerical modeling wasnot determined experimentally, but results from forging practice were used for many years.For the free forging of steels and alloys, friction coefficient values between 0.24 and 0.46 are assumed.

The following boundary conditions were assumed during the numerical tests: Indirect heat transfer between material and tool –"Steel Hot Medium.tef"; Friction coefficient between material and tools–"High.tff";and Heat transfer coefficient between material and environment—"Air. tef". The heat transfer coefficient connecting the tools and the zirconium alloy was assumed to be α = 10,000 W/m^{2}K and the heat transfer coefficient between the zirconium alloy and the environment to be equal to λ = 10 W/m^{2}K. The environment temperature was 25 °C, and the tool temperature was 250 °C. The starting temperature was 950 °C. In all forging steps, a relative 35% crumple was assumed. The speed of the upper tool was v = 8 mm/s, while the lower tool was assumed to be stationary.

Table 1 shows the values of the boundary and initial conditions used in the numerical model.

Table 1. Values of the boundary and initial conditions used in the numerical model.

Parameter Name	Parameter Value
Heat transfer coefficient material-anvils	10,000 W/m^{2}K
Heat transfer coefficient material-environment	10 W/m^{2}K
Relative reduction	35%
Upper anvil speed	8 mm/s
Friction coefficient	0.3
Starting material temperature	950 °C
Starting anvil temperature	250 °C
Environmental temperature	25 °C
Rotary in elongation operation	90 °C

In the FORGE computer program, the diffusion model does not exist. The inference of foundry voids closing is based on the values of hydrostatic pressure and the temperature of the rods being elongated. Therefore, in the forging process, the aim is to achieve the maximum possible values of hydrostatic pressure within the foundry voids, as well as to maintain a high temperature close to that at the beginning of forging. Through the process of closing the foundry voids, there are high values of hydrostatic pressure on their right and left sides, while there is no hydrostatic pressure on their top and bottom sides (positive values of average stress) [13].

Zr alloy bars with a diameter of 100 mm and a length of l = 50 mm were deformed in two compositions of tools in four forging transitions.

The contour and size of the tools used in the investigation are shown in Figures 9 and 10, respectively.

This work describes a graph of the forging operation for flat trapezoidal tools (Figure 9). During in first step, flat trapezoidal tools were used, in which a bar heated to a temperature of 950 °C was deformed with a relative reduction of 35%, and then the bar was rotated 90°. In the next step, with the bar using the same anvils, relative reduction of 35% was again applied, and it was rotated 90° clockwise. Before the third step, the bar was heated to a temperature of 950 °C because, after the second step, the temperature of the bar decreased to 750 °C, and according to the practice of forging Zr alloys, it is impossible to carry out further steps of the forging process. Subsequently, the next steps were already carried out in flat tools with a relative reduction of 35%. After the third pass, the bar was rotated 90° clockwise.

For the rhombic trapezoidal tools (Figure 10), the procedure was analogous to the forging scheme described above.

The geometry and distribution of the modeled foundry voids on the end face of the model Zr alloy bar are shown in Figure 11.

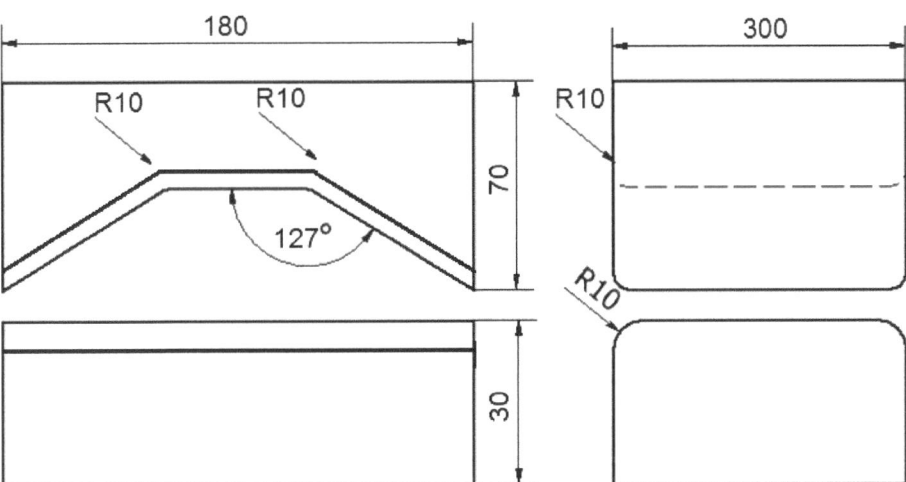

Figure 9. Contour and size of flat trapezoidal tools used for deformation of Zr alloy.

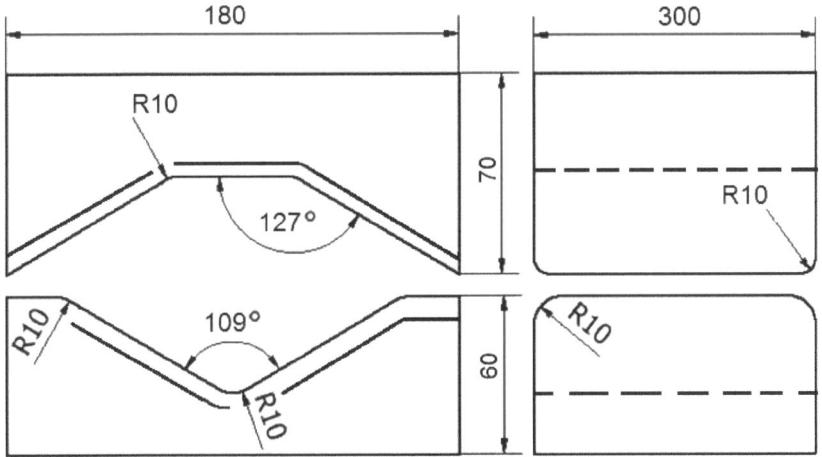

Figure 10. Contour and size of rhombic trapezoidal tools used for deformation of Zr alloy.

The initial model for the forging process was a cylinder with a diameter of 100 mm and a length of 50 mm, in which nine foundry voids were modeled with a length of charge (50 mm); the first axial one had a diameter of 10 mm, and the other eight, with a diameter of 1 mm, were distributed around the circumference of the model charge 25 mm from its axis (Figure 11). The charge modeled in this way corresponded to the shape of an ingot obtained after double remelting in a vacuum in an arc furnace. Artificially modeled holes simulated internal casting discontinuities, such as central porosity and casting voids. Descriptions of these defects can be found in the literature [23–26].

The model inputs and tools were drawn with AutoCad software. The foundry voids were drawn as cylinders inside the model input. Using the various tools in the program, the foundry voids were interpreted by the AC program as holes. The model diagram with artificially simulated foundry voids was exported to the Forge program as a file with the "stl" extension. The same process was performed for all of the tools' compositions. In Forge, using the stl@meshing and volume@meshing tools, a 2D triangular grid was useful to the input and artificially modeled foundry voids, followed by a 3D tetrahedral grid. In total, for the correct simulation of the closure of foundry voids and their good formation, the built-in tools FoldsDetection andSelfContactwere used. Due to the use of the shown tools during the FEM simulation in the places of welded foundry voids, there were no calculation errors or mesh deterioration. During welding, the foundry voids of the nodal points of the tetrahedral elements in close contact connected; thus, the modeled foundry voids were closed. During modeling, there were no errors in the form of interpenetration of nodes of tetrahedral elements around closed foundry voids.

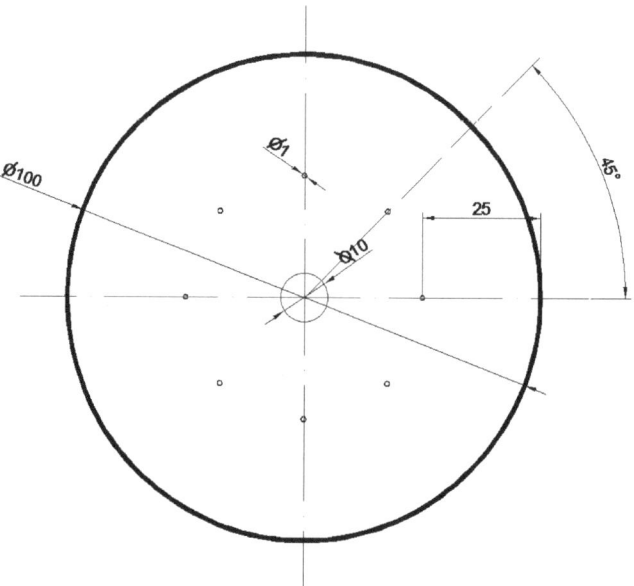

Figure 11. Geometry and distribution of the modeled foundry voids on the end face of the model Zr alloy bar.

On the basis of the simulations of the closing process of foundry voids during the forging of bars made of zirconium alloy in flat and rhombic tools, the volume values of the remaining unwelded foundry voids were determined. The procedure was as follows.After the simulation of the forging process was completed, the stlfile with a deformed bar was exported from the Forge program to the RinoCeros program, where the tetrahedral element grid was separated in this program, and the bar contour elements were removed. After this step, only tetrahedral elements that constituted the contours of the unwelded discontinuities remained. The next step was to merge the previously broken grid of tetrahedral elements. In the last step, using the physical parameters tool in the RinoCeros program, the whole volume of foundry voids for individual forging process was obtained.

After the bar forging process was carried out in two anvil compositions, the values of the hydrostatic pressure at the nodal points of the tetrahedra, constituting the contour of the unwelded foundry voids, were determined as the arithmetic means of the occurring stresses.The same procedure was performed for the value of the effective strain. All the results shown in the article are subject to some errors resulting from the need to analyze the numerical and data and the results obtained.

6. Analysis of Distribution of Hydrostatic Pressure Values During Forging Process

The test effects concerning the hydrostatic pressure distribution during the zirconium alloy bar extension operation in two tool compositions are shown in Figures 12–27.

6.1. Analysis of the Distribution of Hydrostatic Pressure Values During the Forging Process in Flat Trapezoidal Anvils

Figures 12–19 present the distributions of hydrostatic pressure values obtained during the numerical computation of elongating Zr alloy bars in flat trapezoidal anvils.

The data in Figures 12 and 13 show that the axial foundry voids with a diameter of 10 mm were welded in the first forging step, indicating that the use of flat trapezoidal tools results in obtaining the appropriate characteristics of the stress distribution in the axial zones of the deformed rod, with a positive effect on the closing discontinuities. The value of hydrostatic pressure in the axis of the forged bar was 53 MPa. This outcome is especially important because, in the first steps of the forging process, the temperature of the deformed bar does not change, which facilitates welding of the discontinuities. It is also worth noting that, after the casting process for ingot molds, the discontinuities in the axis of the ingot have the largest volume, and they are difficult to weld in the further stages of the elongation process. Therefore, it is very important that the compressive stresses of the highest possible values occur in these zones. The use of a flat trapezoidal anvil in the starting bar forging process also favors the process of welding foundry voids in the areas under the operation of the lower flat tool because, as shown by the hydrostatic pressure distribution (Figure 12), its values are within the range of 26–53 MPa. On the other hand, the forging process in flat trapezoidal anvils does not favor the welding of foundry voids located in places under the action of the upper trapezoidal form. The values of compressive stresses in this place range from 5 to 25 MPa. There were three unwelded foundry voids left.

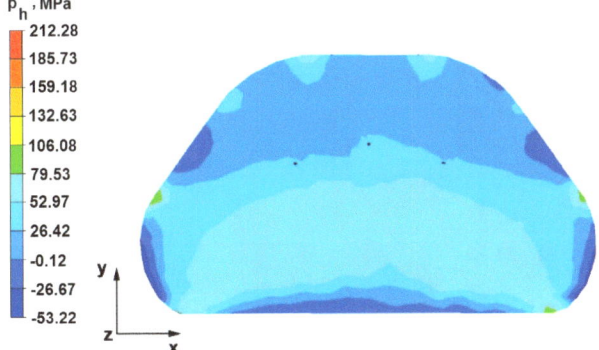

Figure 12. Hydrostatic pressure on the cross-section of a forged bar in the first step with a 35% crumple.

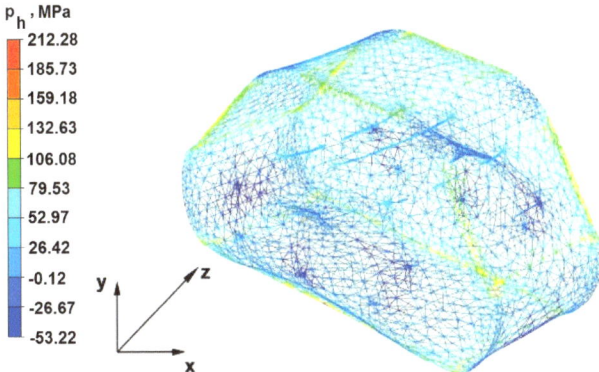

Figure 13. Hydrostatic pressuredistribution with a view of unclosed defects in the volume of a forged bar in the first step with a 35% crumple.

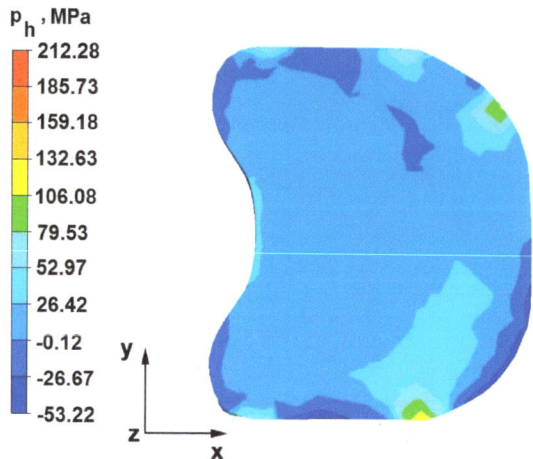

Figure 14. Hydrostatic pressure on the cross-section of a forged bar in the second step after turning through an angle of 90° with a 35% crumple.

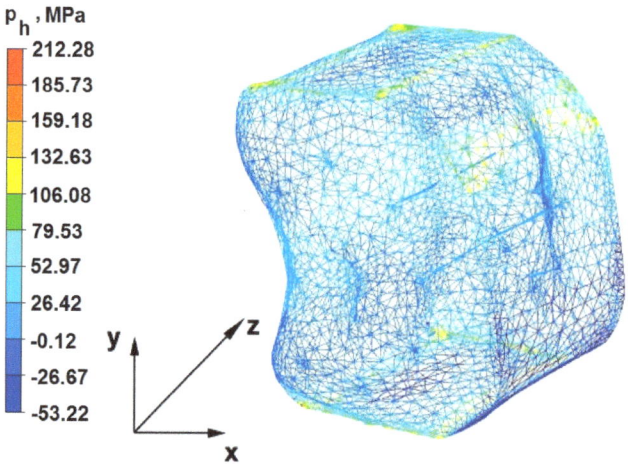

Figure 15. Hydrostatic pressuredistribution with a view of unclosed defects in the volume of a forged rod in the second transition after a rotation of 90° with 35% crumple.

Based on the test results shown in Figures 14 and 15, it can be concluded that, in the predominant volume of the deformed rod, adecreasein the value of the hydrostatic pressure from 53 MPa to a value close to zero was noted, proving its absence. The lack of hydrostatic pressure does not favor the welding of the foundry voids. The only area where the hydrostatic pressure was still present (its value was 26 MPa) was the area on the right, impacted by the lower anvil. Two consecutive discontinuities did not fully weld in the second forging step due to a lack of hydrostatic pressure in the place of their occurrence, and the value of the average stress was 0.12 MPa there; i.e., it was tensile stress, unfavorable for the welding of the discontinuities.

Figures 16 and 17 show the test results obtained after deformation in the third forging step, where the flat trapezoidal tools were replaced with flat tools, and the deformed bar was heated to the starting temperature. The data analysis shows that no unwelded foundry voids were observed in the volume of the deformed bar. In the central deformation zone of the rod, the hydrostatic pressure was 26 MPa. Outside of this zone, there was no hydrostatic pressure.

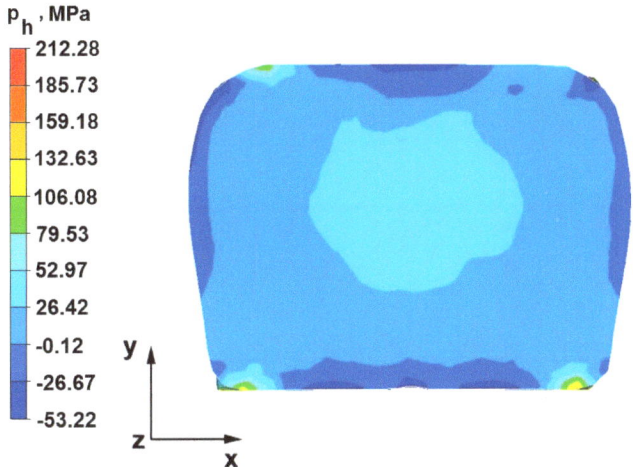

Figure 16. Hydrostatic pressure on the cross-section of a forged bar in the third step after turning through an angle of 90° with a 35% crumple (change to flat anvils).

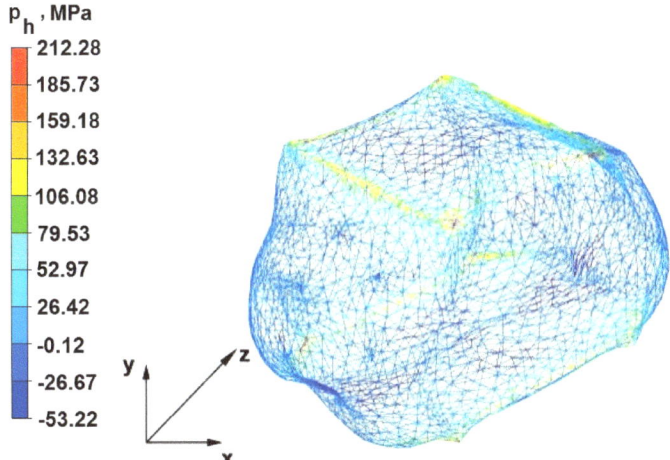

Figure 17. Hydrostatic pressuredistribution with a view of unclosed defects in the volume of a forged rod in the third transition after a rotation of 90° with 35% crumple(change to flat anvils).

The data in Figures 18 and 19 show that, in the fourth and last elongation step in the entire volume of the deformed rod, no hydrostatic pressure was recorded outside the corner areas. There were positive mean stress values in the range of 0.12–53 MPa. The introduction of flat anvils on the last stages of rod formation, after using the trapezoidal flat anvil to introduce high hydrostatic pressure, did not bring the intended effect. During the forging with flat anvils, positive values of average stresses in most areas of the forged rod were observed, preventing the welding of discontinuities, especially those with large initial dimensions.

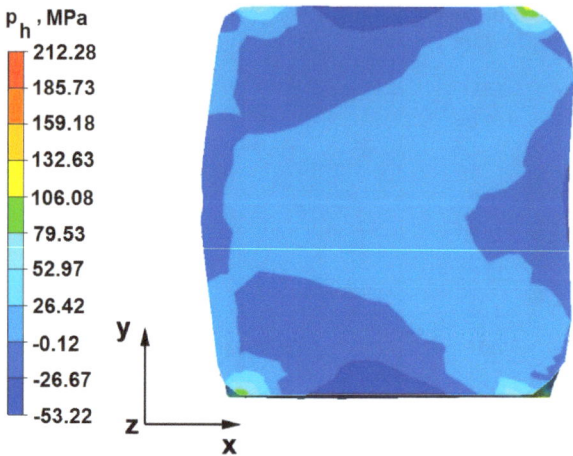

Figure 18. Hydrostatic pressure on the cross-section of a forged bar in the fourth step after turning through an angle of 90° with a 35% crumple (change to flat anvils).

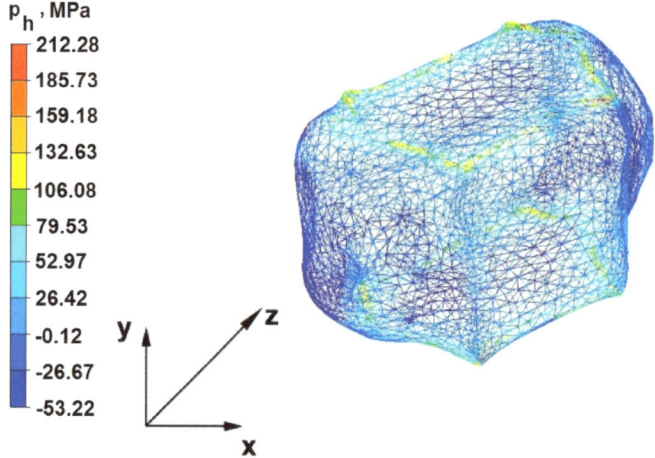

Figure 19. Hydrostatic pressuredistribution with a view of unclosed defects in the volume of a forged rod in thorough transition after a rotation of 90° with 35% crumple (change to flat anvils).

6.2. Analysis of the Distribution of Hydrostatic Pressure Values in the Forging Process in the Rhombic Trapezoidal Tools

Figures 20–27 show the distribution of hydrostatic pressure values obtained in the numerical computation of the elongation of the bar from the zirconium alloy in rhombic trapezoidal tools.

The data in Figures 20 and 21 show that, during the implementation of the initial stages of the elongation using rhombic trapezoidal anvils, the axial discontinuity, as well as the discontinuity lying in the zone of impact of the lower rhombic anvil, is not completely welded. In the place of axial discontinuity occurrence, the hydrostatic pressure was 33 MPa, while in the place of the second occurrence ofunclosed foundry voids, it was 2 MPa. The pressure value obtained in this place was too low for the process of closing the foundry voids. The value of the hydrostatic pressure of 64 MPa is sufficient to weld the discontinuities in the corners of the deformed rod along the x axis.

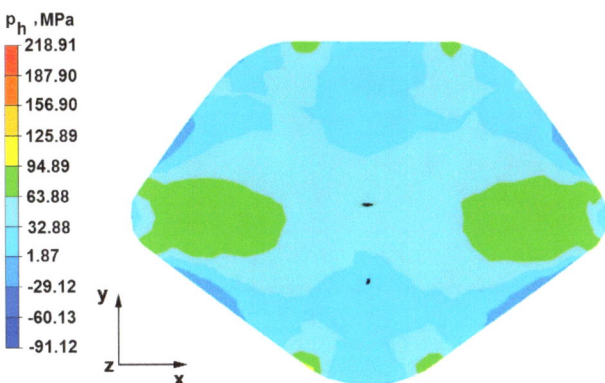

Figure 20. Hydrostatic pressure on the cross-section of a forged bar in the first step with a 35% crumple.

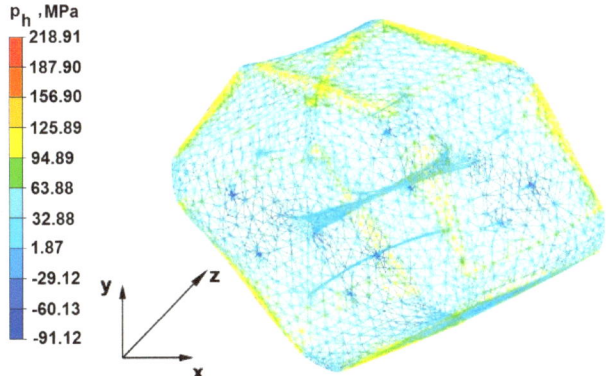

Figure 21. Hydrostatic pressure distribution, along with the view of unclosed defects in the volume of a forged bar in the first step with a 35% crumple.

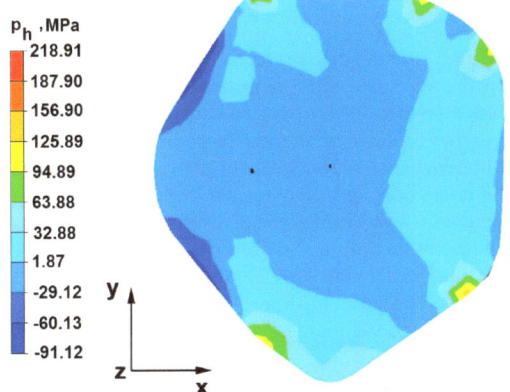

Figure 22. Hydrostatic pressure on the cross-section of a forged bar in the second pass after turning through an angle of 90° with a 35% crumple.

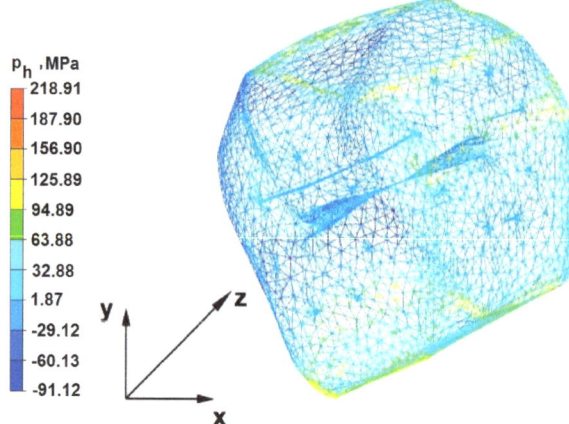

Figure 23. Hydrostatic pressure distribution along with the view of unclosed defects in the volume of a forged rod in the second pass after turning through an angle of 90° with a 35% crumple.

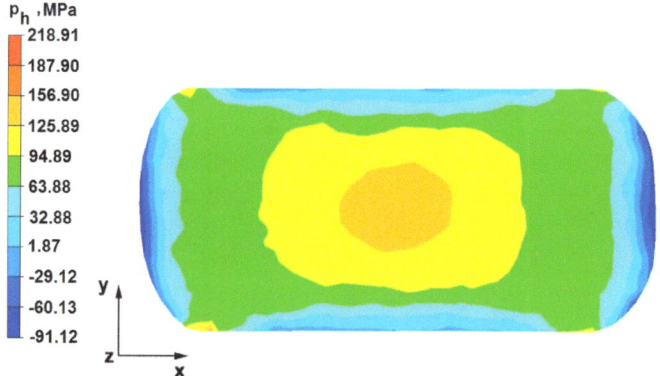

Figure 24. Hydrostatic pressure on the cross-section of a forged bar in the third pass after turning through an angle of 90° with a 35% crumple(change to flat anvils).

By analyzing the data presented in Figures 22 and 23, it can be concluded that, in the majority of the volume of the deformed bar in the next forging step, no hydrostatic pressure occurred; therefore, the foundry voids remaining after the first forging step were also not closed in the second step. In these places, no hydrostatic pressure existed, and the values of positive mean stress ranged from 29 to 60 MPa. This outcome proves that, in this area, tensile stresses occurred, blocking the closing process of foundry voids.

Figures 24 and 25 show the distribution of hydrostatic pressure on the cross-section of a deformed zirconium alloy rod previously heated to the initial forging temperature after the third step, made with flat tools with a 35% crumple. No unclosing foundry voids were observed because, in almost the entire volume of the deformed bar, the hydrostatic pressure achieved values of 64–126 MPa, creating favorable conditions for the closing of foundry voids.

Based on the data in Figures 26 and 27, it can be seen that, after the fourth, final step in the entire volume of the deformed bar, the value of the hydrostatic pressure was sufficiently large, influencing the closing of the foundry voids. The values of hydrostatic pressure occurring in the entire volume of the deformed bar ranged from 33 to 64 MPa.

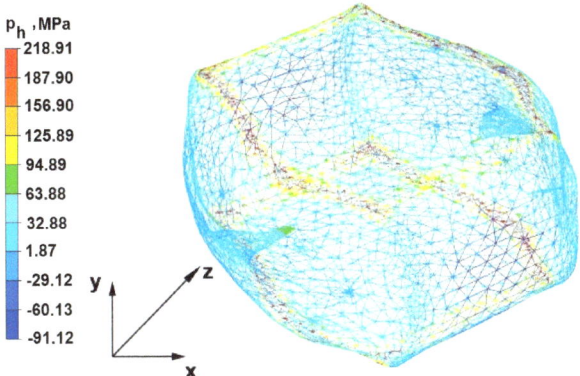

Figure 25. Hydrostatic pressure distribution, along with the view of unclosed defects in the volume of a forged rod in the third transition after turning through an angle of 90° with a 35% crumple(change to flat anvils).

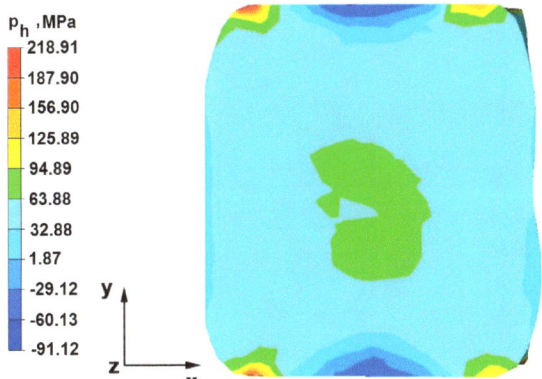

Figure 26. Hydrostatic pressure on the cross-section of a forged bar in the fourth pass after turning through an angle of 90° with a 35% crumple(change to flat anvils).

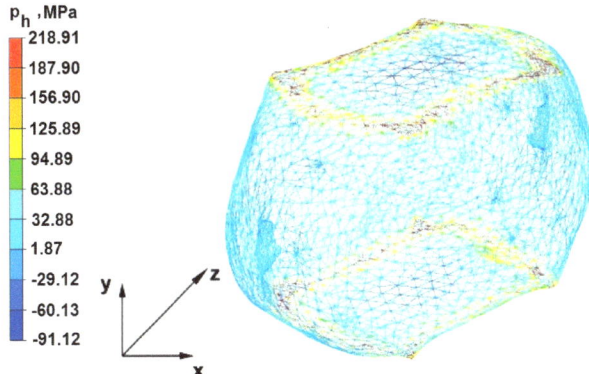

Figure 27. Hydrostatic pressure distribution, along with a view of unclosed defects in the volume of a forged bar in the fourth transition after turning through an angle of 90° with a 35% crumple(change to flat anvils).

7. The Distribution of Deformation Intensity During The Forging Process

To obtain the deliberate characteristics of the deformations that would facilitate closing of the foundry voids, the authors proposed that the forging process should be carried out in rhombic trapezoidal and flat trapezoidal anvils, allowing them to control the direction, orientation, and magnitude of the vectors of friction and the pressure forces acting in the deformation area and to select the appropriate values of the main technological parameters of the forging process.

The effects of the investigation of the distribution of hydrostatic pressure in the elongation of the zirconium alloy bar in flat, trapezoidal, andrhombic-trapezoidal anvils are presented in Figures 28–43.

7.1. The Distribution of the Deformation Intensity Values in the Forging Process in Flat Trapezoidal Anvils

Figures 28–35 show the effective strain distributions obtained during numerical computation of zirconium alloy bar elongation in flat trapezoidal anvils.

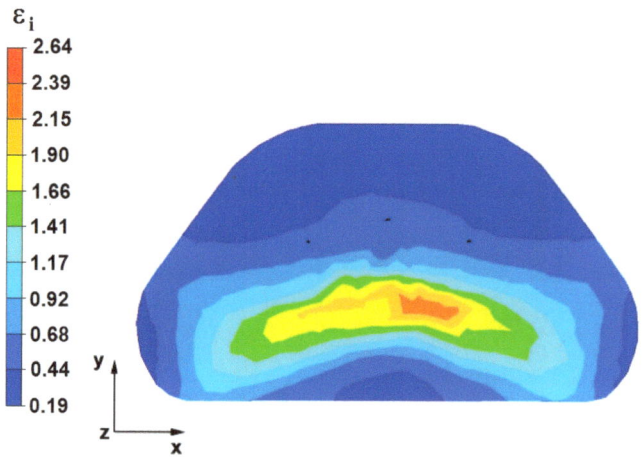

Figure 28. The effective strain on the cross-section of the forged bar in the first step with a 35% crumple.

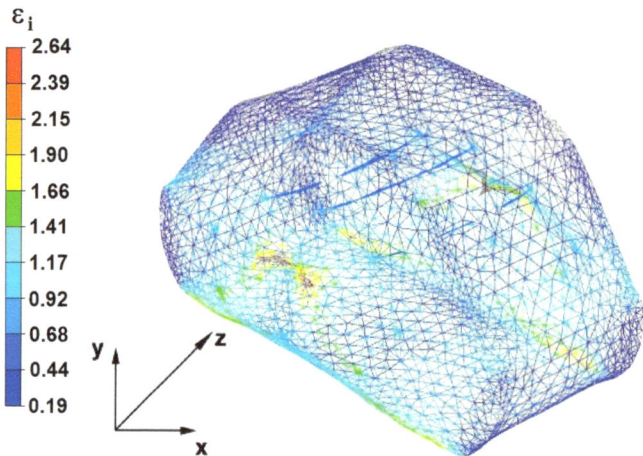

Figure 29. Distribution of the effective strain, along with a view of the unclosed defects in the volume of the forged bar in the first step with a 35% crumple.

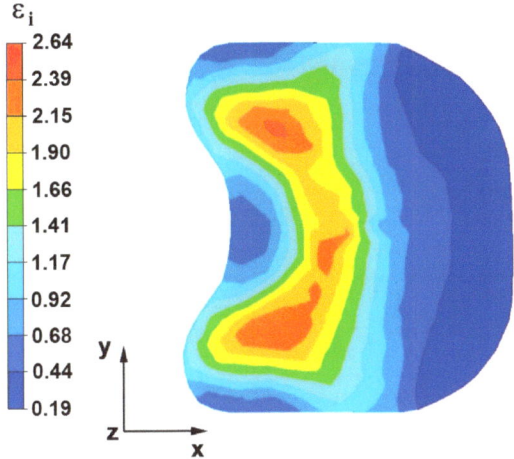

Figure 30. The effective strain on the cross-section of the forged bar in the second step after turning through an angle of 90° with a 35% crumple.

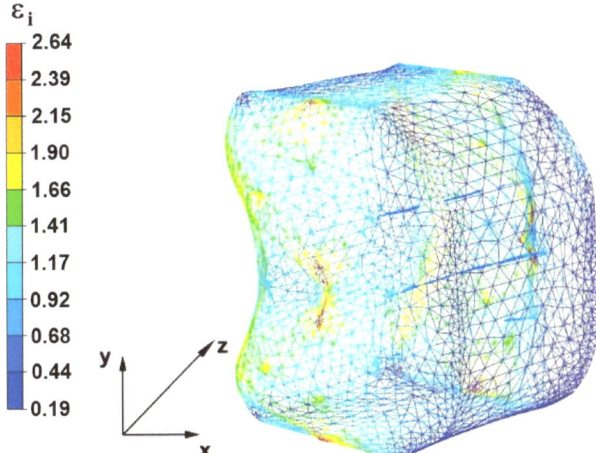

Figure 31. Distribution of the effective strain, along with a view of the unclosed defects in the volume of the forged bar in the second step after turning through an angle of 90° with a 35% crumple.

By analyzing the data in Figures 28 and 29, it can be seen that the axial discontinuity, as well as the foundry voids under the direct influence of the lower flat anvil, was completely closed after the first forging step. A favorable state of deformation developed in the metal located in the deformation basin due to the appropriate directions and orientations of frictional forces and the pressure forces resulting from the impact of the employed surfaces of the shaped tools on the deformed rod. Such a distribution of forces caused the occurrence of large deformations in the middle and lower places of the deformed bar, leading to the effective strain values ranging from 1.17 to 2.64. Only the discontinuities in the areas of the metal impacted by the upper trapezoidal anvil remained completely unwelded. The values of the effective strain were too small there to allow for complete welding of the discontinuities, as they amounted to 0.19–0.44.

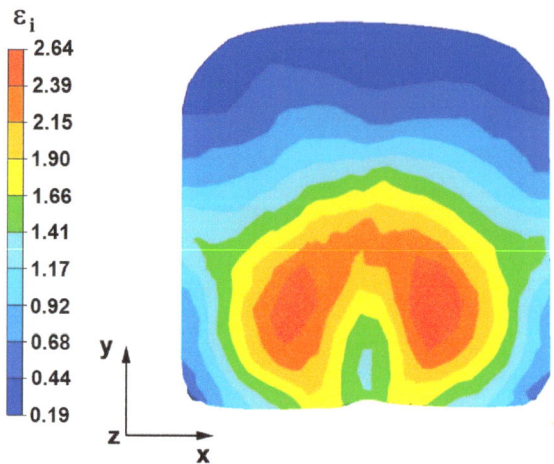

Figure 32. The effective strain on the cross-section of the forged bar in the third step after turning through an angle of 90° with a 35% crumple(change to flat anvils).

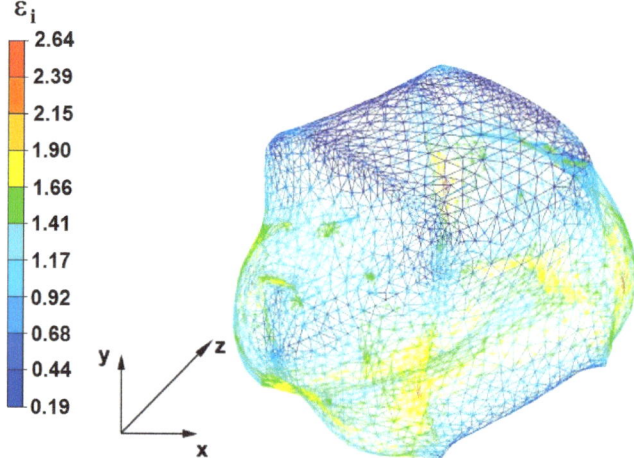

Figure 33. Distribution of the effective strain, along with a view of the unclosed defects in the volume of the forged bar in the third step after turning through an angle of 90° with a 35% crumple(change to flat anvils).

The data in Figures 30 and 31 show that, after the second step, the foundry voids remaining after the first step were still not completely closed because, in the places of their occurrence, the values of the effective strain were too small and amounted to 0.19–0.44. On the other hand, in the area on the left side of the deformed bar, the effective strain values were favorable for the process of closing the foundry voids, as they were in the range of 1.41–2.64.

Figures 32 and 33 show the effective strain values after the third step and after changing the shaped tools to flat tools, before heating the forged bar to the starting temperature and its rotation by an angle of 90°. The data presented in Figures 34 and 35 show that the foundry voids remaining after the second forging step were completely closed. The values of the effective strain in the middle and lower places of the deformed bar were high and ranged from 1.41 to 2.64. On the other hand, the distribution of the effective strain values, which were in the range of 0.19–0.92 and thus unfavorable for the welding of discontinuities, was observed in the zone of the bar impacted by the upper anvil.

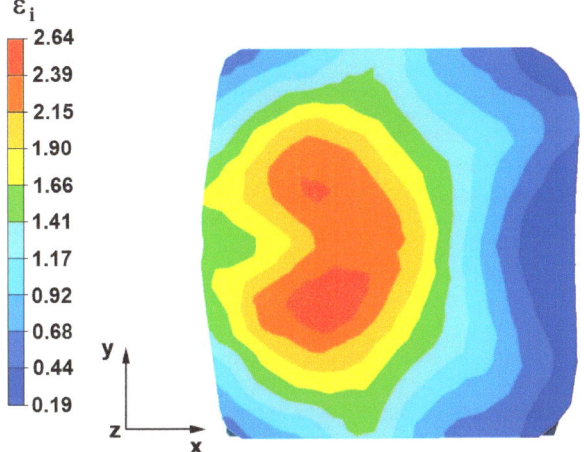

Figure 34. The effective strain on the cross-section of the forged bar in the fourth step after turning through an angle of 90° with a 35% crumple(change to flat anvils).

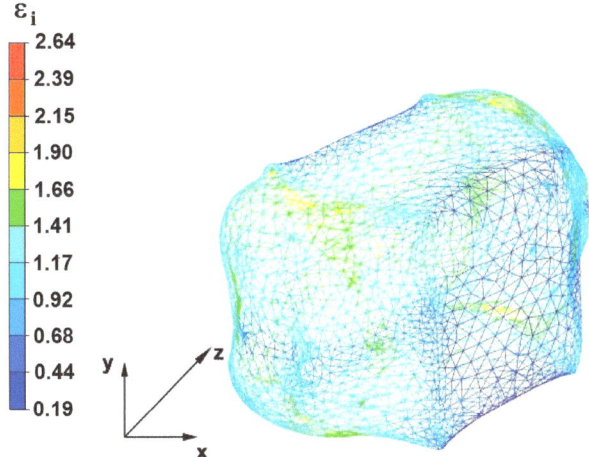

Figure 35. Distribution of the effective strain, along with a view of the unclosed defects in the volume of the forged bar in the fourth step after turning through an angle of 90° with a 35% crumple(change to flat anvils).

By analyzing the data in Figures 34 and 35, it can be concluded that the values of the effective strain in the last (fourth) step slightly increased. Apart from the small zone to the right of the y axis of the deformed bar, these values were large and ranged between 1.17 and 2.64.

7.2. The Distribution of the Deformation Intensity Values in the Forging Process in the Rhombic Trapezoidal Tools

Figures 36–43 show the effective strain distributions obtained in numerical computations of the zirconium alloy bar with rhombic trapezoidal tools.

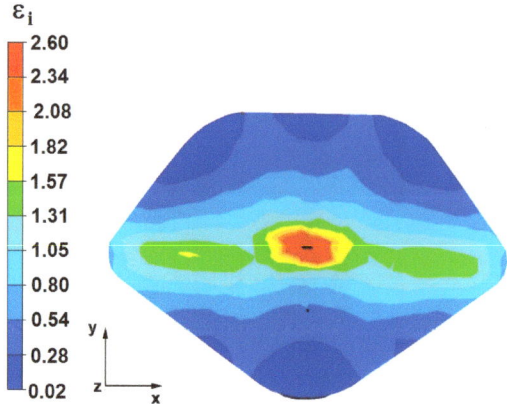

Figure 36. The effective strain on the cross-section of a forged bar in the first step with a 35% crumple.

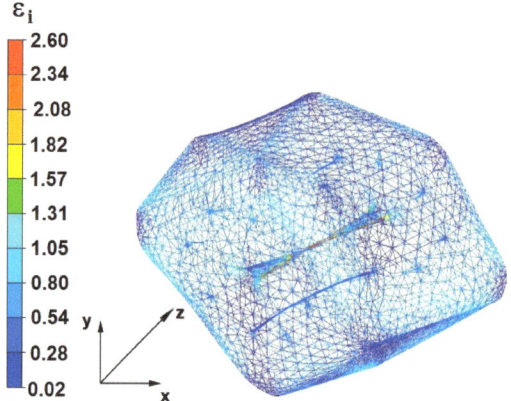

Figure 37. Deformation effective strain, along with the view of non-closed defects in the volume of the forged bar in the first step with a 35% crumple.

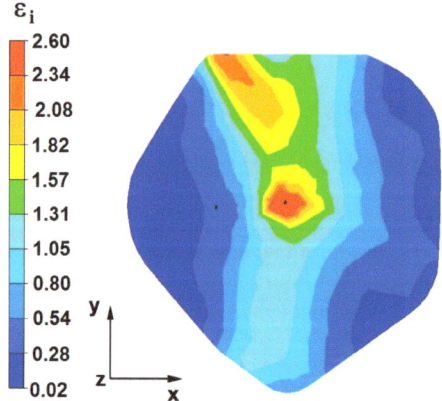

Figure 38. The effective strain on the cross-section of a forged bar in the second step after turning through an angle of 90° with a 35% crumple.

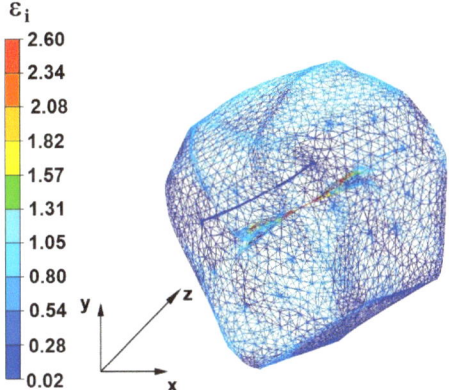

Figure 39. Deformation effective strain, along with the view of non-closed defects in the volume of the forged bar in the second step after turning through an angle of 90° with a 35% crumple.

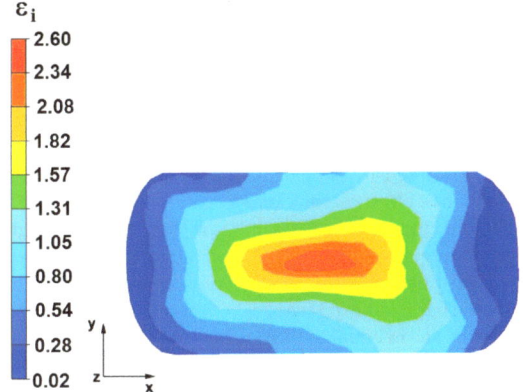

Figure 40. The effective strain on the cross-section of a forged bar in the third step after turning through an angle of 90° with a 35% crumple (change to flat anvils).

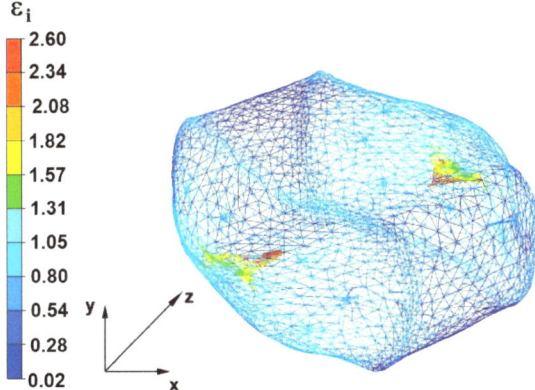

Figure 41. Deformation effective strain, along with the view of non-closed defects in the volume of the forged bar in the third step after turning through an angle of 90° with a 35% crumple (change to flat anvils).

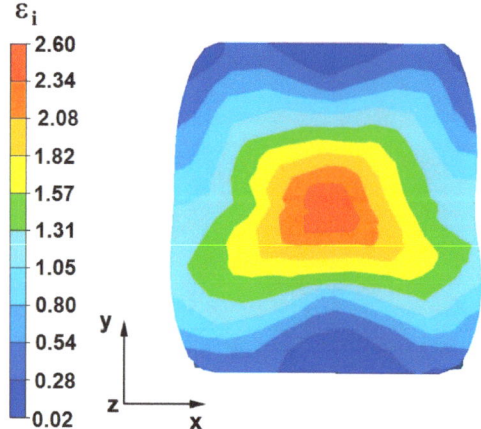

Figure 42. The effective strain on the cross-section of a forged bar in the fourth step after turning through an angle of 90° with a 35% crumple (change to flat anvils).

By analyzing the data presented in Figures 36 and 37, it can be seen that, in the central areas of the deformed bar, the value of the deformation intensity was 2.60, which was insufficient for the complete closing of the axial foundry voids present there. In the lower part of the bar, which was deformed after the first step, unclosed foundry voids were also observed because values of the effective strain in that zone were small, ranging from 0.28 to 0.54. The low intensity of deformations concentrated in the region of foundry voids did not have a positive effect on the closing process of foundry voids.

The data in Figures 38 and 39 show that, due to the high value of the effective strain in the second forging transition, whichwas equal to 2.60, the axial foundry voids were partially welded. Additionally, the foundry voids, now located in the left part of the deformed bar because the bar had been turned 90°, was not closed due to the presence of a small value of the effective strain equal to 0.28.

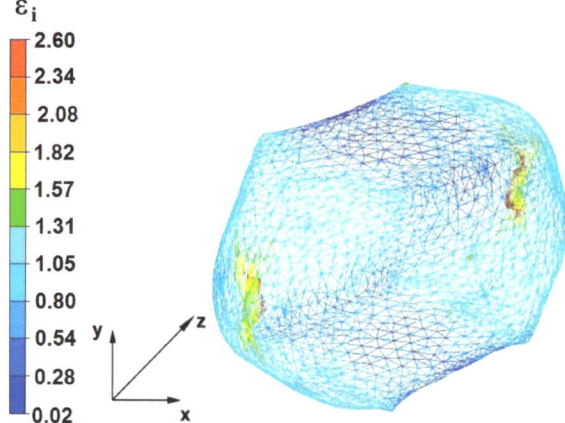

Figure 43. Deformation effective strain, along with the view of non-closed defects in the volume of the forged bar in the fourth step after turning through an angle of 90° with a 35% crumple (change to flat anvils).

From the data presented in Figures 40 and 41, which present the distribution of the effective strain values after the third step, no modeled foundry voids were observed. In most of the cross-sectional zone of the deformed bar, there was a favorable distribution of the effective strain values within the range of 1.05–2.60. Additionally, based on the data presented in Figures 42 and 43, it can be concluded that, on the cross-sectional area of the bar deformed in the fourth step, the values of the effective strain were large (in the range of 1.05–2.60) and facilitated the closure of foundry voids. The exception was the area at the contact surface of the alloy with the anvil, in which the effective strain values were small and fell within the range of 0.02–0.28. However, internal foundry voids never occur in the near surface zones.

8. The Influence of the Anvil Shape on Changes in the Volume of Discontinuities, Hydrostatic Pressure, and the Effective Strain in the Zr Alloy in the First Two Steps

Figures 44–46 show diagrams of the total volume of unclosed foundry voids, mean values of hydrostatic pressure, and arithmetic mean values of the effective strain occurring around unclosed foundry voids for the anvils analyzed in the article, divided into the first and second steps.

To determine the influence of the deformation basin shape on the closing of internal foundry voids, the authors also presented the research results in [18].

By analyzing the data presented in Figures 44–46, the authors try to answer the question of how the shape of the hitting surfaces of the tools affects the closing of internal foundry voids.

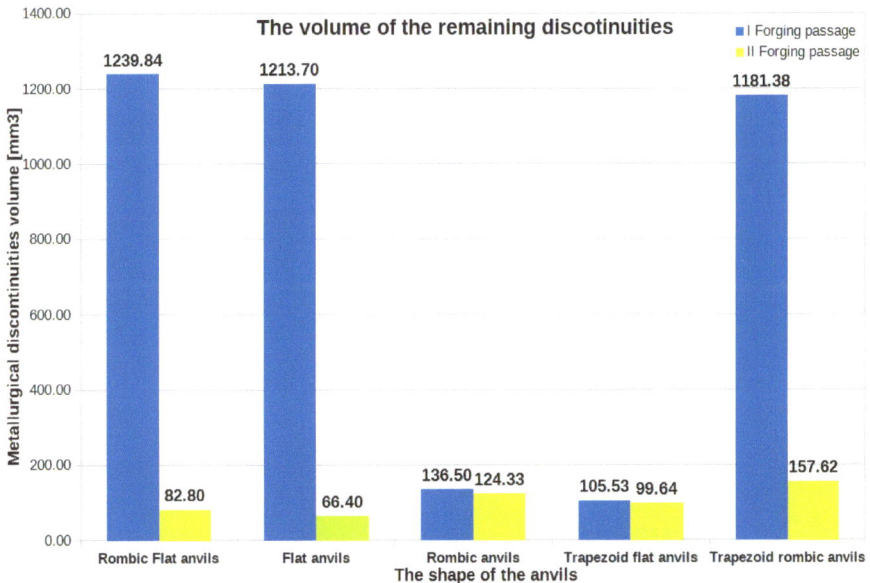

Figure 44. The total volume of unclosed foundry voids in the volume of a forged Zr alloy bar for two forging steps in relation to individual tools' compositions.

By analyzing the data presented in Figure 44, it can be seen that the best results in terms of the closing of foundry voids during the forging process of Zr alloy bars after the first and second steps were obtained for rhombic–flat and flat tools [18]. After the second step, the values of the total volumes of unclosed foundry voids ranged from 66.40 mm^3 to 82.80 mm^3, which constituted only 1.5% to 2% of the volume of the initially modeled discontinuities. It is worth noting that the total starting value of the volume of all modeled foundry voids was 4239 mm^3. The conducted research shows that the closing of foundry voids can also be performed with the use of rhombic trapezoidal and flat trapezoidal anvils. For these tools, after the first step, an observed reduction in the total volume of foundry voids amounted to 105.53 mm³ and 1181.38 mm³, respectively. After the second step for these tools, a decrease in the total value of the volume of unclosed foundry voids was noted. The volumes of unclosed foundry voids were 99.64 mm³ and 157.62 mm³, respectively, which accounted for 2.3% and 3.7% of the volume of the initially modeled foundry voids.

Figure 45 also presents data on the value of hydrostatic pressure obtained after forging processeswith flat and shaped tools presented in [18] to compare the hydrostatic pressure values obtained for different anvils. The data presented in Figure 45 show that, for the assembly of flat–trapezoidaland rhombic–trapezoidal anvils, the value of the hydrostatic pressure after the first step was similar and amounted to 26 MPa and 17 MPa, respectively. After the second step, a decrease in this value was observed for both the first and the second anvils' composition, while for the rhombic–trapezoidal anvils, the decrease was about 16 MPa.

Based on the analysis of the data shown in Figure 45, it can be seen that the highest value of hydrostatic pressure after the first and second steps was obtained for the rhombic–flat and flat tools. The high value of the hydrostatic pressure around the foundry voids is not unambiguous in relation to the total value of the closed foundry voids (Figure 44).

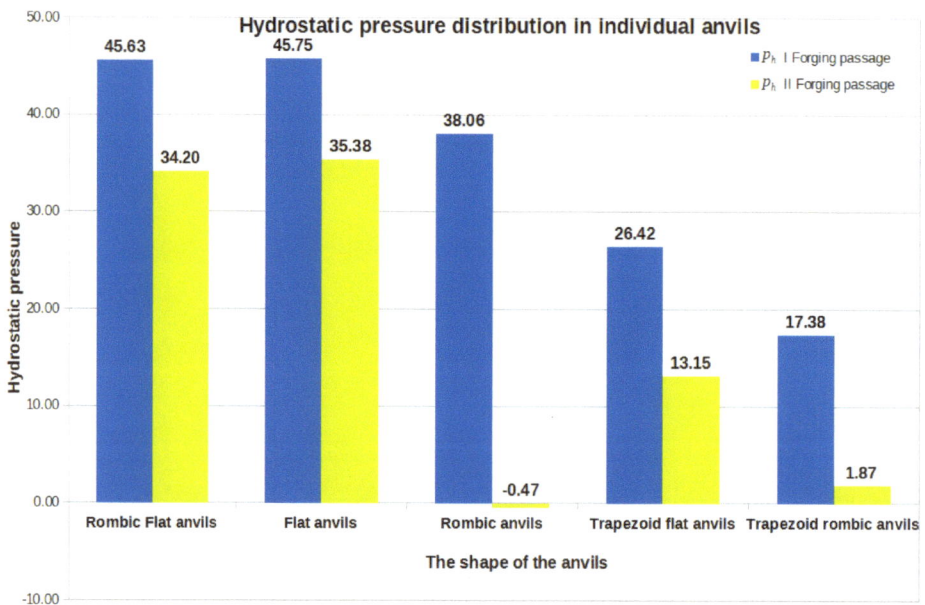

Figure 45. The arithmetic mean value of the hydrostatic pressure around the unclosed foundry voids in relation to the individual tools' compositions.

Figure 46 compares the data on the value of the effective strain obtained after the forging process in flat and shaped tools presented in [18]. The study of the information in Figure 46 shows that the most favorable arithmetic mean value of the effective strain after two steps was obtained for flat tools and was equal to 1.70. The arithmetic mean values of the effective strain calculated for the bar forging in the remaining tool compositions, except for the rhombic–trapezoidal tools, had much lower values. In the implementation of the first two steps with the rhombic–trapezoidal anvils, the arithmetic mean value of the deformation intensity after the first transition was 1.44, and after the second transition, it was 1.31. These values were, however, slightly lower than the values obtained for the deformation with flat anvils. This slightly smaller difference resulted in significant inhibition of the welding process of metallurgical discontinuities after the second forging transition. For trapezoidal anvils, the worst results were obtained in terms of closing of the foundry voids (Figure 44).

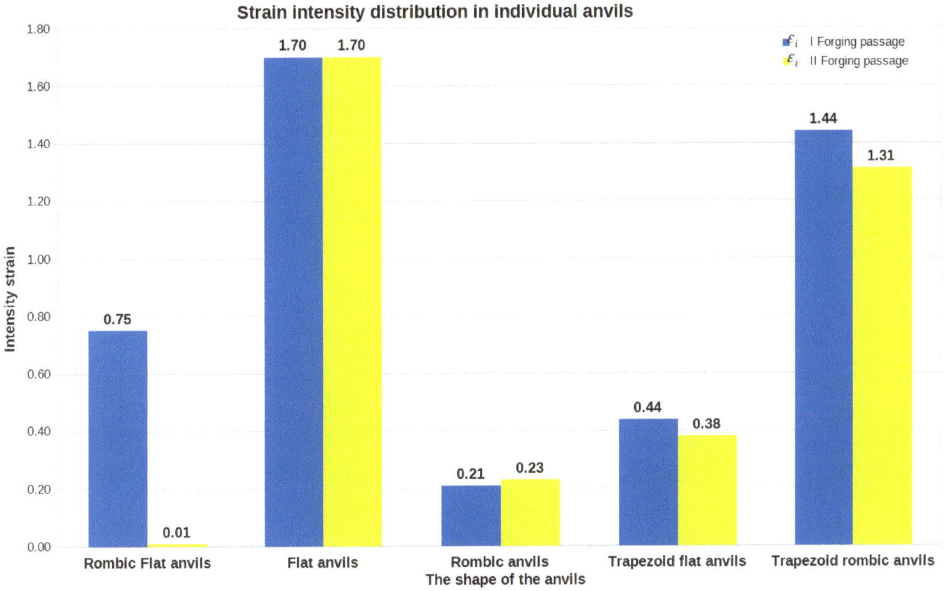

Strain intensity distribution in individual anvils

Anvil shape	I Forging passage	II Forging passage
Rombic Flat anvils	0.75	0.01
Flat anvils	1.70	1.70
Rombic anvils	0.21	0.23
Trapezoid flat anvils	0.44	0.38
Trapezoid rombic anvils	1.44	1.31

Figure 46. The arithmetic mean value of the deformation intensity occurring around unclosed foundry voids in relation to individual tool compositions.

Considering the operations of the forging processes of Zr alloy with five different tool compositions, it can be stated that it is favorable, in terms of closed foundry voids, to maintain the effective strain in the area of the closed foundry voids at the level of 1.70 and higher and the hydrostatic pressure at the level of 45 MPa and higher.

9. Conclusions

Based on the investigation of the results of the conducted research, the following conclusions were drawn:

1. The use of different hitting surfaces of the tools to close internal foundry voids significantly affects their closing;
2. The most important factor for the closing of internal foundry voids in elongation is not only the value of hydrostatic pressure, which, depending on the anvil's shape used, should be between 53 and 126 MPa, but, above all, the shape of the deformation basin, which influences the distribution of effective strain;
3. The highest values of the effective strain, in the range of 1.31–1.44, were obtained during the operation of the forging process of zirconium alloy bars with rhombic–trapezoidal tools, providing good material processing and at the same time contributing to obtaining better mechanical properties of the finished product;
4. Closing of all modeled metallurgical discontinuities for both anvil compositions was achieved in the third step;
5. The greatest axial discontinuity was closed with the flat–trapezoidal anvils;
6. When carrying out the elongation operation to weld foundry voids, the highest possible values of relative reduction in the range of 30–40% should be used because high values of reduction affect the formation of stresses and compressive deformations in the local zones of the deformed bar that favor the closing of foundry voids.

After the analysis of the forging process operation to weld the internal foundry voids occurring in Zr alloy ingots, the following guidelines for the forging technology were proposed.

- For the initial stages of the elongation operation, anvils with different working surfaces should be used, based on which it is possible to weld the largest (in terms of volume) metallurgical discontinuities in the rod axis. In addition, the use of these anvils results in good material processing and allows for obtaining a homogeneous distribution of mechanical properties in the starting volume of the deformed bar.
- After the first two steps of the forging operation, additional heating of the bar should be applied.
- For the final stages of forging, it seems reasonable to use flat tools due to the higher values of the effective strain and hydrostatic pressure in the volume of the deformed bar.
- In all forging transitions, irrespective of the anvils used, the highest possible values of relative reduction should be used.

The authors intend to verify the obtained modeling results under laboratory conditions in the near future.

Author Contributions: Conceptualization, G.B., A.K. and A.A.; methodology, G.B. and S.S.; software, K.O.; formal analysis, K.O. and A.N.; writing—review and editing A.K. and S.S.; funding acquisition, S.S. All authors have read and agreed to the published version of the manuscript.

Funding: Czestochowa University of Technology, ul. J.H. Dąbrowskiego 69, 42-201 Częstochowa.

Institutional Review Board Statement: Not applicable.

Informed Consent Statement: Not applicable.

Data Availability Statement: Not applicable.

Conflicts of Interest: The authors declare no conflict of interest.

References

1. Melechow, R.; Tubielewicz, K. *Materiały Stosowane W Energetyce Jądrowej [Materials Used in Nuclear Energy]*; Politechnika Częstochowska, seria Monografie Nr 86; Wydawnictwo Politechniki Częstochowskiej: Częstochowa, Poland, 2002; p. 229, ISBN 9788371931635/8371931638.
2. Zajmovskij, A.S.; Nikulina, A.V.; Reshetnikov, N.G.; CirkonievyeSplavy, V. *Yadernoj Ehnergetike [Zirconium Alloys in Nuclear Power]*; Energoatomizdat: Moscow, Russia, 1994; p. 256, ISBN 5-283-03767-3.
3. Azhazha, V.M.; V'yugov, P.N.; Lavrinenko, S.D. *Cirkonij I Ego Splavy: Tekhnologii Proizvodstva, Oblasti Primeneniya [Zirconium and Its Alloys: Production Technologies, Applications]*; NNC HFTI: Kharkiv, Ukraine, 1998; p. 89.
4. Alekseev, A.; Goryachev, A.; Izhutov, A.L. Eksperimental'noeizuchenie v reaktore MIR povedeniya TVELOV VVER—1000 v rezhimah RIA i LOCA [Experimental Study of the Behavior of VVER—1000 Fuel Rods in the MIR Reactor in the RIA and LOCA Modes]. In Proceedings of the X Rossijskaya Konfereciya Poreatornomu Materialovedeniyu, Dimitrovgrad, Russia, 27–31 May 2013; pp. 140–151.
5. Markelov, V.A.; Novikov, V.V.; Nikulina, A.V.; Shishov, V.N.; Peregud, M.M.; Konkov, V.F.; Tselishchev, A.V.; Shikov, A.K.; Kabanov, A.A.; Bocharov, O.V.; et al. Sostojanierazrabotkiiosvoenijacirkonievyhsplavovdljatvjelovi TVS aktivnyhzonjadernyhvodoohlazhdaemyhreaktorov v obespechenieperspektivnyhtoplivnyhciklovikonkurentosposobnostinamirovomrynke. Voprosyatomnojnaukiitehniki. Materialoved. *I Novye Mater.* **2006**, *N2*, 63–72.
6. Brusnicyn, S.V.; Loginov, Y.U.N.; Mysik, R.K. *DefektySlitkovChernyh I CvetnyhSplavov, PrednaznachennyhDlyaPlasticheskojDeformacii [Defects in Ingots of Ferrous and Non—Ferrous Alloys Intended for Plastic Deformation]*; UGTU-UPI: Ekaterinburg, Russia, 2007; p. 167, ISBN 978-5-321-01228-4.
7. Shved, F.I. *SlitokVakuumnogoDugovogoPereplava [Vacuum Arc Remelting Ingot]*; Tat'yanyLur'e: Chelyabinsk, Russia, 2009; p. 428.
8. Geisler, A.; Morteza, S.; Morin, J.B.; Loucif, A.; Jahazi, M. Void closure during open die forging of large size martensitic stainless-steel ingots: An experimental-analytical-numerical study. *Int. J. Mater. Form.* **2023**, *16*, 11. [CrossRef]
9. Kakimoto, H.; Arikawa, T.; Takahashi, Y.; Tanaka, T.; Imaida, Y. Development of forging process design to close internal voids. *J. Mater. Process. Technol.* **2010**, *210*, 415–422. [CrossRef]
10. Chen, K.; Yang, Y.; Shao, G.; Liu, K. Strain function analysis method for void closure in the forging process of the large—sized steel ingot. *Comput. Master. Sci.* **2012**, *51*, 72–77. [CrossRef]
11. Niu, L.; Zhang, Q. A void closure model based on hydrostatic integration and the Lode parameter for additive manufacturing AlSi10Mg. *J. Manuf. Process.* **2022**, *73*, 235–247. [CrossRef]

12. Li, X.; Wu, X.C.; Zhang, Z. Analysis of the relationship between void direction and closure efficiency in forging process. *Adv. Mater. Res.* **2011**, *291–294*, 246–250. [CrossRef]
13. Saby, M.; Bouchard, P.O.; Bernacki, M. Void closure criteria for hot metal forming. *J. Manuf. Process.* **2015**, *19*, 239–250. [CrossRef]
14. Harris, N.; Shahriari, D.; Jahazi, M. Development of a fast converging material specific void closure model during ingot forging. *J. Manuf. Process.* **2017**, *26*, 131–141. [CrossRef]
15. Banaszek, G.; Bajor, T.; Kawałek, A.; Knapiński, M. Modeling of the Closure of Metallurgical Defects in the Magnesium Alloy Die Forging Process. *Materials* **2022**, *15*, 7465. [CrossRef] [PubMed]
16. Banaszek, G.; Ozhmegov, K.; Kawalek, A.; Sawicki, S.; Magzhanov, M.; Arbuz, A. Investigation of the influence of hot forging parameters on the closing conditions of internal metallurgical defects in zirconium alloy ingots. *Materials* **2023**, *16*, 1427. [CrossRef] [PubMed]
17. Dyja, H.; Kawałek, A.; Ozhmegov, K.V.; Sawicki, S. The thermomechanical conditions of open die forging of zirconium alloy ingots determined by rheological tests. *Metalurgija* **2020**, *59*, 39–42.
18. Dyja, H.; Gałkin, A.; Knapiński, M. *Reologia Metali Odkształcanych Plastycznie [The Rheology of Plastically Deformed Metals]*; Politechnika Częstochowska, seria Monografie Nr 190; Wydawnictwo Politechniki Częstochowskiej: Częstochowa, Poland, 2010; p. 371, ISBN 978-83-7193-471-1.
19. Henzel, A.; Shpittel, T. *Raschetjen Ergosilovyh Parametrov V Processah Obrabotki Metallov Davleniem: Spravochnik [Calculation of Energy—Power Parameters in Metal Forming Processes. Directory]*; Metallurgija: Moscow, Russia, 1982; p. 360.
20. Gerasimovich, A.S. *Approximation of Dependencies by Polynomials*; Ves' SergievPasad: SergievPosad, Russia, 2013; p. 104, ISBN 978-5-91582-043-1.
21. *Forge 2008 3D Forging Simulation Software: Rheology and Interfaces Module, FPDBase Version 1.3*; Transvalor S.A.: Sophia Antipolis, France, 2008.
22. Kawalek, A.; Rapalska-Nowakowska, J.; Dyja, H.; Koczurkiewicz, B. Physical and numerical modelling of heat treatment the precipitation—hardening complex—phase steel (CP). *Metalurgija* **2013**, *52*, 23–26.
23. Falencki, Z. *Analysis of Cast Defects*; AGH: Cracow, Poland, 1997; p. 141.
24. Elbel, T.; Kralova, Y.; Hampl, J. Expert System for Analysis of Casting Defects–ESVOD. *Arch. Foundry Eng.* **2015**, *15*, 17–20. [CrossRef]
25. Sika, R.; Rogalewicz, M.; Kroma, A.; Ignaszak, Z. Open Atlas of Defects as a Supporting Knowledge Base for Cast Iron Defects Analysis. *Arch. Foundry Eng.* **2020**, *20*, 55–60. [CrossRef]
26. Mehta, N.D.; Gohil, A.V.; Doshi, S.J. Innovative Support System for Casting Defect Analysis—A Need of Time. *Mater. Today Proc.* **2018**, *5*, 4156–4161.

Disclaimer/Publisher's Note: The statements, opinions and data contained in all publications are solely those of the individual author(s) and contributor(s) and not of MDPI and/or the editor(s). MDPI and/or the editor(s) disclaim responsibility for any injury to people or property resulting from any ideas, methods, instructions or products referred to in the content.

Article

The Optimized Homogenization Process of Cast 7Mo Super Austenitic Stainless Steel

Runze Zhang [1], Jinshan He [1,*], Shiguang Xu [1], Fucheng Zhang [2] and Xitao Wang [1,3,*]

1. Collaborative Innovation Center of Steel Technology, University of Science and Technology Beijing, Beijing 100083, China
2. State Key Laboratory of Metastable Materials Science and Technology, Yanshan University, Qinhuangdao 066004, China
3. Shandong Provincial Key Laboratory for High Strength Lightweight Metallic Materials, Advanced Materials Institute, Qilu University of Technology (Shandong Academy of Science), Jinan 250353, China
* Correspondence: hejinshan@ustb.edu.cn (J.H.); xtwang@ustb.edu.cn (X.W.)

Citation: Zhang, R.; He, J.; Xu, S.; Zhang, F.; Wang, X. The Optimized Homogenization Process of Cast 7Mo Super Austenitic Stainless Steel. *Materials* 2023, *16*, 3438. https://doi.org/10.3390/ma16093438

Academic Editors: Janusz Tomczak, Konrad Laber, Beata Leszczyńska-Madej, Grażyna Mrówka-Nowotnik and Magdalena Barbara Jabłońska

Received: 3 April 2023
Revised: 19 April 2023
Accepted: 21 April 2023
Published: 28 April 2023

Copyright: © 2023 by the authors. Licensee MDPI, Basel, Switzerland. This article is an open access article distributed under the terms and conditions of the Creative Commons Attribution (CC BY) license (https://creativecommons.org/licenses/by/4.0/).

Abstract: Super austenitic stainless steels are expected to replace expensive alloys in harsh environments due to their superior corrosion resistance and mechanical properties. However, the ultra-high alloy contents drive serious segregation in cast steels, where the σ phase is difficult to eliminate. In this study, the microstructural evolution of 7Mo super austenitic stainless steels under different homogenization methods was investigated. The results showed that after isothermal treatment for 30 h at 1250 °C, the σ phase in steels dissolved, while the remelting morphologies appeared at the phase boundaries. Therefore, the stepped solution heat treatment was further conducted to optimize the homogenized microstructure. The samples were heated up to 1220 °C, 1235 °C and 1250 °C with a slow heating rate, and held at these temperatures for 2 h, respectively. The elemental segregation was greatly reduced without incipient remelting and the σ phase was eventually reduced to less than 0.6%. A prolonged incubation below the dissolution temperature will lead to a spontaneous compositional adjustment of the eutectic σ phase, resulting in uphill diffusion of Cr and Mn, and reducing the homogenization efficiency of ISHT, which is avoided by SSHT. The hardness reduced from 228~236 Hv to 220~232 Hv by adopting the cooling process of "furnace cooling + water quench". In addition, the study noticed that increasing the Ce content or decreasing the Mn content can both refine the homogenized grain size and accelerate diffusion processes. This study provides a theoretical and experimental basis for the process and composition optimization of super austenitic stainless steels.

Keywords: super austenitic stainless steel; homogenization; Cerium; Manganese

1. Introduction

7Mo super austenitic stainless steel (SASS) is the highest-alloyed austenitic stainless steel with more than 50% alloying elements, including molybdenum and nickel. It possesses stronger corrosion resistance (PREN $\geq$ 40) and mechanical properties than common stainless steels [1]. Compared with nickel-based alloys, 7Mo SASS has significant advantages of cost-effectiveness and rivaled-performance, which can partially replace the application of expensive alloys in harsh environments [2]. However, the formation of the σ phase is difficult to control due to the strong tendency of elemental segregation. The σ phase is a brittle phase that arises as a eutectic phase in SASSs [3]. An excessive σ phase will cause cracking during processing [4,5] and reduce corrosion resistance [6]. In order to reduce the influence of the σ phase on hot processing and application, the σ phase can be reduced by regulating the alloy composition [3,7–9] or by adopting homogenization treatment [10]. However, the dissolution temperature of the σ phase increases with the increasing alloying contents in steels, to pose a challenge in the homogenization process [11].

The traditional homogenization process for steels is the isothermal solution heat treatment (ISHT), with the main processes of heating → holding → rapid cooling. In 304 stainless steel (18Cr-8Ni), homogenization can be achieved by holding the steel at 1050~1150 °C for 0.17 h [12,13]. The first-generation SASS 904 L (20Cr-23Ni-4Mo) needs to be held at 1100~1150 °C for 4~8 h. Until 654 SMO (25Cr-20Ni-7Mo), where the temperature is increased to 1280 °C and it need to be held for more than 30 h [14,15]. At the same time, as the Mo contents increase further, the dissolution temperature of the eutectic σ phase gradually increases, accompanied by the lower liquidus temperature of alloys [16,17]. Therefore, higher-alloy contents providing excellent overall performance pose a significant challenge to subsequent thermal processing.

The stepped solution heat treatment (SSHT) is widely used in superalloys, which can effectively increase the melting point of inter-dendrites by holding the steel slightly below the overburning temperature [18]. For this method, a slow heating rate is used as part of the homogenization [19]. Due to the significant differences between SASSs and common stainless steels, SASSs have similar characteristics with the superalloys, including: (1) high alloying contents (>50%) with serious element segregation; (2) the eutectic phase with high dissolution temperature; (3) the heat treatment window is extremely narrowed. For these reasons, optimizing the homogenization parameters of SASSs based on the typical SSHT is possible and has a prospect of application.

Alloying elements can also influence the homogenization efficiency by changing the as-cast microstructure. For example, alloying modifier can increase crystal defects to accelerate the atomic diffusion, while the addition of refractory elements with a slow diffusion rate will greatly increase the homogenization time [17]. Typical refiners in steels are rare earth elements, which have been confirmed as helpful elements in reducing the second phase [7,20–22]. The austenite-forming element Mn has also been studied in high-alloyed steels for reducing the element segregation tendency [7,23]. However, the Mn content in SASSs is controlled below 3 wt.% mostly, and the application of rare earth elements in SASSs is still in the initial stage. The effect of Ce and Mn elements on SASSs still needs to be explored.

In this study, in order to reduce the segregation and the σ phase, based on the 654SMO, we developed 7Mo SASSs by adding Ce elements and raising Mn elements. Thermodynamic calculations show the dissolution temperature of the σ phase in 7Mo SASSs over 1270 °C, while the remelting temperature is about 1250 °C. Therefore, it is very difficult to dissolve the σ phase completely. To develop a suitable homogenization process—by adopting ISHT and SSHT, exploring the evolution of the second phase and element segregations—we find that SSHT can shorten the homogenization time effectively and optimize the microstructure. The effect of Ce and Mn elements on the homogenization process was also explored. This study applies the SSHT innovatively to austenitic stainless steels to provide theoretical guidance for the production and development of 7Mo SASSs. It also establishes the foundation for the composition optimization of super austenitic stainless steels.

2. Materials and Methods

The 7Mo SASS ingots were all provided by Taiyuan Iron & Steel Group Co. LTD of China. Test steels for each composition were smelted in a vacuum induction furnace (VIM). The smelting process is shown in Figure 1: pouring 15% liquid steel into the VIM, feeding rare earth wire with full stirring, then pouring the remaining liquid steel and solidifying. The sampling location is shown in Figure 1d, from the edge to 2/3R of the center. The chemical composition of 7Mo SASSs in the experiments are shown in Table 1. The nitrogen content of the samples was determined by oxygen, nitrogen and hydrogen analyzers (TCH-600). The content of rare earth Ce in the steel was measured by inductively coupled plasma mass spectrometry (ICP-MS, TQ-ICP-MS)). Other alloying elements in the steel were detected by inductively coupled plasma emission spectrometry (ICP-OES, Optima 5300 DV).

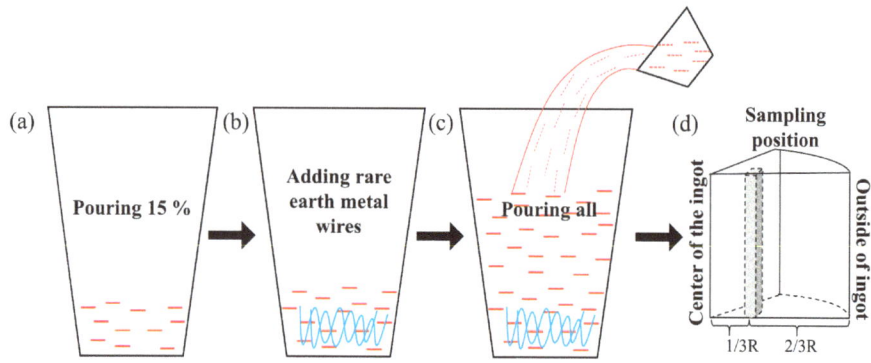

Figure 1. (**a–c**) 7Mo super austenitic stainless steel smelting process; (**d**) Sampling location.

Table 1. Chemical composition of 7Mo SASSs (wt.%).

	Cr	Mo	Ni	Mn	N	Cu	Si	Ce	Fe
6Mn-Ce	25.012	6.771	18.891	5.733	0.459	0.405	0.091	0.024	bal.
6Mn	25.182	6.892	18.819	6.028	0.418	0.412	0.108	0.001	bal.
3Mn	25.321	7.088	19.052	3.124	0.438	0.443	0.113	0.001	bal.

Each of the three samples was cut into 10 mm × 10 mm × 10 mm. The actual ingot we used is shown in Figure 2a. All homogenization processes were performed in a muffle furnace (KSL-1400X) with a refractory brick door as shown in Figure 2b. In order to prevent high-temperature oxidation, the samples were protected by argon gas at high temperature. To minimize the temperature fluctuation, B-type thermocouples were used to control the temperature.

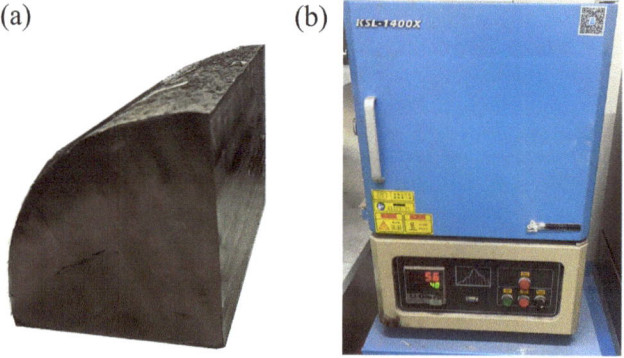

Figure 2. (**a**) the ingot for the experiment; (**b**) the muffle furnace (KSL-1400X).

The parameters of the ISHT and SSHT are given in Figure 3. For ISHT, the samples were homogenized at 1250 °C and held for 30 h, followed by water quench. For SSHT, the samples were heated at 4 K/min to 1200 °C with the furnace held for 2 h, followed by heating at 2 K/min to 1235 °C held for 2 h, then heated at 2 K/min to 1250 °C for 2 h. At each stage, the "furnace cooling + water quench" was used to obtain the homogenized samples.

Before observation, samples were polished with 50 #~2000 # sandpaper and mechanically polished with 1.5~2.5 diamond polishing paste. The distribution of the second phase was observed by field emission scanning electron microscopy (SEM, SM-7001F) with a backscattering probe (BSE). As shown in Figure 4, the σ phase in the polished sample appears bright-white under SEM-BSE, and the morphology of inter-dendrites can be observed at the same time. X-ray diffraction (XRD, Rigaku TTR3) was used for the identification of phases with a step of 0.02°, a scan time of 3 s per step, and a scan angle of 30°~120°. After the homogenization, the grain boundaries were observed and counted by an optical microscope (OM, Axio Imager M2m). The electrolytic erosion had parameters of 10% ethanol hydrochloric acid at 7 V for 40 s.

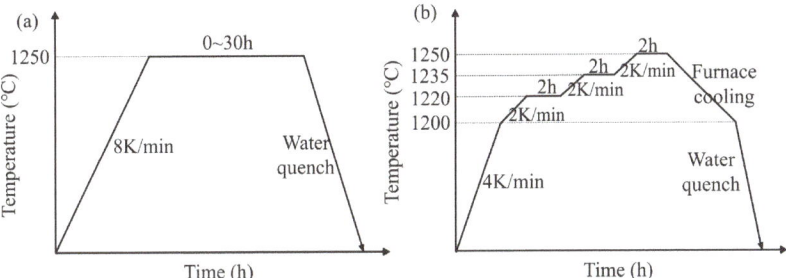

Figure 3. (a) ISHT parameters; (b) SSHT parameters.

The electron probe X-ray micro-analyzer (EPMA, model JXA-8530F) was used for measuring the distribution of elements and the samples were polished after grinding. Since the dendritic morphology after homogenization is inconspicuous, the element distribution can be determined using the dot-matrix method by EPMA [24], as shown in the Figure 4. The homogenized samples were observed at the same magnification as cast samples. A large number of points were selected within the area, and each point was detached and classified.

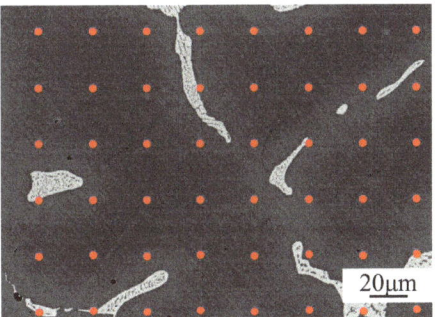

Figure 4. Schematic diagram of component distribution in dot-matrix method: a large number of points are evenly selected on the sample as shown by the red dots.

Vickers hardness (VTD512) was used to test the hardness of each sample. Before the test, they were grounded and polished. Twenty points were randomly selected in each sample center, and the load was set as 500 g and kept for 15 s. Finally, the average value of the tests were taken.

All thermodynamic calculations in this study were performed using the Thermocalc software 2023a with TCFE10 database. DICTRA dynamics calculation was used to calculate the distribution of elements at the inter-dendrite and the σ/γ phase interface during homogenization. The database was MOBFE4.

To solve the homogenization problem, the single-phase model was adopted. Because of the symmetry of the as-cast microstructure, the three-dimensional diffusion process of alloy was simplified into one dimensional problem by using the plate model. According to the average size of equiaxed crystal measured by experiments, the minimum diffusion unit is taken as 1/2 of the size of equiaxed crystal. Considering the accuracy and efficiency of the simulation, the diffusion element was divided into 50 nodes. The initial composition distribution was defined based on the results of the EPMA tests. Finally, the time and temperature conditions of the simulation homogenization were set.

In order to describe the distribution of elements in the front of the interphase boundary during the dissolution process, the plate model was used to divide two-phase regions. The minimum diffusion unit of the σ phase was 5 μm, and the minimum diffusion unit of austenite was 1/2 length of isometric crystal size of several steels. The initial composition distribution was defined according to the EPMA results. The σ phase diffusion element was divided into 10 nodes and the austenitic diffusion element into 50 nodes.

3. Results

3.1. As-Cast Microstructure

Figure 5 shows the microstructures of 6Mn-Ce (Figure 5a), 6Mn (Figure 5b) and 3Mn (Figure 5c) under SEM-BSE. The second phase appears bright-white in color with a coral-like morphology distributed at inter-dendrites, showing a typical eutectic morphology and a high-level element enrichment. The phases were analyzed by XRD as shown in Figure 5d–f. It can be seen that the as-cast sample consisted of the σ + γ phases, which is consistent with previous studies [25]. By comparing the XRD pattern of the three steels, there are multiple crystalline σ phase diffraction peaks in the 6Mn-Ce and 3Mn steels, which should be caused by the finer austenite dendrites and more dispersed σ phase. Meanwhile, the (111) peak of austenite in 6Mn-Ce was significantly wider, and this phenomenon is mainly due to the lattice distortion caused by the increased Mn and Ce content in the steel. The volume fractions of the σ phase and average grain size were counted in Figure 5g,h. The volume fractions of the σ phase in 6Mn-Ce, 6Mn and 3Mn steels were 5.9%, 10.5% and 12.0%, respectively. 6Mn-Ce, 6Mn -0.001Ce and 3Mn had an average isometric crystal size of 110.2 μm, 124.3 μm and 106.2 μm, respectively. It is obvious that increasing the Ce content or reducing the Mn content is beneficial for refining the microstructure.

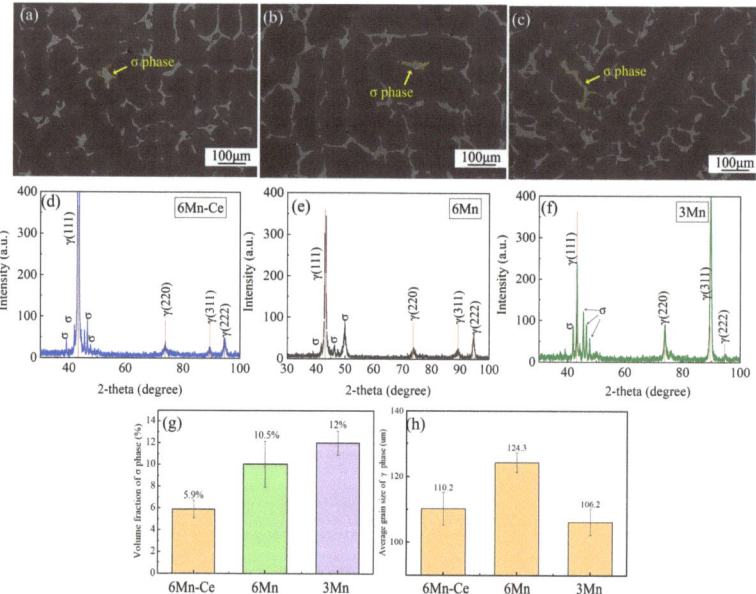

Figure 5. (**a**–**c**) the microstructure of 6Mn-Ce, 6Mn and 3Mn under SEM-BSE; (**d**–**f**) the XRD results of 6Mn-Ce, 6Mn and 3Mn; (**g**,**h**) summary of the volume fraction of σ phase and the average size of isotropic crystals.

3.2. Determination of the Heating Temperature

The phase transition temperature is crucial to guiding the heat treatment and thermal processing. The equilibrium phase diagrams of 6Mn-Ce (black line), 6Mn (red line) and 3Mn (blue line) obtained by the Thermo-calc thermodynamic calculation are shown in Figure 6. The predicted temperatures are concluded in Table 2. T_σ is the temperature at which the σ phase dissolved completely. T_δ and T_L are the lowest temperatures at which the δ phase and the liquid appeared, respectively. In both 6Mn-Ce and 6Mn, the σ phase dissolved at 1267 °C, while the δ phase formed at 1268 °C and remained stable until the liquid formed. It can be seen that for 6Mn SASSs with different Ce contents, the single-austenite region was only 1 °C, and the temperature difference between T_L and T_σ was only 59 °C. For 3Mn steel, the σ phase was completely dissolved at 1279 °C, and the δ phase appeared at 1285 °C until the liquid phase formed, where the single-austenite temperature range was also only 6 °C and the temperature window between T_L and T_σ was 59 °C. Therefore, increasing the Mn content narrows the single-austenite region of 7Mo SASSs, while Ce has little effect on it.

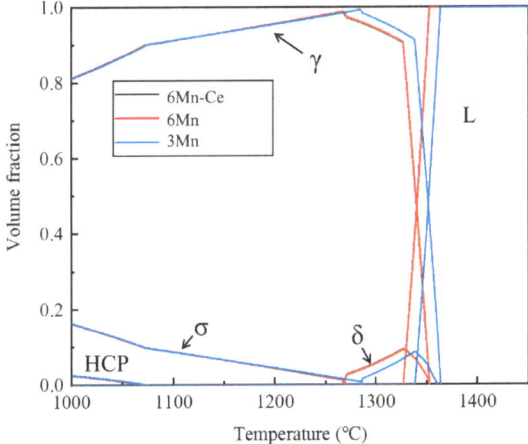

Figure 6. The equilibrium diagrams of 6Mn-Ce, 6Mn and 3Mn.

Table 2. Temperatures of completely dissolved σ phase (T_σ); temperatures at which the δ phase (T_δ) and the liquid (T_L) formed.

	T_σ (°C)	T_δ (°C)	T_L (°C)
6Mn-Ce	1267.48	1268.00	1326.21
6Mn	1267.82	1268.30	1326.18
3Mn	1279.10	1285.84	1338.00

The Gulliver-Scheil calculation [26] was further used to predict the melting point between dendrites based on the EPMA results, and is denoted as T_{LL} in Table 3. As can be noticed, the maximum temperature for the homogenization of 6Mn steel with different Ce content should be lower than 1250 °C, while the ultimate temperature for 3Mn steel was 1260 °C. According to Figure 6, the volume fractions of the σ phase at 1250 °C under equilibrium in 6Mn-Ce, 6Mn and 3Mn were 0.206%, 0.199% and 0.0215%, respectively. A study showed that the corrosion resistance of steels is hardly affected when the volume fraction of the σ phase is below 0.6% [27]. In order to facilitate the comparison of the effects of Ce and Mn, 1250 °C was chosen as the extreme temperature for the homogenization process.

Table 3. Calculation of the liquidus temperature (T_{LL}).

wt.%	Cr	Mo	Mn	Ni	Fe	Ce	T_{LL} (°C)
6Mn-Ce	25.66	6.62	6.42	16.16	42.00	0.02	1250.74
6Mn	25.05	8.55	6.43	17.01	39.69	0.001	1248.30
3Mn	25.94	7.51	3.68	17.84	41.68	0.001	1264.38

3.3. The Process of ISHT and SSHT

Figure 7a–c shows the microstructural evolution of 6Mn-C, 6Mn and 3Mn during the ISHT. According to the morphological observation, the σ phase in the three steels gradually transforms from a coral-like shape to a regular shape, and the dendritic morphology also disappears with the extension of time. After holding for 30 h, the σ phase in the three steels almost completely dissolved and equiaxed crystals formed. Further analysis of the microstructure at 1250 °C as shown in Figure 7d–f, which contains a large number of regular-shape holes, appeared at the interface of the σ phase. From Figure 7e, a ring exists around these holes, which is a typical remelting morphology. Figure 7f shows a remolten ball with a unique patterned surface in the homogenized sample. The appearance of these defects indicates that incipient remelting occurred after the ISHT [28].

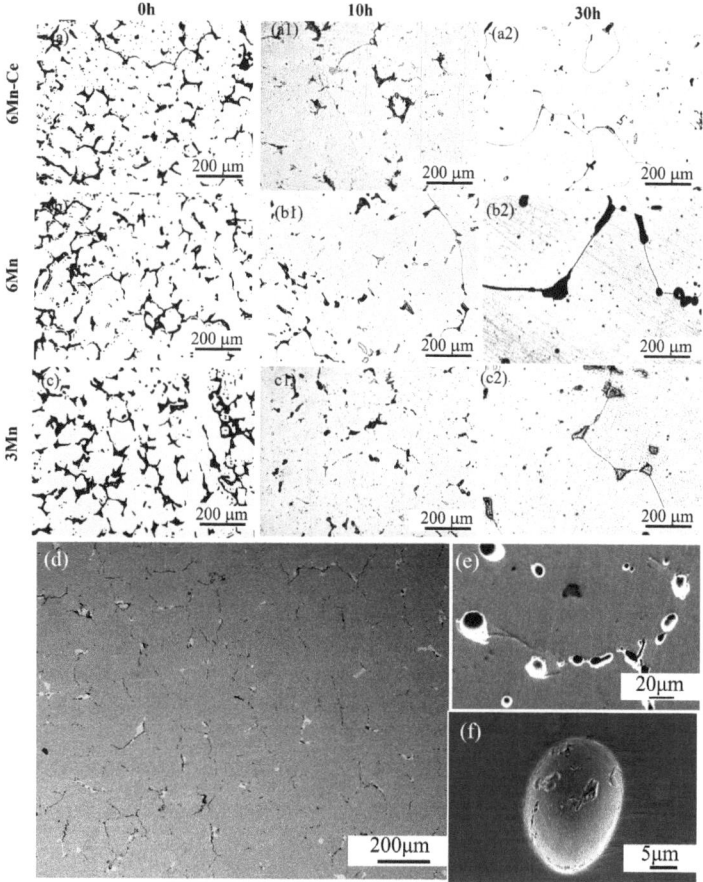

Figure 7. The microstructural evolution of (a–a2) 6Mn-Ce, (b–b2) 6Mn and (c–c2) 3Mn during ISHT; (d,e) remelting holes after ISHT; (f) the remolten ball.

The microstructure evolution during SSHT was subsequently observed, as shown in Figure 8. After the 1st-SSHT, a large number of the σ phase remained in all samples, and the size of the γ phase inside the coral-like σ phase expanded, as shown in Figure 8a1′–c1′. Until the completion of the 3rd-SSHT, the σ phase was almost dissolved in 6Mn-Ce and 3Mn steels. Further observation of the 6Mn steel after the 3rd-SSHT without remelting holes, as shown in Figure 8b3′, indicated that SSHT can effectively alleviate the remelting of the alloy.

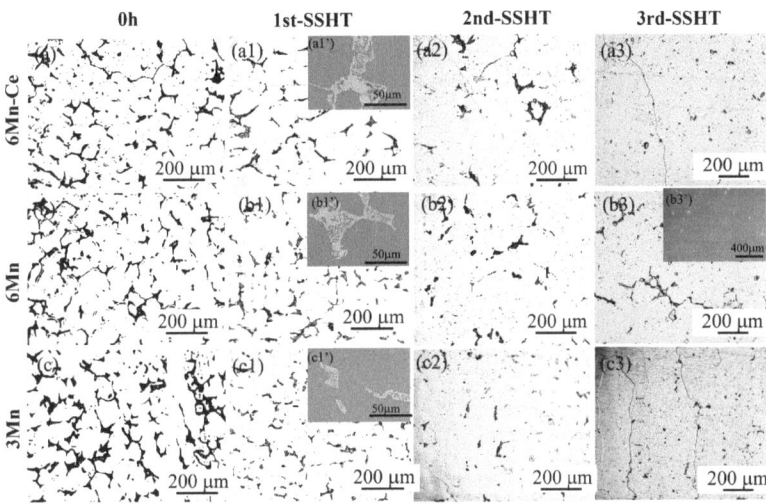

Figure 8. Microstructure evolution of steels during SSHT: (**a**–**a3**) 6Mn-Ce, (**b**–**b3**) 6Mn and (**c**–**c3**) 3Mn; (**a1′,b1′,c1′**) show the eutectic σ phase in three steels without remelting after 1 st-SSHT; (**b3′**) indicates that the 6Mn steel with the lowest melting point remained a small amount of σ phase without remelting after 3rd-SSHT.

3.4. Dissolution of σ Phase and Diffusion of Elements during ISHT and SSHT

Quantitative statistics were conducted based on the evolution of the σ phase during homogenization, as shown in Figure 9. After 7 h of homogenization (ISHT for 7 h, 1st-SSHT finished), the dissolution rate of the σ phase in 3Mn steels was faster in ISHT, with the residual σ phase about 2.03%, 4.31% and 5.82% in 6Mn-Ce, 6Mn and 3Mn, respectively; meanwhile, the residual σ phase in 6Mn-Ce, 6Mn and 3Mn were 3.56%, 4.48% and 4.74% after SSHT, respectively. After about 9.5 h of homogenization (9.5 h for ISHT and 2nd-SSHT end), 1.16%, 3.01% and 3.88% of the σ phase in 6Mn-Ce, 6Mn and 3Mn remained after ISHT, respectively. While after the 2nd-SSHT, 6Mn-Ce, 6Mn and 3Mn remained the volume fractions of the σ phase of 2.83%, 3.88% and 3.89%, respectively. After 12~13 h of homogenization (ISHT for 12~13 h, end of 3rd-SSHT), the σ phase in 6Mn-Ce, 6Mn and 3Mn was reduced to 0.51%, 2.23% and 2.89% after ISHT, respectively; and the σ phase in 6Mn-Ce, 6Mn and 3Mn was reduced to 0.26%, 2.40% and 0.55% after SSHT, respectively. Therefore, during the ISHT, the dissolution rate of the σ phase first accelerated and then gradually smoothed. While during the SSHT, the dissolution rate was slow at the beginning and then accelerated by higher temperatures. Since the corrosion resistance was almost unaffected when the σ phase was below 0.6% [27], both 6Mn-Ce and 3Mn achieved the homogenization effect under the SSHT.

The elemental distribution during homogenization is illustrated in Figure 10. The red curve represents elemental change during the ISHT, and the black curve is for the SSHT. Figure 10a–c is the 6Mn-Ce, 6Mn and 3Mn steels, respectively. The segregation coefficient is $K = K_L/K_S$, where K_L is the alloy contents of inter-dendrites and K_S is the element distribution at the dendrites. In general, the further the K value deviates from 1, the greater the segregation. When the 0.95 < K < 1.1, the homogenization is basically completed.

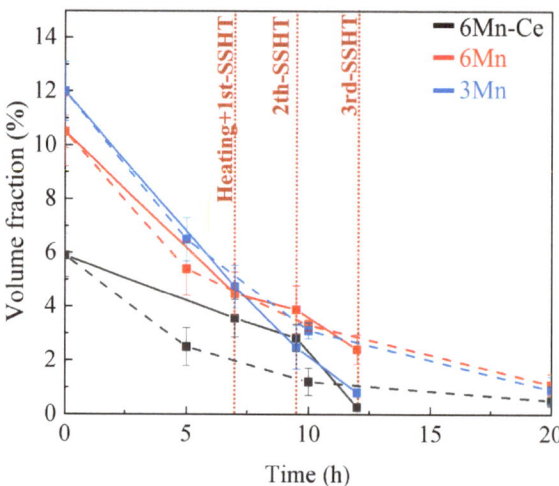

Figure 9. The volume fraction of σ phase under different homogenization processes: the dashed and solid lines are the changes of the volume fraction of σ phase during ISHT and SSHT processes, respectively.

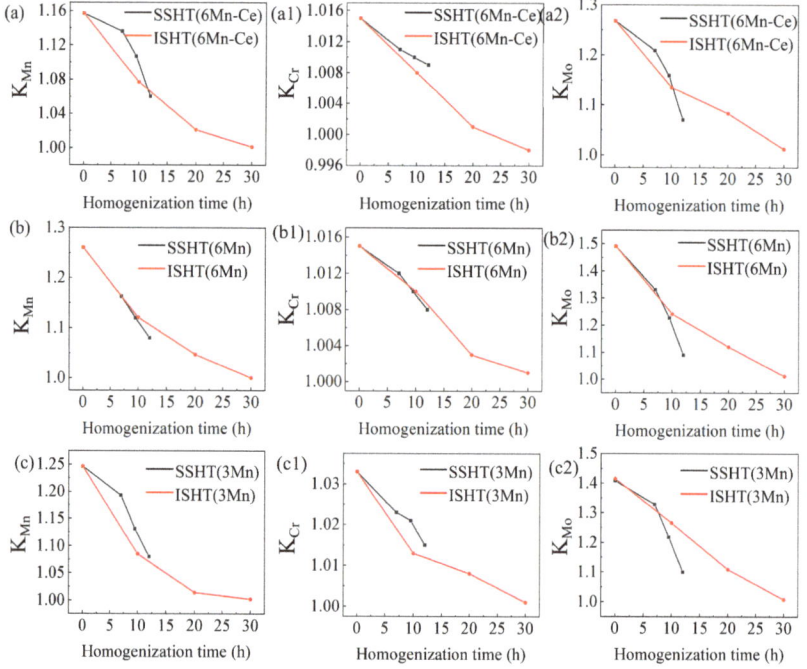

Figure 10. The K value of Mn, Cr and Mo in (**a–a2**) 6Mn-Ce, (**b–b2**) 6Mn and (**c–c2**) 3Mn under ISHT and SSHT.

In the ISHT, the K of each element gradually decreased with the extension of time, and the diffusion rates of Mn, Cr and Mo were faster in the early stage, followed by a smooth trend. Until 30 h, the K value of Cr, Mo and Mn elements reached K ≤ 1.02 which means that the homogenization of the three steels had been completed. For the SSHT, the homogenization rate was relatively slow in the early stage and accelerated when the temperature rises to 1235 °C. After the 3rd-SSHT, the K of all elements of 6Mn-Ce and 3Mn decreased significantly to K < 1.1, reaching the homogenization standard.

3.5. Hardness of the Homogenized Sample

Figure 11a shows the grain size of 6Mn-Ce, 6Mn and 3Mn after different homogenization processes. After adopting the SSHT, the average austenite grain size of 6Mn-Ce reduced from 1.21 mm to 1.14 mm; the average austenite grain size of 6Mn reduced from 1.65 mm to 1.45 mm; and 3Mn reduced from 1.32 mm to 1.10 mm. This is mainly due to the reduction in holding time at the high temperature. The hardness of the three steels after two homogenization processes was investigated using Vickers hardness, as shown in Figure 11b. It can be seen that using the SSHT can slightly reduce the microstructure hardness, mainly because the "furnace cooling + water quench" reduces the thermal stress rather than the water cooling directly. Due to the amounts of casting defects, such as element segregation and inclusions, furnace cooling was used to reduce the temperature difference, preventing the stress distortion caused by inclusions. On the other hand, the solute elements were fully diffused to avoid internal stress.

The effect of Ce and Mn on the homogenized steels can be noticed according to Figure 11; increasing the Ce content or decreasing the Mn content can reduce the grain size and increase the average hardness.

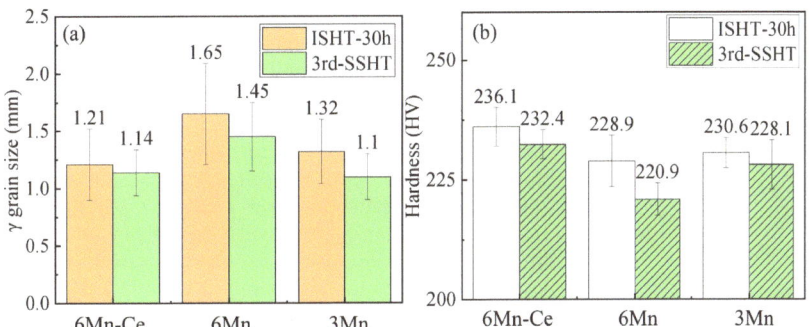

Figure 11. (**a**) the average grain size; and (**b**) the average hardness of 6Mn-Ce, 6Mn and 3Mn after ISHT and SSHT.

4. Discussion

By investigating the evolution of the σ phase and element segregation of cast 7Mo SASSs with different Ce and Mn contents during the ISHT and SSHT, the main conclusions are as follows: (1) the SSHT process improves the homogenization efficiency of 6Mn-Ce and 3Mn steels and optimizes the homogenized microstructure; (2) increasing the Ce content or decreasing the Mn content both contribute to a faster homogenization process and improve the properties of the homogenized microstructure. Therefore, the discussion will be further developed based on the above points.

Firstly, the advantages of SSHT are discussed. According to Figure 7d, most of the remelting holes exist at the phase interface. A previous study [29] showed that the dissolved atoms accumulate at the phase interface because of a fast heating rate, which leads to a lower liquidus temperature, triggering overburning. Regarding the remelting region as a semi-solid melting pool, the elements will be segregated at the phase interface as time prolongs [30,31]. At the same time, the liquidus temperature will be lower than 1250 °C, according to Table 3, when the atoms accumulate at phase interfaces. Once the overburning occurs, the σ phase will be formed again from the liquid pool "L + γ + σ" after cooling, making the σ phase hard to be dissolved at the final stage.

Moreover, DICTRA was used to calculate the elemental changes at the front of the σ phase interface during the isothermal process, as shown in Figure 12. In the three steels, enrichment of Cr and Mn elements appears at the front of the interface of the σ phase, as shown by the arrow in Figure 12. With the extension of holding time, Mn content in the σ phase decreases, while the Cr element accumulates in the σ phase. For the uphill diffusion within the eutectic phase, it is considered that due to the chemical composition of the eutectic phase deviating from the equilibrium, the elements in the phase will be adjusted spontaneously to lead to the uphill diffusion of alloying elements under long-term isothermal treatment and reduce the homogenization efficiency [32]. The variation of the σ phase in the equilibrium composition of 6Mn-Ce steel with temperature is shown in Figure 12g. In order to ensure the phase stability of the intermetallic compound, the σ phase was adjusted in proportion to the composition element content before reaching the melting point. The SSHT can exactly avoid the long-term isothermal process at a certain temperature and avoid the occurrence of uphill diffusion.

Therefore, for materials with a narrow heat treatment temperature range, the SSHT is more suitable. The slow heating rate can effectively reduce the temperature fluctuation of the furnace and increase the diffusion coefficient of the atoms, which accelerates the diffusion of elements that are produced by the dissolved σ phase [19]. The application of graded insulation is intended to increase the liquidus temperature step by step. This method is also verified in CMSX-10 [33], where the graded insulation helps to increase the liquidus temperature by 25 °C.

The effects of Ce and Mn elements on the homogenization process can be obtained from Figures 7 and 8. According to the slope of the dissolution curve of the σ phase in 6Mn-Ce, 6Mn and 3Mn, the dissolution rate of the σ phase obeys the relationship of 6Mn-Ce > 3Mn > 6Mn, which shows the dissolution of the σ phase is accelerated by increasing the Ce or Mn content. A study denoted that decreasing dendrite spacing or reducing the initial concentration can both accelerate the diffusion process [34]. The addition of Ce promotes the formation of rare-earth inclusions in SASSs [3,8]. These inclusions can refine the dendrites to reduce the second phase, which also can play a role in pinning grain boundaries at high temperatures to increase the average hardness. However, excess Mn contents will participate in the formation of Ce-O-Mn inclusions with little effect on heterogeneous nucleation, resulting in coarse dendrites [7], leading to a lower dissolution rate. Meanwhile, according to Figure 8, the dissolution rate of Cr and Mo elements slows down at the initial stage of homogenization when the Mn content is increased from 3 wt.% to 5 wt.%. According to the previous study, the Mn element has the effect of promoting the dendritic segregation of Cr and Mo elements, which is the main reason for the lower dissolution rate.

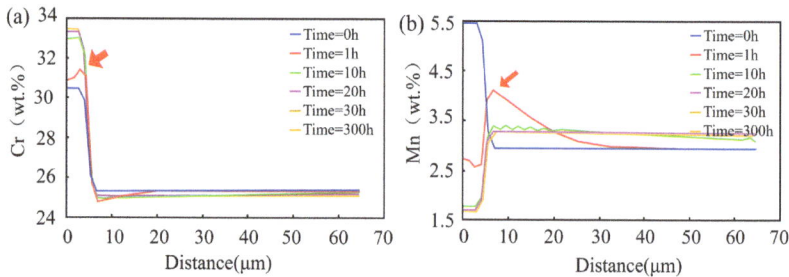

Figure 12. Cont.

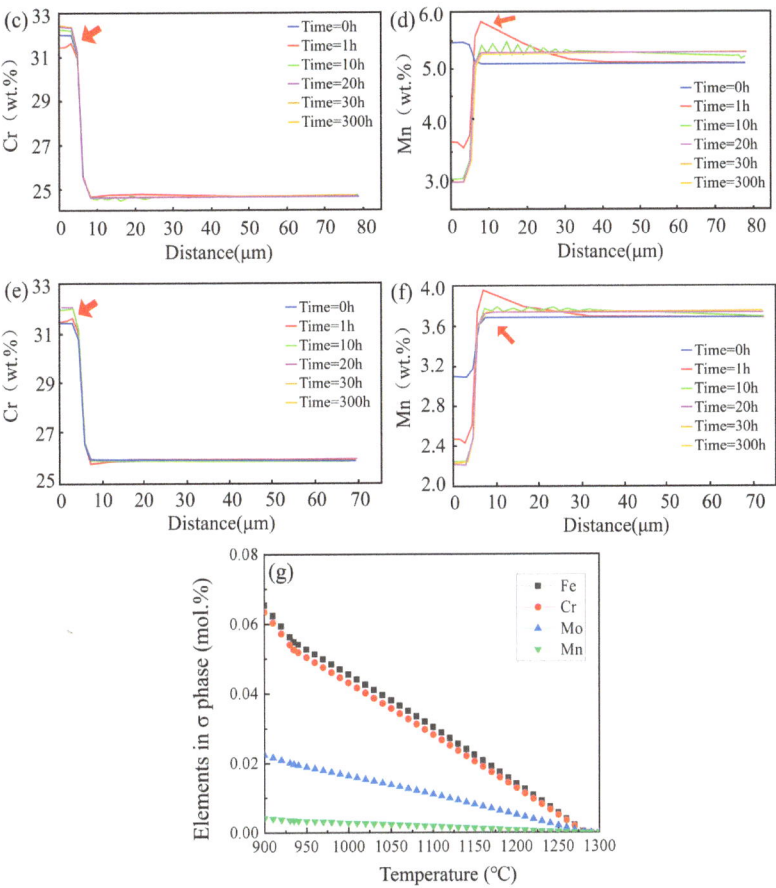

Figure 12. Changes in element distribution at the front of the interface of σ phase during isothermal homogenization: (**a,b**) 6Mn-Ce; (**c,d**) 6Mn; (**e,f**) 3Mn, after 1h ISHT, all elements are enriched at the phase interface, as shown by the arrow; (**g**) variation of σ phase equilibrium composition with temperature in 6Mn-Ce steel.

5. Conclusions

In this study, the microstructural evolution of 7Mo super austenitic stainless steels with different Ce and Mn contents during homogenization were investigated, and the effect of alloy elements and homogenization parameters on the dissolution process were explored by studying dissolution kinetics and hardness, with the following results:

(1) Increasing the Ce element can refine the cast microstructure and reduce the σ phase from 10.5% to 5.9%, which speeds up the homogenization process and increases the average hardness of the homogenized microstructure.

(2) Raising the Mn element promotes severe element segregation and coarse dendrites in the cast microstructures, which slows down the dissolution rate of atoms. Homogenized grains are also enlarged, with smaller hardness after the Mn contents increase by 2 wt.%.

(3) The stepped solution heat treatment can make the cast 7Mo SASS meet the homogenization standard efficiently, shorten the homogenization time from 30 h to 12~13 h, as well as avoid the incipient remelting.

Author Contributions: Conceptualization, J.H., X.W. and F.Z.; methodology, R.Z., S.X. and F.Z.; formal analysis, R.Z.; investigation, R.Z. and S.X.; writing—original draft, R.Z.; writing—review and editing, R.Z., J.H. and X.W.; supervision, J.H. and X.W. All authors have read and agreed to the published version of the manuscript.

Funding: This work was supported by the National Natural Science Foundation of China (No. U1810207) and the Innovation Pilot Project for the Fusion of Science, Education and Industry (International Cooperation) from the Qilu University of Technology (No. 2020KJC-GH03).

Institutional Review Board Statement: Not applicable.

Informed Consent Statement: Not applicable.

Data Availability Statement: All data are reported in the article.

Conflicts of Interest: The authors declare no conflict of interest.

References

1. Wallen, B.; Liljas, M.; Stenvall, P. AVESTA 654SMO™–A new nitrogen-enhanced superaustenitic stainless steel. *Mater. Corros.* **1993**, *44*, 83–88. [CrossRef]
2. Olsson, J.; Wasielewska, W. Applications and experience with a Superaustenitic 7Mo stainless steel in hostile environments. *Mater. Corrosion* **1997**, *48*, 791–798. [CrossRef]
3. Zhang, S.; Yu, J.; Li, H.; Jiang, Z.; Geng, Y.; Feng, H.; Zhang, B.; Zhu, H. Refinement mechanism of cerium addition on solidification structure and sigma phase of super austenitic stainless steel S32654. *J. Mater. Sci. Technol.* **2021**, *102*, 105–114. [CrossRef]
4. Kalandyk, E.B.; Zapala, E.R.; Palka, P. Effect of Isothermal Holding at 750 degrees C and 900 degrees C on Microstructure and Properties of Cast Duplex Stainless Steel Containing 24% Cr-5% Ni-2.5% Mo-2.5% Cu. *Materials* **2022**, *15*, 8569. [CrossRef] [PubMed]
5. Wang, H.; Wang, A.; Li, C.; Yu, X.; Xie, J.; Liu, C. Effect of Secondary-Phase Precipitation on Mechanical Properties and Corrosion Resistance of 00Cr27Ni7Mo5N Hyper-Duplex Stainless Steel during Solution Treatment. *Materials* **2022**, *15*, 7533. [CrossRef]
6. Hosseini, V.A.; Karlsson, L.; Wessman, S.; Fuertes, N. Effect of Sigma Phase Morphology on the Degradation of Properties in a Super Duplex Stainless Steel. *Materials* **2018**, *11*, 933. [CrossRef] [PubMed]
7. Zhang, R.; He, J.; Xu, S.; Zhang, F.; Wang, X. The roles of Ce and Mn on solidification behaviors and mechanical properties of 7Mo super austenitic stainless steel. *J. Mater. Res. Technol.* **2023**, *22*, 1238–1249. [CrossRef]
8. Wang, Q.; Wang, L.; Zhang, W.; Li, J.; Chou, K. Effect of cerium on the austenitic nucleation and growth of high-Mo austenitic stainless steel. *Metall. Mater. Trans. B* **2020**, *51B*, 1773–1783. [CrossRef]
9. Yu, J.; Zhang, S.; Li, H.; Jiang, Z.; Feng, H.; Xu, P.; Han, P. Influence mechanism of boron segregation on the microstructure evolution and hot ductility of super austenitic stainless steel S32654. *J. Mater. Sci. Technol.* **2022**, *112*, 184–194. [CrossRef]
10. Biezma, V.M.; Martin, U.; Linhardt, P.; Ress, J.; Rodríguez, C.M.; Bastidas, D. Non-destructive techniques for the detection of sigma phase in duplex stainless steel: A comprehensive review. *Eng. Fail. Anal.* **2021**, *122*, 105227. [CrossRef]
11. Hsieh, C.C.; Wu, W. Overview of intermetallic sigma phase precipitation in stainless steels. *ISRN Metall.* **2012**, *2012*, 1–16. [CrossRef]
12. Nasir, N.I. The effect of heat treatment on the mechanical properties of stainless steel type 304. *Int. J. Sci. Eng. Res.* **2015**, *3*, 87–93.
13. Vermilyea, D.; Tedmon, C.; Broecker, D.E. Some effects of heat treatment variables on the sensitization of type 304 stainless steel. *Corrosion* **1975**, *31*, 140–142. [CrossRef]
14. Gao, J.; Fan, S.; Zhang, S.; Jiang, Z.; Li, H. Segregation behavior and homogenization process of novel super austenitic stainless steel 654SMO. *Iron Steel* **2018**, *53*, 83–89.
15. Li, Z.; Wei, Z.; Qi, W. Hot deformation behavior of super austenitic stainless steel. *Iron Steel* **2017**, *52*, 72.
16. Su, X.; Xu, Q.; Wang, R.; Xu, Z.; Liu, S.; Liu, B. Microstructural evolution and compositional homogenization of a low Re-bearing Ni-based single crystal superalloy during through progression of heat treatment. *Mater. Des.* **2018**, *141*, 296–322. [CrossRef]
17. Liu, J.L.; Meng, J.; Yu, J.; Zhou, Y.; Sun, X.F. Influence of solidification conditions and alloying elements Re and Ti on micropores formed during homogenization of Ni base single crystal superalloy. *J. Alloys Compd.* **2018**, *746*, 428–434. [CrossRef]
18. Zhang, Y.; Liu, L.; Huang, T.; Yue, Q.; Sun, D.; Zhang, J.; Yang, W.; Su, H.; Fu, H. Investigation on a ramp solution heat treatment for a third generation nickel-based single crystal superalloy. *J. Alloys Compd.* **2017**, *723*, 922–929. [CrossRef]
19. Tancret, F. Thermo-Calc and Dictra simulation of constitutional liquation of gamma prime (γ') during welding of Ni base superalloys. *Comput. Mater. Sci.* **2007**, *41*, 13–19. [CrossRef]
20. Yang, Q.; Cui, Z.; Liao, B.; Yao, M. Effect of rare earth elements on growth dynamics of austenite in 60CrMnMo steel. *J. Rare Earths* **1999**, *17*, 46–48.
21. Lou, D.; Cui, K.; Grong, O.; Akselsen, O.M. *Effect of Rare Earth Metals and Calcium in Solid Phase Transformation Microstructure of Low Alloy Steels*; International Symposium on Microslloyed Steels: Columbus, OH, USA, 2002.
22. Guo, F.; Zheng, C.; Wang, P.; Li, D.; Li, Y. Effects of Rare Earth on Austenite-Ferrite Phase Transformation in a Low-Carbon Fe-C Alloy. *Acta Metall. Sin.* **2023**, *36*, 6. [CrossRef]

23. Raghavan, V. Effect of manganese on the stability of austenite in Fe-Cr-Ni alloys. *Metall. &Mater. Trans. A* **1995**, *26*, 237–242.
24. Flemgins, M.C.; Barone, R.V.; Brody, H.D. Microsegregation in iron-base alloys. *J. Iron Steel Inst.* **1970**, *208*, 371–380.
25. Phillips, N.S.L.; Chumbley, L.S.; Gleeson, B. Phase transformations in cast superaustenitic stainless steels. *J. Mater. Eng. Perform.* **2009**, *18*, 1285–1293. [CrossRef]
26. Sundman, O.B.; Ansara, I. *The Gulliver–Scheil Method for the Calculation of Solidification Paths*; Woodhead Publishing: Sawston, UK, 2008.
27. Kurosu, S.; Nomura, N.; Chiba, A. Effect of σ phase in Co-29Cr-6Mo alloy on corrosion behavior in saline solution. *Mater. Trans.* **2006**, *47*, 1961–1964. [CrossRef]
28. Andilab, B.; Ravindran, C.; Dogan, N.; Lombardi, A.; Byczynski, G. In-situ analysis of incipient melting of Al_2Cu in a novel high strength Al-Cu casting alloy using laser scanning confocal microscopy. *Mater. Charact.* **2020**, *159*, 110064. [CrossRef]
29. Lombardi, A.; Mu, W.; Ravindran, C.; Dogan, N.; Barati, M. Influence of Al_2Cu morphology on the incipient melting characteristics in B206 Al alloy. *J. Alloy. Compd.* **2018**, *747*, 131–139. [CrossRef]
30. Yan, Y.; Shan, W.; Hao, Y.; Li, Y. Microstructure and composition evolution of magnesium alloy MB15 during semi-solid isothermal heat treatment. *Mater. Sci. Technol.* **2005**, *49*, 652–655.
31. Chen, T.J.; Ma, Y.; Hao, Y.; Lu, S.; Sun, J. Structural evolution of ZA27 alloy during semi-solid isothermal heat treatment. *Trans. Nonferrous Met. Soc. China* **2001**, *1*, 98–102.
32. Hegde, S.R.; Kearsey, R.M.; Beddoes, J.C. Designing homogenization–solution heat treatments for single crystal superalloys. *Mater. Sci. Eng. A* **2010**, *527*, 5528–5538. [CrossRef]
33. Fuchs, G.E. Solution heat treatment response of a third generation single crystal Ni-base superalloy. *Mater. Sci. Eng. A* **2001**, *300*, 52–60. [CrossRef]
34. Cardona, M.; Fulde, P.; Klitzing, K.v.; Merlin, R.; Queisser, H.J.; Stormer, H. *Diffusion in Solids*; Springer Series in Solid-State Sciences; Springer: Berlin/Heidelberg, Germany, 2007.

Disclaimer/Publisher's Note: The statements, opinions and data contained in all publications are solely those of the individual author(s) and contributor(s) and not of MDPI and/or the editor(s). MDPI and/or the editor(s) disclaim responsibility for any injury to people or property resulting from any ideas, methods, instructions or products referred to in the content.

Article

Innovative Methodology for Physical Modelling of Multi-Pass Wire Rod Rolling with the Use of a Variable Strain Scheme

Konrad Błażej Laber

Department of Metallurgy and Metal Technology, Faculty of Production Engineering and Materials Technology, Czestochowa University of Technology, 19 Armii Krajowej Ave., 42-200 Czestochowa, Poland; konrad.laber@pcz.pl; Tel.: +48-34-325-07-97

Abstract: This paper presents the results of physical modelling of the process of multi-pass rolling of a wire rod with controlled, multi-stage cooling. The main goal of this study was to verify the possibility of using a torsion plastometer, which allows conducting tests on multi-sequence torsion, tensile, compression and in the so-called complex strain state to physically replicate the actual technological process. The advantage of the research methodology proposed in this paper in relation to work published so far, is its ability to replicate the entire deformation cycle while precisely preserving the temperature of the deformed material during individual stages of the reproduced technological process and its ability to quickly and accurately determine selected mechanical properties during a static tensile test. Changes in the most important parameters of the process (strain, strain rate, temperature, and yield stress) were analyzed for each variant. After physical modelling, the material was subjected to metallographic and hardness tests. Then, on the basis of mathematical models and using measurements of the average grain size, chemical composition, and hardness, the yield strength, ultimate tensile strength, and plasticity reserve were determined. The scope of the tests also included determining selected mechanical properties during a static tensile test. The obtained results were verified by comparing to results obtained under industrial conditions. The best variant was a variant consisting of physically replicating the rolling process in a bar rolling mill as multi-sequence non-free torsion; the rolling process in an NTM block (no twist mill) as non-free continuous torsion, with the total strain equal to the actual strain occurring at this stage of the technological process; and the rolling process in an RSM block (reducing and sizing mill) as tension, while maintaining the total strain value in this block. The differences between the most important mechanical parameters determined during a static tensile test of a wire rod under industrial conditions and the material after physical modelling were 1.5% for yield strength, approximately 6.1% for ultimate tensile strength, and approximately 4.1% for the relative reduction of the area in the fracture and plasticity reserve.

Keywords: physical modelling; wire rod rolling; variable strain state; hot torsion test; metallographic tests; mechanical properties; cold upsetting steel

Citation: Laber, K.B. Innovative Methodology for Physical Modelling of Multi-Pass Wire Rod Rolling with the Use of a Variable Strain Scheme. *Materials* **2023**, *16*, 578. https://doi.org/10.3390/ma16020578

Academic Editor: Chih-Chun Hsieh

Received: 4 December 2022
Revised: 20 December 2022
Accepted: 3 January 2023
Published: 6 January 2023

Copyright: © 2023 by the author. Licensee MDPI, Basel, Switzerland. This article is an open access article distributed under the terms and conditions of the Creative Commons Attribution (CC BY) license (https://creativecommons.org/licenses/by/4.0/).

1. Introduction

The wire rod rolling process in modern rolling mills is characterized by high dynamics with the linear speeds of the rolled band reaching values of 120 m/s or more [1]. Such a high speed of the rolled band combined with short intervals between individual deformations results in the processes occurring in the material itself also becoming dynamic. The strain rate of the material often exceeds the value of 2000 s^{-1}, which significantly hinders the physical modelling of such dynamic processes with the use of available research apparatus. Providing a required strain value, strain rate, and temperature during physical modelling that is similar to those occurring during the actual process ensures the high accuracy of obtained results [2]. These parameters have a direct impact on the shape and nature of the changes in the yield stress of the tested material, and thus the shaping of the microstructure and properties of the finished product [3,4]. In a situation where it is impossible to ensure any of the above-mentioned parameters are at the required level, one solution may be to adopt certain simplification assumptions. During the following stage, it should be verified whether such assumptions can be applied and how this affects the final result. Using certain simplifying assumptions depends on, among others, the specificity of the analyzed process and the tested material [2].

The most commonly used methods for the physical modelling of rolling processes are the compression test [5–9] and the torsion test [10–14]. These methods, despite the dynamic development of the laboratory equipment, have some limitations.

Most of the work concerning the physical modelling of the rolling processes published so far focuses on the assumption that the microstructure and properties of the final product are mainly determined by the few final passes, and the impact of the earlier stages of the technological process is less important [15]. As shown by the results of the research presented, inter alia, in the works [16–21], such a simplification is acceptable. Therefore, physical modelling studies are usually carried out only for the final stage of the analyzed technological process, often during compression tests by using the GLEEBLE metallurgical processes simulator [16–21]. The biggest advantage of this research methodology consists of the possibility of obtaining large values of strain rate, which significantly affects the level of yield stress. On the other hand, the biggest disadvantage is the limited value of the total strain value (about 1.2) and sometimes the limitations in the scope of precisely controlling the process of accelerated (controlled) cooling, which is important in the case of modelling processes which include the so-called multi-stage interoperational cooling. Moreover, after physical modelling, the material often cannot be used to determine mechanical properties directly during a tensile test, and determining the mechanical properties takes place indirectly based on measuring the grain size, hardness, and chemical composition [17,20,21], which is time-consuming and may include some error.

However, during physical modelling of the rolling process during a torsion test using a torsion plastometer, a limitation may be the low value of the strain rate. Apart from the standard free or non-free torsion tests, modern torsion plastometers enable testing in the course of multi-sequence torsion (including alternating), tensile, and compression tests, and in the so-called complex strain state test (simultaneous torsion with compression or simultaneous torsion with tensile tests). Furthermore, they enable precise temperature control during individual stages of the reproduced technological process, and a sole torsion test allows for much greater deformation values than a compression test.

Most of the works published so far on the physical modelling of real technological processes in torsion tests refer to the rolling of sheet metal bars [12,22–25], stepped shafts [26], or service pipes [10]. Therefore, it is reasonable to conduct research on solving problems related to the physical modelling of wire rod rolling processes in modern rolling mills, which are characterized by high linear velocities of the rolled strand with the use of modern torsion plastometers, enabling testing of the variable strain state.

This paper presents the results of physical modelling of the process of multi-pass wire rod rolling with controlled, multi-stage cooling. The main objective of the study was to verify the possibility of using a torsion plastometer, enabling testing with the use of a variable strain scheme to physically reproduce the technological process of rolling a wire rod.

The advantage of the research methodology proposed in the paper in reference to the works published so far is the possibility of reproducing the entire strain cycle (with a total strain value of 14.32), precisely preserving the temperature value of the deformed material during individual stages of the reproduced technological process, and the possibility of quickly and accurately determining selected mechanical properties during a static tensile test directly from the material after physical modelling.

The tests presented in the paper were carried out on several variants with the use of the STD 812 torsion plastometer [27], which, in addition to standard free or non-free torsion tests, allows conducting tests during multi-sequence torsion (also alternating), tensile, and compression tests, and also in the so-called complex strain state test (simultaneous torsion with compression or simultaneous torsion with tensile test).

For each variant, changes in the most important process parameters (strain, strain rate, temperature, and yield stress) were analyzed in detail. After physical modelling, each time the material was subjected to metallographic tests at characteristic points in order to assess the microstructure of the tested steel and the average ferrite grain size. In addition, Vickers hardness tests were also carried out. Then, based on the available mathematical models and using the measurements of the average ferrite grain size, chemical composition, and hardness [28,29], the yield strength, ultimate tensile strength, and plasticity reserve were determined. The scope of research carried out as part of the work also included determining selected mechanical properties directly in the static tensile test. The obtained results were verified under industrial conditions.

Basing on the obtained research results, it has been determined that the best variant is a variant consisting of physically reproducing the rolling process in a bar rolling mill as multi-sequence non-free torsion; the rolling process in an NTM block (No-Twist Mill) as non-free continuous torsion, with the total strain equal to the actual strain occurring at this stage of the technological process; and the physical modelling of the rolling process in an RSM block (reducing and sizing mill) as a tension, while maintaining the total strain value in this block. The differences between the most important mechanical parameters determined during a wire rod static tensile test under industrial conditions and during tensile tests of material after physical modelling were 1.5% for the yield strength, approximately 6.1% for the ultimate tensile strength, and approximately 4.1% for the relative reduction of area at fracture and the plasticity reserve.

According to the authors, such high accuracy between the research results obtained under industrial and experimental conditions results mainly from inducing a high level of yield stress during the final stage of physical modelling of the analyzed process (RSM block), by changing the deformation state from torsion to tension. The level of yield stress obtained in such a manner is close to the value of the yield stress occurring in the actual technological process, which is crucial for activating the microstructure rebuilding processes.

2. Materials and Methods

2.1. Materials

The research presented in the paper was carried out for low-carbon cold upsetting steel of the 20MnB4 grade, with a chemical composition in accordance with PN-EN 10263-4:2004 (Table 1) [17,21,30].

Table 1. Chemical composition of 20MnB4 steel [17,21,30].

Steel Grade	Steel Number	Melt Analysis, Mass%						
		C	Si	Mn	P_{max}, S_{max}	Cr	Cu_{max}	B
20MnB4	1.5525	0.18–0.23	≤0.30	0.90–1.20	0.025	≤0.30	0.25	0.0008–0.005

2.2. Methods

Physical modelling tests were carried out for the entire production cycle of rolling a wire rod with a final diameter of 5.5 mm, for an exemplary combined-type rolling mill (a combination of a bar rolling mill and a wire rod rolling mill) (Figure 1). The rolling process in a continuous rolling mill took place over 17 passes, while rolling in a wire rod rolling mill took place in 2 blocks: a No-Twist Mill (NTM) 10-rolling block, and a reducing and sizing mill (RSM) 4-rolling block.

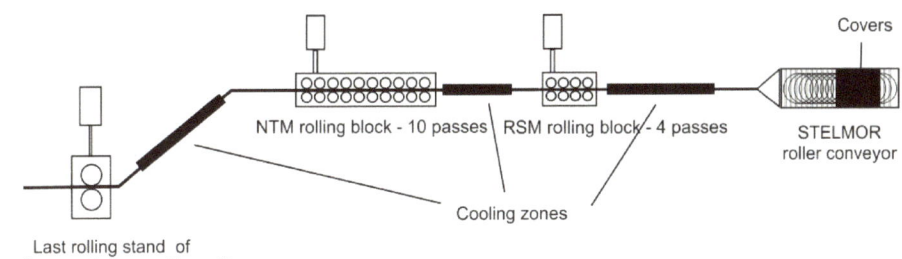

Figure 1. General scheme of the analyzed combined-type rolling mill [17]. In order to obtain a finished product with an even fine-grained ferritic–perlithic microstructure without a clear band structure, the final stage of deformation should take place in the austenitic range, when its temperature is 30–80 °C higher than the initial temperature of the austenite transformation, Ar_3 [31–35]. For 20MnB4 steel, the Ar_3 temperature is 780 °C. In the case of low-carbon and low-alloy steels, which are intended for further cold plastic processing, the most advantageous temperature for forming coils is a temperature of about 850–900 °C. Such a method of laying the coils provides an increased plasticity of the metal, beneficial for the cold drawing process, and allows a decrease in the recrystallizing annealing time after the drawing process [32].

However, during tests it is necessary to take into consideration the temperature increase caused by the deformation of the material in the RSM block with a high strain rate, which for the tested steel grade was about 50 °C. Therefore, during the tests, the temperature of the rolled band in the RSM block was 850 °C. After the rolling process, the material was cooled in two stages. During the first stage of the controlled cooling process, the 20MnB4 steel was cooled from the rolling end temperature to 500 °C at a cooling rate of approx. 10 °C/s, while during the second stage of cooling, the tested material was cooled to 200 °C at a cooling rate of 1 °C/s.

Several different tests were carried out in this paper with the use of the STD 812 torsion plastometer [27], which, in addition to standard free or non-free torsion tests, allows conducting tests during multi-sequence torsion (also alternating), tensile, and compression tests, and in the so-called complex strain state test (simultaneous torsion with compression or simultaneous torsion with tensile test). For the tests, circular samples with dimensions of diameter, d, of 9 mm and length, l, of 2 mm were used (Figure 2). An S-type thermocouple (PtRh10-Pt) welded to the side surface of the sample was used to register and control changes in temperature. The general view of the test chamber and the main parameters of the device are presented in Figure 3.

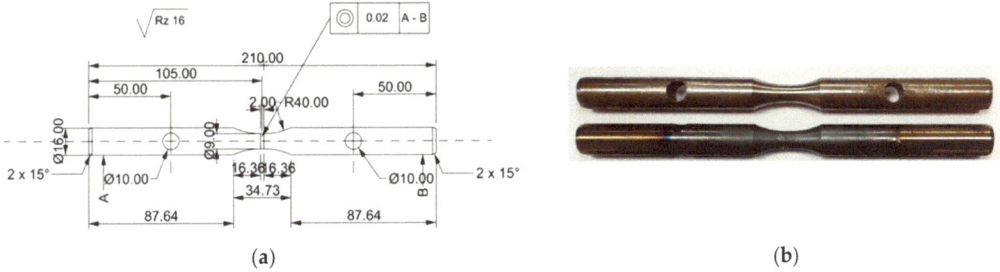

Figure 2. Samples for physical modelling tests: (**a**) technical specification and (**b**) general view of 20MnB4 steel samples before and after physical modelling.

Heating:	induction
Maximum testing temperature:	1500 °C
Heating and cooling rate:	up to 100 K/s
Minimum time between strains:	60 ms
Medium: Vacuum 10^{-4} mbar, neutral gas, air	
Torsion	
Rotational velocity:	up to 500 rpm
Number of rotations:	up to 30
Torque:	up to 50 Nm
Strain rate:	up to 60 s^{-1}
Tension and compression	
Change in length:	ca. 15 mm
Strain rate:	up to 30 mm/s
Strain force:	up to 25kN
Strain rate:	up to 1.0 s^{-1}
True strain: dependent on specimen dimensions	
The device allows for torsion with tension or compression	

(a)   (b)

Figure 3. STD 812 torsion plastometer: (**a**) device chamber: 1—specimen, 2—holders, 3—thermocouples type S, 4—induction solenoid, 5—cooling system jets, 6—pyrometer, 7—sensors for laser measurement of specimen diameter and (**b**) basic specification [2,27].

In order to determine the actual strain value, the relation (1) was used, the actual strain rate was determined based on the relation (2), while the yield stress was calculated according to the Formula (3) [36,37]:

$$\varepsilon = \frac{2 \cdot \pi \cdot r \cdot N}{\sqrt{3} \cdot L} \quad (1)$$

$$\dot{\varepsilon} = \frac{2 \cdot \pi \cdot r \cdot \dot{N}}{\sqrt{3} \cdot 60 \cdot L} \quad (2)$$

$$\sigma_p = \frac{\sqrt{3} \cdot 3M}{2\pi r^3} \quad (3)$$

where r is the sample radius, L is the sample length, N is the number of sample twists (revolutions), $\dot{N}$ is the torsion speed (rpm), and M is the torque.

Research concerning the physical modelling of the wire rod rolling process was carried out in three variants, differing mainly in the method of applying deformation.

Variant 1 (V1): Physical modelling of the analyzed process in this variant was carried out during non-free torsion tests, as a cycle of 31 individual deformations at a certain temperature with specified intervals between successive deformations, and taking into account multi-stage controlled cooling during the individual stages of the technological process:

- *17 individual deformations as non-free torsion, representing the rolling process in a continuous bar rolling mill;*
- *10 individual deformations as non-free torsion, representing the rolling process in an NTM block of a wire rod rolling mill;*
- *4 individual deformations as non-free torsion, representing the rolling process in an RSM block of a wire rod rolling mill.*

Variant 2 (V2): The physical modelling of the analyzed process in this variant was carried out in non-free torsion tests, as a cycle of 19 deformations at a certain temperature with specified intervals between following deformation cycles, and taking into account multi-stage controlled cooling at individual stages of the technological process:

- 17 individual deformations as non-free torsion, representing the rolling process in a continuous bar rolling mill;
- 1 deformation reproducing the rolling process at an NTM block of a wire rod rolling mill, as non-free torsion, with a total strain value equal to the actual strain value occurring at this stage of the technological process;
- 1 deformation reproducing the rolling process at an RSM block of a wire rod rolling mill, as non-free torsion, with a total strain value equal to the actual strain value occurring in this block.

Variant 3 (V3): The physical modelling of the analyzed process in this variant was carried out during non-free torsion and tensile tests, as a cycle of 19 deformations at a specific temperature with specified intervals between successive deformation cycles, and taking into account multi-stage controlled cooling during individual stages of the technological process:

- 17 individual deformations as non-free torsion, representing the rolling process in a continuous bar rolling mill,
- 1 deformation reproducing the rolling process in an NTM block of a wire rod rolling mill, as non-free torsion, with a total strain value equal to the actual strain value occurring at this stage of the technological process and,
- 1 deformation reproducing the rolling process in an RSM block of a wire rod rolling mill, as tension, while maintaining the total strain value in this block.

The general diagram of thermo-mechanical treatment, reproducing the entire rolling process of a 20 MnB4 steel wire rod with a diameter of 5.5 mm, is shown in Figure 4.

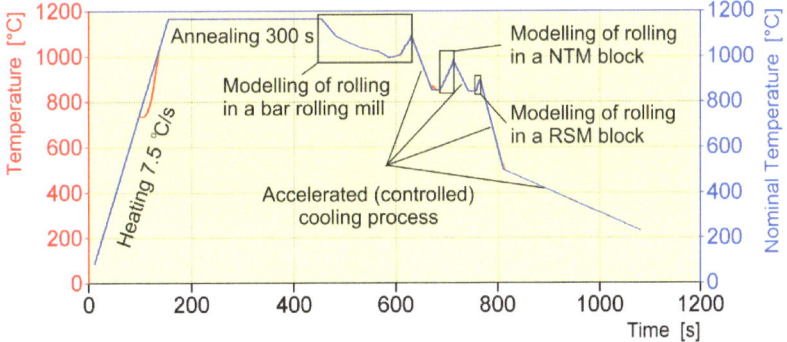

Figure 4. General model of thermo-mechanical treatment representing the entire rolling process of a 20MnB4 steel wire rod with a diameter of 5.5 mm.

Before the deformation process, the material was heated to a temperature of 1165 °C, corresponding to the temperature in the equalizing zone of a heating furnace (under industrial conditions). Then, in order to unify the temperature distribution in the entire sample working zone, the 20MnB4 steel was heated for 300 s. The following stage consisted of cooling for 30 s to a temperature of 1086 °C, replicating the cooling of the band during its transport from the furnace to the first rolling pass stand. Then, the tested samples were deformed over 17 cycles with strain parameters in accordance with Table 2, replicating the rolling process in a continuous rolling mill. The following stage consisted of accelerated cooling to 851 °C, corresponding to the band temperature before the NTM block of a rolling mill. During the following stage of physical modelling, the rolling process in an NTM block was reproduced (Table 3). Then, the process of accelerated cooling between NTM and RSM blocks was modelled, with cooling to a temperature of 845 °C, corresponding to the band temperature before the RSM block. The next stage of physical modelling consisted of reproducing the rolling process in an RSM block (Table 3). During the final stage, the tested steel grade was cooled in a controlled manner to 500 °C with an average cooling rate of about 10 °C/s, and then to 200 °C with a cooling rate of 1 °C/s.

Table 2. Parameters of the deformation process during physical modelling of rolling a round 20MnB4 steel rod in a medium continuous rolling mill, for rolling a 5.5 mm diameter wire rod (research variants: V1, V2, and V3) [17,21].

Pass Number	Temperature, T (°C)	Strain, ε (-)	Strain Rate, $\dot{\varepsilon}$ (s^{-1})	Yield Stress (Torsion), σ_p (MPa)	Break Time After Deformation, t (s)
1	1086/1084	0.18	0.16	29.3	26.47
2	1057/1056	0.39	0.35	42.9	19.89
3	1037/1035	0.28	0.39	60.5	29.98
4	1023/1022	0.59	0.96	69.9	11.33
5	1010/1008	0.46	1.15	74.9	8.91
6	995/993	0.50	2.02/1.15	79.8	6.13
7	997/996	0.45	2.45/1.15	84.9	11.65
8	1005/1004	0.48	4.71/1.15	72.2	3.35
9	1009/1009	0.44	5.57/1.15	90.0	2.63
10	1022/1022	0.54	10.39/1.15	90.7	1.85
11	1030/1029	0.48	12.07/1.15	93.6	3.09
12	1049/1048	0.50	20.53/1.15	73.5	2.28
13	1052/1049	0.51	24.74/1.15	84.8	3.18
14	1069/1068	0.50	46.34/1.15	69.9	1.35
15	1072/1070	0.41	47.13/1.15	83.0	1.11
16	1087/1080	0.51	79.93/1.15	80.1	0.90
17	1091/1092	0.31	70.63/1.15	78.9	8.52/55.0

Table 3. Parameters of the deformation process during physical modelling of rolling a 20MnB4 steel wire rod with a diameter of 5.5 mm in NTM and RSM blocks of a wire rod rolling mill (test variant V1) [17,21].

Pass Number	Temperature, T (°C)	Strain, ε (-)	Strain Rate, $\dot{\varepsilon}$ (s^{-1})	Yield Stress (Torsion), σ_p (MPa)	Break Time after Deformation, t (s)
			NTM		
18	851/849	0.49	156.02/1.15	106.4	0.091/2.0
19	860/860	0.51	171.25/1.15	141.8	0.074/2.0
20	867/866	0.56	276.33/1.15	147.7	0.058/2.0
21	883/883	0.54	303.93/1.15	143.3	0.048/2.0
22	892/890	0.56	477.46/1.15	145.3	0.037/2.0
23	908/908	0.53	584.28/1.15	139.1	0.032/2.0
24	918/918	0.62	991.51/1.15	131.4	0.024/3.0
25	941/942	0.57	1042.10/1.15	106.3	0.020/3.0
26	956/955	0.62	1753.46/1.15	97.40	0.015/3.0
27	982/983	0.56	1809.67/1.15	86.5	0.82/45.0
			RSM		
28	845/842	0.53	2368.05/1.15	96.0	0.012/3.0
29	873/874	0.48	2275.43/1.15	126.9	0.007/3.0
30	894/893	0.13	1853.11/1.15	111.9	0.004/0.9
31	895/894	0.10	1680.68/1.15	42.0	

After physical modelling, changes in the most important parameters of the process (strain, strain rate, temperature, and yield stress) were analyzed in detail.

During the second stage of work, samples were cut out from the material for metallographic tests after physical modelling using an EDM 32 wire electro-driller, and then metallographic microsections (nitrile etching) were prepared (Figure 5).

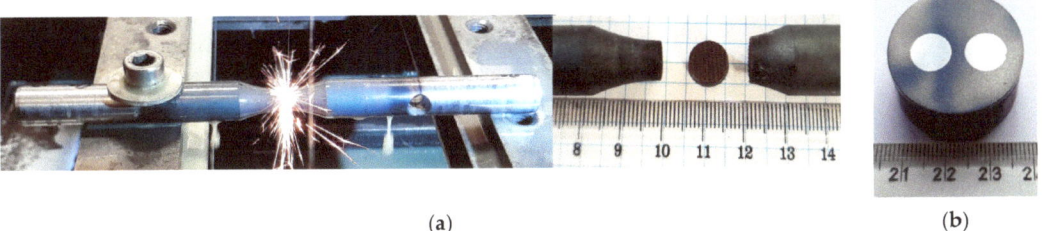

(a)   (b)

Figure 5. Method of sampling for metallographic tests from material after physical modelling: (a) general view and (b) sample metallographic microsections.

The locations of characteristic points where metallographic analyses and hardness measurements were carried out are presented in Figure 6 (r = radius).

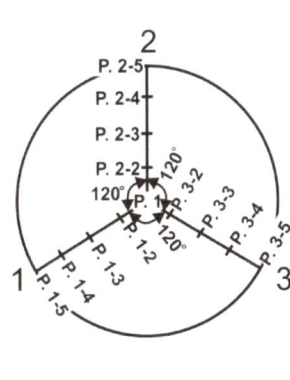

Points	Distance V 1, V 2		V 3
P 1	sample center	sample center	sample center
P 2	0.3 r	1.35 mm	1.20 mm
P 3	0.5 r	2.25 mm	2.00 mm
P 4	0.7 r	3.15 mm	2.81 mm
P 5	1.0 r	4.50 mm	4.01 mm

The differences in distance (in mm) for variants V 1, V 2 and V3 are resulted of different strain scheme between them. V 1 and V 2—non- free torsion (diameter is constans durig the experiment), V3—tension in the last stage of physical modelling (RSM rolling block) and reduction of diameter.

(a)   (b)

Figure 6. Cross-section of the material after physical modelling of the wire rod rolling process including marked measuring points: (a) general view and (b) distance table.

An assessment of the microstructure of the tested steel was carried out using light microscopy (Nikon Eclipse MA-200 microscope with NIS-Elements software) [38]. The average ferrite grain size was determined using the perpendicular secant method [39]. The hardness tests were carried out using the Vickers method with the use of a Future-Tech FM-700 microhardness meter (load 1000 gf, time 5 s).

In the following stage, on the basis of available mathematical models and using measurements of average ferrite grain sizes, chemical composition, and hardness [28,29], the yield strength (YS), ultimate tensile strength (UTS), and plasticity reserve were determined.

$$YS = \frac{HV}{0.378} - 123, \qquad (4)$$

$$UTS = \frac{HV}{0.352} + 70, \qquad (5)$$

$$YS = 62.6 + 26(\%Mn) + 60(\%Si) + 759(\%P) + 213(\%Cu) + 3286(\%N) + \frac{19.7}{\sqrt{\frac{D_\alpha}{1000}}}, \quad (6)$$

$$UTS = 165 + 54(\%Mn) + 100(\%Si) + 652(\%P) + 473(\%Ni) + 635(\%C) + 2173(\%N) + \frac{11}{\sqrt{\frac{D_\alpha}{1000}}}, \quad (7)$$

where HV is the Vickers hardness; %Mn, %Si, %P, %Cu, %N, %Ni, and %C are the content in mass percent of manganese, silicon, phosphorus, copper, nitrogen, nickel, and carbon, respectively, in the steel; and D_α is the ferrite grain size in μm.

The scope of research carried out in terms of the work also included determining selected mechanical properties directly during the static tensile test using a Zwick Z/100 strength machine [38].

During the final stage of the work, the obtained research results were verified under industrial conditions.

3. Results and Discussion

3.1. Analysis of the Main Parameters of the Deformation Process

The general course of temperature and plasticizing yield stress during the physical modelling of the rolling process of 20MnB4 steel round bars in a medium continuous rolling mill is shown in Figure 7. This stage of physical modelling was the same for all three variants analyzed in the work.

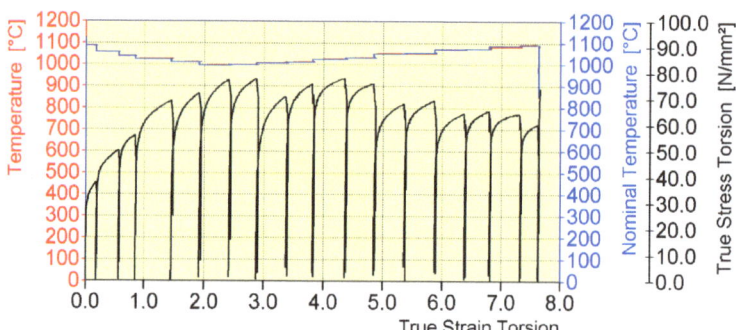

Figure 7. General course of temperature (nominal and obtained) and yield stress during physical modelling of rolling round 20MnB4 steel bars in a medium continuous rolling mill, for rolling a 5.5 mm diameter wire rod (test variants: V1, V2, and V3).

The most important parameters of the deformation process during the physical modelling of rolling of round bars made of 20MnB4 steel in a medium continuous rolling mill, which have been tested, determined, and verified in previous works [17,21], are presented in Table 2. The "/" symbol is followed by parameters that can be achieved by the STD 812 torsion plastometer, taking into account its technical parameters mainly inertia.

When analyzing the data presented in Figure 7 and in Table 2, a decrease in the temperature of the band during the initial stage of the deformation process (deformations 1–7) and a simultaneous increase in the value of the yield stress to 84.9 MPa (for pass No. 7) were determined. This resulted from the long intervals between the deformations and the low strain rate value. Analyzing the remaining deformations replicating the rolling process in a continuous rolling mill (No. 8–17), it is possible to notice a gradual increase in the temperature of the deformed steel, caused, among others, by an increase in the rolling speed, and thus shorter intervals between subsequent deformations. There is a general downward trend in the yield stress to a value of 78.9 MPa (pass No. 17). Comparing the values of the nominal (set) temperature and that obtained during the physical modelling of the rolling process of 20MnB4 steel in a bar rolling mill, a high accuracy for this analyzed parameter was found. The biggest error was between the set temperature value and the obtained value in pass No. 16, at 0.6%.

Analyzing the data in Table 2, it can be noted that starting from pass No. 6, it was not possible to maintain the required strain rate value. This was due to the inertia of the device. The time needed to reach the required strain rate value was about 0.25 s. Using higher strain rate values than 1.15 s^{-1}, with values of individual deformations of 0.5–0.6, resulted in exceeding the strain value. According to published data, among others, in the paper [2], despite a great difference between the desired value of strain rate and the value achievable by the torsion plastometer, the differences in the values of yield stress did not exceed 6.3%, which did not have a significant impact on the microstructure and mechanical properties of the tested material after this stage of the rolling process [2].

The minimum time between deformations achievable by a torsion plastometer was 0.9 s. After the last deformation in the rod rolling mill (No. 17), in accordance with the industrial conditions, the material was cooled to a temperature of about 851 °C, required before the first deformation in an NTM block of a wire rod rolling mill (Table 3). Accelerated cooling was carried out using argon supplied by special nozzles to the central part of the sample area (Figure 3a, in the case of slow cooling rate, the device simultaneously cools down and heats the tested material to ensure the required temperature value). The accelerated cooling time was increased from about 8.5 s to about 55 s (cooling over 40 s and holding 15 s (Table 2)). Increasing this time and holding at a temperature of about 851 °C was necessary due to the inductive method of heating the samples and the shape of the inductive exciter itself (Figure 3a). Using a cooling time in accordance with industrial conditions (approx. 8.5 s) after reaching the required temperature (851 °C) and shutting down the cooling system resulted in a rapid increase in temperature in the middle part of the samples (Figure 8a). This increase was caused by heat conduction towards the central part of the samples from areas with a higher temperature. This resulted in the inability to achieve the required temperature value during the initial stage of the deformation process in an NTM block. The difference between the nominal temperature value and the desired value was 4.7% (Figure 8a).

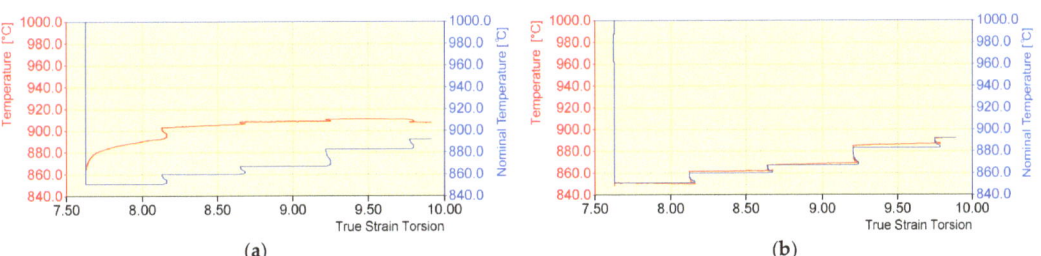

Figure 8. Example temperature distribution (nominal and obtained) during the initial stage of the deformation process in an NTM block: (**a**) accelerated cooling time before the NTM block (8.5 s) and (**b**) accelerated cooling time before the NTM block (55 s: accelerated cooling for 40 s and holding for 15 s).

The general course of temperature and yield stress during the physical modelling of a 20MnB4 steel wire rod rolling process in NTM and RSM blocks of a wire rod rolling mill for the V1 test variant is shown in Figure 9.

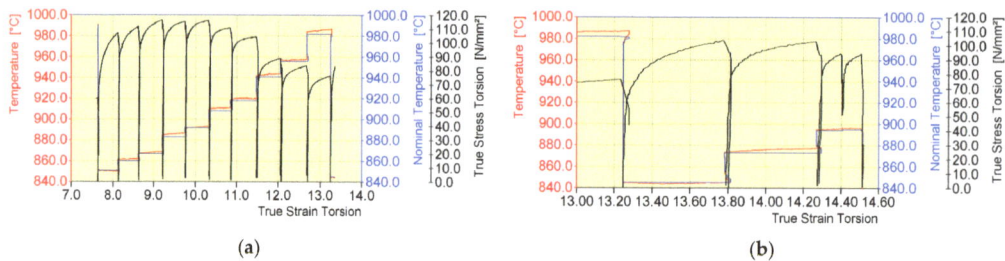

Figure 9. General course of temperature (nominal and obtained) and yield stress during physical modelling of rolling a 20MnB4 steel wire rod with a diameter of 5.5 mm (test variant V1): (**a**) in an NTM block and (**b**) in an RSM block.

The most important parameters of the deformation process during the physical modelling of rolling a 20MnB4 steel wire rod with a diameter of 5.5 mm in NTM and RSM blocks of a wire rod rolling mill according to the V1 test variant are presented in Table 3. As before, the "/" symbol is followed by parameters that can be achieved by the STD 812 torsion plastometer, taking into account its technical parameters.

Analyzing the temperature changes during the deformation reproducing the rolling process in an NTM block (No. 18–27), it is possible to observe a continuous increase in temperature caused mainly by short intervals between successive deformations. Comparing the nominal (set) and obtained temperature values during the physical modelling of the 20MnB4 steel rolling process in an NTM block, a high accuracy of this analyzed parameter was found. The biggest error between the set temperature value and the obtained value occurred for passes No. 18 and 22, at 0.2%.

The course of yield stress during the first five deformation sequences at this stage of the physical modelling process is somewhat unusual. As the temperature increases, the yield stress decreases and then slightly increases. This may result from an uneven temperature distribution after the process of accelerated cooling between the continuous rolling mill and the NTM block. During subsequent cycles of deformations simulating the rolling process in an NTM block, the course of yield stress is typical, i.e., as the temperature of the deformed material increases, the yield stress value of the deformed steel decreases.

The minimum possible time between deformations achieved by the torsion plastometer was 2 s. After the final deformation in an NTM block (No. 27), in accordance with industrial conditions, accelerated material cooling to a temperature of about 845 °C was required before the first deformation in an RSM block of a wire rod rolling mill was applied (Table 3). The accelerated cooling time, similar to the accelerated cooling after a continuous rolling mill (pass No. 17), was increased from approx. 0.82 s to approx. 45 s (cooling over 30 s and holding 15 s to ensure the required temperature values in the RSM block).

Analyzing the temperature changes during the deformation reproducing the rolling process in an RSM block (No. 28–31), it is possible to observe (similar to the case of deformation in the NTM block) a continuous increase in temperature caused mainly by short intervals between successive deformations. By comparing the values of the nominal (set) temperature and that obtained during the physical modelling of the 20MnB4 steel rolling process in an RSM block, a high accuracy of this analyzed parameter was found. The largest error between the set temperature value and the obtained value occurred for pass No. 28 and was 0.4%. Despite an increase in temperature during the first two deformation sequences (No. 28 and 29), the yield stress increased (Table 3), which may result from an uneven distribution of temperature after the process of accelerated cooling between the NTM and RSM blocks. A decrease in the yield stress value was observed for the two final deformation sequences (No. 30 and 31). This resulted mainly from a small strain value of about 0.1. The minimum time between deformations achievable by the torsion plastometer during this stage of physical modelling was 0.9 s.

The general course of temperature and yield stress during the physical modelling of 20MnB4 steel wire rod rolling process in NTM and RSM blocks of a wire rod rolling mill for the V2 test variant is shown in Figure 10. The most important parameters of the deformation process during the physical modelling of rolling a 20MnB4 steel wire rod with a diameter of 5.5 mm in NTM and RSM blocks of a wire rod rolling mill according to the V2 test variant are presented in Table 4. As before, the "/" symbol is followed by parameters that can be achieved by the STD 812 torsion plastometer, taking into account its technical parameters. In this variant, the sequences of individual passes in NTM and RSM blocks of a wire rod rolling mill were replaced with individual deformations as non-free torsion, with a total strain value equal to the actual strain value occurring in these blocks. Such an assumption was tested during research presented, for example, in works [17,40]. Based on the obtained results, it was found that replacing the sequence of the four final deformations with one did not cause a large error in the distribution of yield stress and did not incur large errors in terms of assessing the structural construction of the samples.

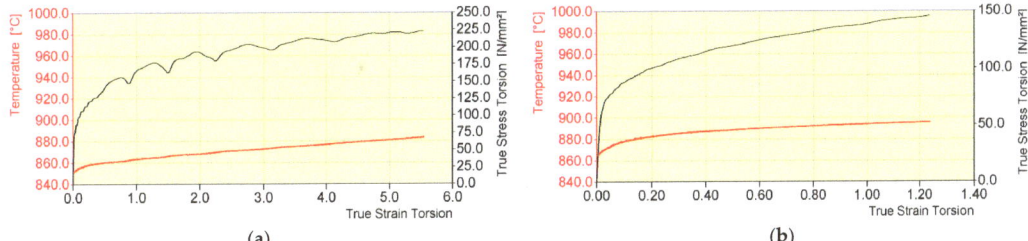

Figure 10. General course of temperature (nominal and obtained) and yield stress during physical modelling of rolling a 20MnB4 steel wire rod with a diameter of 5.5 mm (test variant V2): (**a**) in an NTM block and (**b**) in an RSM block.

Table 4. Parameters of the deformation process during physical modelling of rolling a 20MnB4 steel wire rod with a diameter of 5.5 mm in NTM and RSM blocks of a wire rod rolling mill (test variant V2) [17,21].

Pass Number	Temperature, T (°C)	Strain, ε (-)	Strain Rate, $\dot{\varepsilon}$ (s^{-1})	Yield Stress (Torsion), σ_p (MPa)	Break Time after Deformation, t (s)
			NTM		
18	851–982/851–883	5.56	156.02–1809.67/40	222.8	0.82/6.0
			RSM		
19	845–895/853–895	1.24	2368.05–1680.68/10	145.4	

Replacing the sequence of individual deformations in NTM and RSM blocks of a wire rod rolling mill with single deformations with strain values equal to the actual strain value occurring in these blocks allowed for the use of a higher value of the strain rate compared to the test variant V1. In the case of the NTM block, the strain rate was 40 s^{-1}, whereas in the case of the RSM block it was 10 s^{-1}. In turn, this had a positive impact in that it increased the value of the yield stress. In relation to the V1 test variant, the yield stress in the NTM block increased in value to about 223 MPa, while in the case of the RSM block, the yield stress value increased to approximately 145 MPa. During the physical modelling of the rolling process in NTM and RSM blocks according to the V2 test variant, the increase in temperature in these blocks was also programmed. Taking into account the inductive method of heating the samples and the shape of the inductive exciter itself (Figure 3a), in order to increase the accuracy between the assumed and the obtained temperature, in this variant the speed of accelerated cooling was increased between the continuous rolling mill and the NTM block as well as between the NTM and RSM blocks, and the resistance immediately after accelerated cooling was abandoned. This facilitated an increase in temperature as a result of the rapid thermal conductivity from areas with higher temperatures towards the central part of the samples. The error between the temperature assumed after deformation in the NTM block and the temperature obtained during physical modelling was 10%. In the case of the RSM block, the required temperature increase was achieved.

The general course of temperature and yield stress during the physical modelling of a 20MnB4 steel wire rod rolling process in NTM and RSM blocks of a wire rod rolling mill for the test variant V3 is shown in Figure 11.

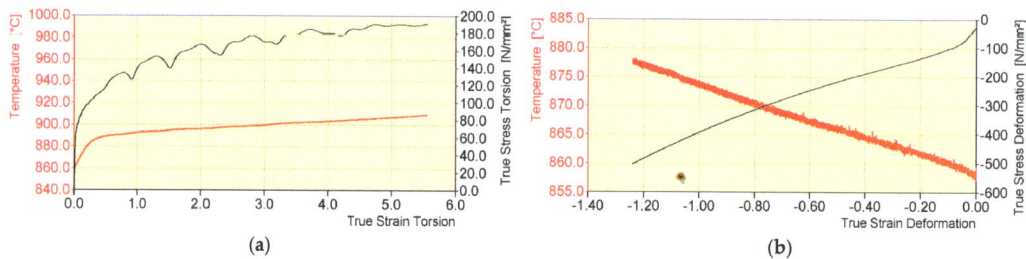

Figure 11. General course of temperature (nominal and obtained) and yield stress during physical modelling of rolling a 20MnB4 steel wire rod with a diameter of 5.5 mm (test variant V3): (**a**) in an NTM block (torsion) and (**b**) in the RSM block (tension).

The most important parameters of the deformation process during the physical modelling of rolling a 20MnB4 steel wire rod with a diameter of 5.5 mm in NTM and RSM blocks of a wire rod rolling mill according to the V3 test variant are presented in Table 5. The "/" symbol is followed by the parameters that can be achieved by the STD 812 torsion plastometer, taking into account its technical parameters.

Table 5. Parameters of the deformation process during physical modelling of rolling a 20MnB4 steel wire rod with a diameter of 5.5 mm in NTM and RSM blocks of a wire rod rolling mill (test variant V3) [17,21].

Pass Number	Temperature, T (°C)	Strain, (-)	Strain Rate, $\dot{\varepsilon}$ (s^{-1})	Yield Stress, σ_p (MPa)	Break Time after Deformation, t (s)
		NTM (torsion)			
18	851–982/856–909	5.56	156.02–1809.67/40	191.0	0.82/6.0
		RSM (tension)			
19	845–895/858–878	1.24	2368.05–1680.68/10	503.0	

In this variant, the sequences of individual passes in NTM and RSM blocks of a wire rod rolling mill were replaced by single deformations: one deformation reproducing the rolling process in an NTM block of a wire rod rolling mill as non-free torsion, with the total strain value equal to the actual strain value occurring during this stage of the technological process, and one deformation reproducing the rolling process in an RSM block of a wire rod rolling mill as tension, while maintaining the total strain value in this block.

Replacing the sequence of individual deformations in NTM and RSM blocks of a wire rod rolling mill with single deformations with strain values equal to the actual strain value occurring in these blocks (similar to the case of variant V2) allowed using a higher value of the strain rate in relation to the V1 test variant. In the case of the NTM block, the strain rate was 40 s^{-1}, whereas in the case of the RSM block, it was 10 s^{-1}. In turn, this had a positive impact in that there was an increase in the value of the yield stress. The yield stress in the NTM block reached approximately 191 MPa, while in the case of the RSM block, the yield stress increased to approximately 500 MPa, mainly as a result of changing the strain state. Similar to the case of the V2 variant, a temperature increase in the NTM and RSM blocks was also programmed in this variant. In order to increase the accuracy between the assumed and the obtained temperature, in this variant (similar to variant V2) the speed of accelerated cooling was increased between the continuous rolling mill and the NTM block as well as between the NTM and RSM blocks, and the resistance immediately after accelerated cooling was abandoned. This facilitated an increase in temperature as a result of the rapid thermal conductivity from areas with higher temperatures towards the central part of the samples. The error between the temperature assumed after deformation in the NTM block and the temperature obtained during physical modelling was 7.4%. In the case of the RSM block, the error between the temperature assumed after deformation and the temperature obtained during physical modelling was less than 2%.

It was found that the obtained level of yield stress, as a result of replacing the sequence of individual deformations in NTM and RSM blocks with one deformation and changing the strain state in the RSM block, is similar to the value of yield stress occurring in the actual technological process [17], which is crucial for the process of microstructure reconstruction.

The general course of changes in yield stress for all analyzed variants is shown in Figure 12.

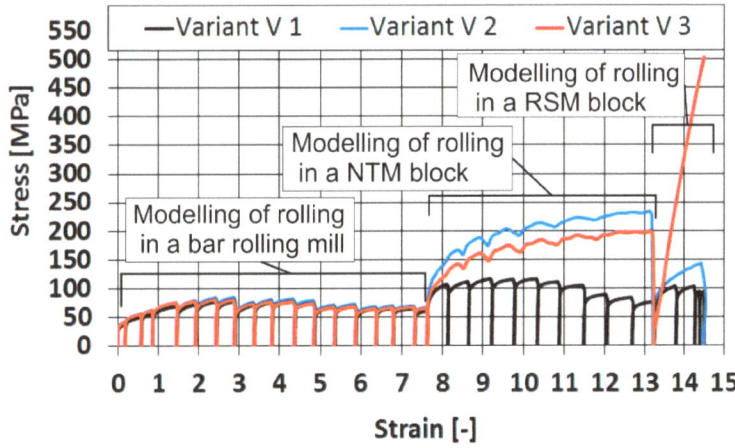

Figure 12. General course of yield stress during physical modelling of rolling a 20MnB4 steel wire rod with a diameter of 5.5 mm (test variant V1, test variant V2, test variant V3.

3.2. Impact of the Used Deformation Process Conditions on the Microstructure of the Tested Material

Examples of the 20MnB4 steel microstructure after the physical modelling process according to the V1–V3 variants are shown in Figures 13–15. The results of measuring the ferrite grain size and the hardness of the tested steel grade are presented in Table 6 and Figure 16.

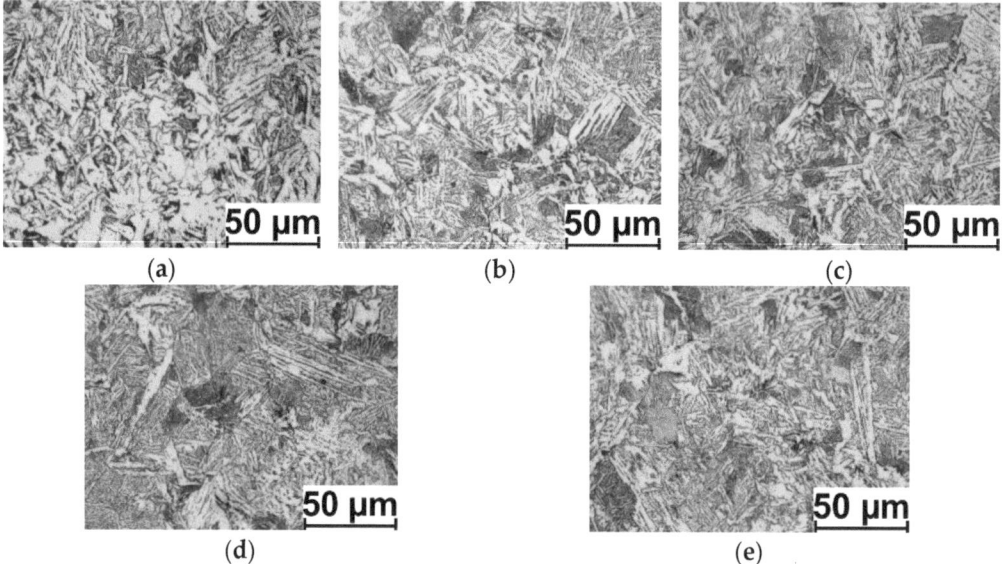

Figure 13. Example microstructures of 20MnB4 steel after physical modelling of a wire rod rolling process for test variant V1 (marking of points in accordance with Figure 5): (**a**) point P. 1, (**b**) point P. 1-2, (**c**) point P. 1-3, (**d**) point P. 1-4 and (**e**) point P. 1-5.

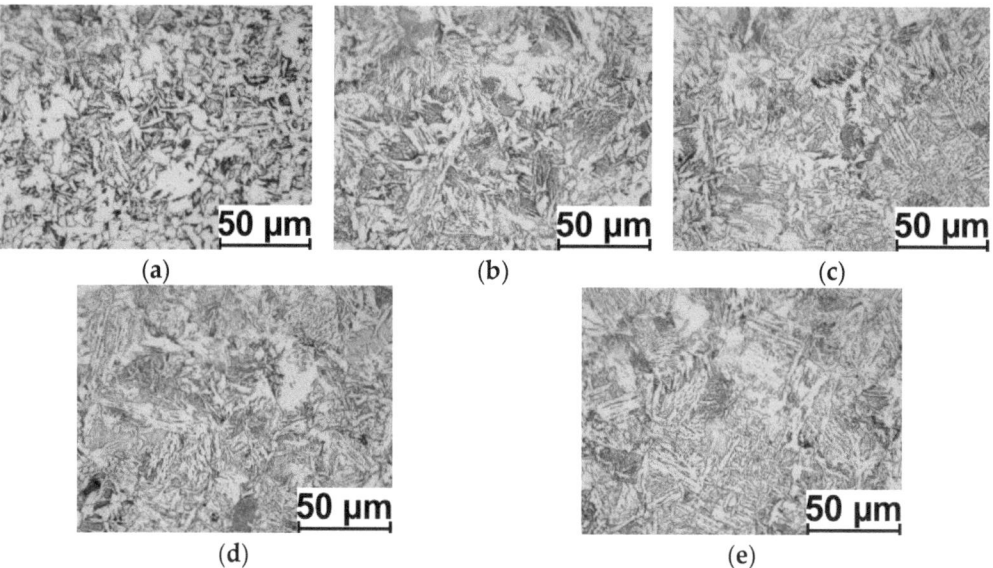

Figure 14. Example microstructures of 20MnB4 steel after physical modelling of a wire rod rolling process for test variant V2 (marking of points in accordance with Figure 5): (**a**) point P. 1, (**b**) point P. 1-2, (**c**) point P. 1-3, (**d**) point P. 1-4, and (**e**) point P. 1-5.

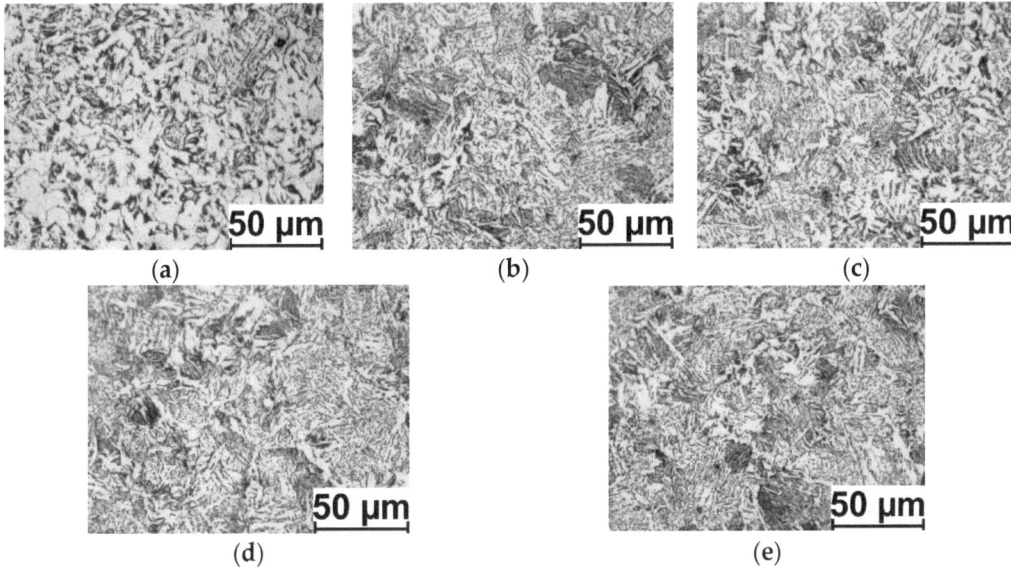

Figure 15. Example microstructures of 20MnB4 steel after physical modelling of the wire rod rolling process for test variant V3 (marking of points in accordance with Figure 5): (**a**) point P. 1, (**b**) point P. 1-2, (**c**) point P. 1-3, (**d**) point P. 1-4, and (**e**) point P. 1-5.

Table 6. Results of measurements of ferrite grain size and hardness of 20MnB4 steel after physical modelling of the rolling process of a 5.5 mm diameter wire rod.

Points (According to Figure 6)	Variant V1		Variant V2		Variant V3	
	Ferrite Grain Size, D_α (µm)	Hardness (HV)	Ferrite Grain Size, D_α (µm)	Hardness (HV)	Ferrite Grain Size, D_α (µm)	Hardness (HV)
P.1	8.14	179.75	7.31	182.37	8.31	158.30
P. 1-2	6.38	206.83	6.73	203.90	7.31	209.20
P. 1-3	6.27	210.70	6.54	209.17	6.99	211.90
P. 1-4	6.17	212.78	6.27	210.77	6.94	215.30
P. 1-5	5.92	216.33	6.13	213.13	6.64	221.48
P.1	8.14	179.75	7.31	182.37	8.31	158.30
P. 2-2	6.46	208.83	6.95	198.43	7.55	204.20
P. 2-3	6.12	209.17	6.85	205.33	7.29	211.40
P. 2-4	5.97	210.97	6.60	206.40	7.21	212.83
P. 2-5	5.85	214.90	5.70	213.43	6.99	220,65
P.1	8.14	179.75	7.31	182.37	8.31	158.30
P. 3-2	6.00	206.03	7.00	207.23	6.93	205.50
P. 3-3	5.84	211.03	6.25	210.20	6.93	209.77
P. 3-4	5.51	215.13	6.18	210.87	6.19	215.50
P. 3-5	5.48	215.97	6.17	213.00	6.16	221.80
Average value	6.43	205.19	6.62	203.26	7.20	202.30

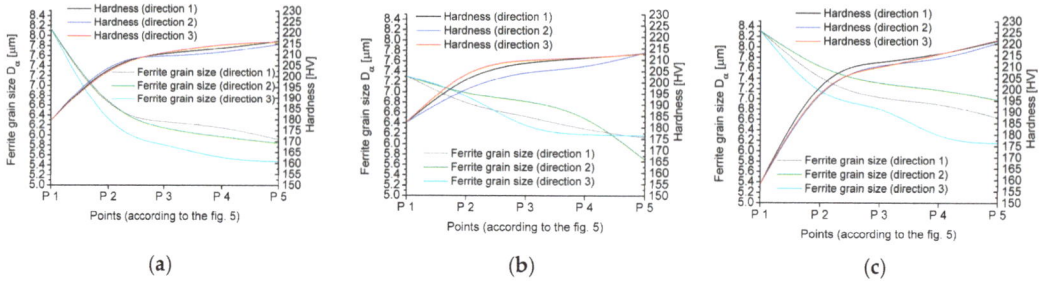

Figure 16. Distribution of the ferrite grain size and hardness of 20MnB4 steel after physical modelling of the rolling process of a wire rod with a diameter of 5.5 mm: (**a**) test variant V1, (**b**) test variant V2, and (**c**) test variant V3.

Based on the analysis of the test results of 20MnB4 steel after physical modelling of the rolling process of a wire rod according to the test variant V1, it was found that the D_α ferrite grain size was between 8.14 and 5.48 µm. The average D_α ferrite grain size was 6.43 µm (Table 6, Figure 16). The hardness of the tested material after physical modelling according to this variant ranged from 179.75 to 216.33 HV (average value 205.19 HV).

The D_α ferrite grain size for the material after physical modelling in accordance with variant V2 was in the range of 7.31 to 5.7 µm. The average D_α ferrite grain size was only slightly larger (comparably) to the value achieved as a result of modelling in accordance with the V1 variant and was at 6.62 µm (Table 6 and Figure 16). This was due to an only slightly larger value of the yield stress during the physical modelling of the deformation process in the RSM block (despite an increase in the yield stress during the physical modelling of the deformation process in the NTM block in this variant in relation to the V1 variant, Figure 12). The hardness of the tested steel after physical modelling according to the technological variant V2 ranged from 182.37 to 213.43 HV (average value 203.26 HV).

Analyzing the test results for variant V3, it was determined that the D_α ferrite grain size of 20MnB4 steel was in the range of 8.31 to 6.16 µm. The average D_α ferrite grain size was 7.20 µm (Table 6, Figure 16).

The increase in the average D_α ferrite size resulted mainly from the much higher value of the yield stress during the physical modelling of the deformation process in the RSM block, mainly due to the change in strain state from non-free torsion to tension (and the increase in the strain rate). Moreover, the value of yield stress during the physical modelling of the deformation process in the NTM block in this variant in relation to the V1 variant was also higher (Figure 12). After physical modelling according to the technological variant V3, the hardness of 20MnB4 steel ranged from 158.30 to 221.80 HV, while the average hardness value was 202.30 HV.

Based on an analysis of the D_α ferrite grain size distribution in a cross-section (along the radius) (Figure 16), it was determined that the largest D_α ferrite grains occurred along the axis of the tested samples, while the smallest D_α ferrite grains occurred in subsurface areas. This is due to the characteristics of the torsion test itself, in which the smallest strain value occurs in the axis of the material subject to torsion, while the largest strain value occurs in the subsurface areas. Based on the data presented in Table 6 and Figure 16, it was determined that there was a simultaneous increase in the hardness of the tested steel along with a decrease in D_α ferrite grain size.

Analyzing the distribution of D_α ferrite grain size in a cross-section of the samples (along the radius), it is possible to observe a relatively high homogeneity in size. This may result from the high value of the total strain value (14.32) and the small length of the sample's measured part (2 mm, total torsion angle of about 635°), whereas by analyzing the grain size at individual measurement points (Figures 13–15) (in accordance with Figure 6), it is possible to notice a relatively large heterogeneity and the acicular shape of grains in certain areas. This may be due to the relatively high cooling rate after the deformation process (10 °C/s). According to the research results presented in paper [17], the applied cooling speed is the cooling speed limit for 20MnB4 steel, and if this value is exceeded, bainitic structures begin to form in the material.

Comparing the ferrite grain size of a wire rod produced in industrial conditions with the values obtained as a result of our physical modelling, it was found that the best results were obtained using physical modelling according to variant V3. The average size of the ferrite grain in the cross-section of the wire rod produced under industrial conditions was 7.52 µm. However, the average size of the ferrite grain in the longitudinal cross-section of the finished product was equal to 8.11 µm. The error between the ferrite grain sizes measured in a wire rod cross-section and obtained as a result of physical modelling according to the technological variant V3 was 4.3%, whereas the error between the ferrite grain sizes measured on the longitudinal cross-section of a wire rod and obtained as a result of physical modelling for this technological variant was 11.2%. On this basis, it can be concluded that the average ferrite grain size in a wire rod obtained under industrial conditions is similar to the grain size obtained in samples after physical modelling of the rolling process according to variant V3.

3.3. The Impact of the Applied Conditions of the Deformation Process on the Selected Mechanical Properties of the Tested Steel

Table 7 presents the results of research concerning selected mechanical properties of the material after physical modelling of the rolling process of a 20MnB4 steel wire rod with a diameter of 5.5 mm calculated using Formulas (4)–(7). These properties were determined from average hardness values and ferrite grain size. Moreover, this table presents the results of selected mechanical properties of the material after physical modelling, determined in the course of a static tensile test. This table also includes the results of selected mechanical properties of 20MnB4 steel obtained after rolling under industrial conditions. Additionally, this table also includes the results of tests obtained in previous studies [17,21], which carried out modelling only in the final stage of the analyzed rolling process (RSM block) in compression tests, using the GLEEBLE 3800 metallurgical process simulator.

A general view of the 20MnB4 steel samples after physical modelling of the process of rolling a wire rod during testing mechanical properties in a static tensile test is presented in Figure 17a. Examples of samples before and after the tests are shown in Figure 17b, while examples of tensile curves are shown in Figure 17c.

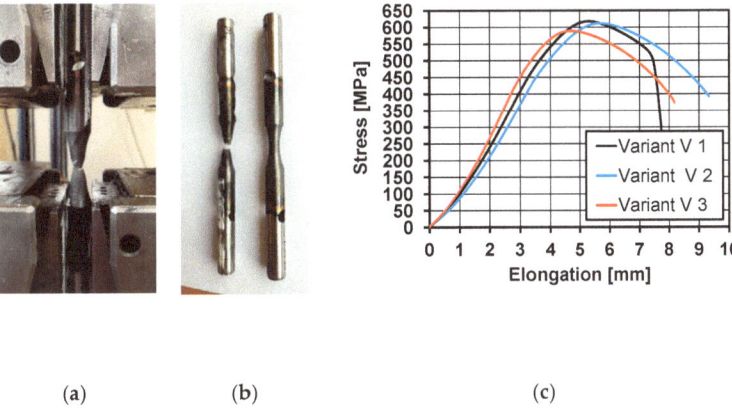

(a)　　　　　　(b)　　　　　　(c)

Figure 17. Sample results from testing the mechanical properties of the material after physical modelling of rolling a wire rod: (**a**) general view of a 20MnB4 steel sample after physical modelling, during determination of mechanical properties in a static tensile test, (**b**) general view of samples before and after the test, and (**c**) sample tensile curves of the tested material after physical modelling.

Table 7. Measurement results of selected mechanical properties of 20MnB4 steel.

Research Variant of Physical Modelling	Mechanical Properties									
	Calculated Using Formulas:								Static Tensile Test		
	(4)	(5)	(4), (5)	(6)	(7)	(6), (7)	Yield Strength, YS (MPa)	Ultimate Tensile Strength, UTS (MPa)	YS/UTS	Relative Reduction of Area at Fracture, Z (%)
	Yield Strength, YS (MPa)	Ultimate Tensile Strength, UTS (MPa)	YS/UTS	Yield Strength, YS (MPa)	Ultimate Tensile Strength, UTS (MPa)	YS/UTS				
Variant V1	420	653	0.64	426	566	0.75	429	605	0.71	62.03
Variant V2	415	648	0.64	422	564	0.75	422	598	0.71	57.49
Variant V3	412	645	0.64	412	558	0.74	418	591	0.71	66.93
Physical modelling in compression test [17,21]	401	632	0.63	397	550	0.72				
Industrial research results							412	557	0.74	69.80

Based on an analysis of the results of mechanical properties tests (Table 7), it was determined that applying the modifications to the physical modelling of the wire rod rolling process resulted in a decrease in the yield strength and ultimate tensile strength of the tested steel, regardless of the applied analytical Formulas (4)–(7). This was also confirmed by the results of static tensile tests. No changes in values were observed with regard to the plasticity reserve. Due to the small length of the test area of the used samples (2 mm) and the inability to install an accurate extensometer, the relative elongation was not tested during the static tensile test. Analyzing the obtained test results concerning mechanical properties, it was found that the best research variant allowing to obtain a high correspondence between the mechanical properties determined in industrial research and for the material after physical modelling is the research variant V3. The differences between the most important mechanical parameters determined during a static tensile test of a wire rod under industrial conditions and material after physical modelling were 1.5% for the yield strength, approximately 6.1% for the ultimate tensile strength, and approximately 4.1% for the relative reduction of area at fracture and plasticity reserve. The high consistency between the mechanical properties determined in industrial research and for the material after physical modelling according to variant V3 was also confirmed by the results of calculations of selected mechanical properties using Equations (4)–(7).

When comparing the obtained results of mechanical properties tests (static tensile test for variant V3) and those presented in earlier works [17,21] with the results of industrial research (Table 7), it was found that the lowest accuracy was provided by Equations (4) and (5). The differences between the analyzed mechanical properties determined using these formulas and the values obtained in industrial conditions were 2.7% for the yield strength, approximately 13.5% for the ultimate tensile strength, and approximately 14.9% for the plasticity reserve. This may be due to the fact that these relations have been developed only for steel with a specific chemical composition. Moreover, these dependencies are based only on hardness measurements, which can be burdened with a certain error. A much higher accuracy was obtained using Equations (6) and (7) and a static tensile test. The differences between the analyzed mechanical properties determined using formulas (6) and (7) and the values obtained in industrial conditions were 3.6% for the yield strength, approximately 1.3% for the ultimate tensile strength, and approximately 2.7% for the plasticity reserve. The differences between the most important mechanical parameters determined during a static tensile test of a wire rod under industrial conditions and material after physical modelling were 1.5% for the yield strength, approximately 6.1% for the ultimate tensile strength, and approximately 4.1% for the relative reduction of area at fracture and plasticity reserve. Results obtained using these two test methods are comparable. The greater accuracy of results of mechanical properties tests obtained using Equations (6) and (7) may result from the fact that these relations take into account the chemical composition of a particular steel grade and ferrite grain size. An advantage of the research methodology proposed in this paper in relation to the works published so far is that it can quickly and accurately determine selected mechanical properties during a static tensile test (considering the results obtained using Equations (6) and (7)). The static tensile test also makes it possible to determine the relative reduction of area at fracture Z, which is also an important parameter when assessing whether the steel can be processed by cold plastic processing.

4. Directions of Further Research

In the future, we have planned to carry out the physical modelling of the analyzed process with the use of a complex deformation scheme (simultaneous tension and torsion) during the modelling of the rolling stages in NTM and RSM blocks of a rolling mill. The purpose of these tests is to create a strong deformation (neck) location in the deformed samples. Based on the results of many experimental and theoretical studies, it has been proven that the deformation speed in the neck during a tensile test is higher than the average speed calculated on the basis of changes in the measurement length of samples [41]. Taking advantage of this phenomenon will result in the obtained deformation speed being at a level at which, according to the literature data [32], the plasticizing stress of the tested material will not show significant changes.

An increase in the deformation speed in the analyzed process should positively affect the stress and activation of microstructure reconstruction processes and consequently increase the accuracy of the results of physical modelling (ferrite grain size and mechanical properties) in relation to the results of industrial tests. However, this will require designing a new geometry of samples, as the current geometry leads to a deformation location outside the research area, which disqualifies it for further analysis (Figure 18). The location of deformation outside the research area is also caused, in the analyzed case, by the dynamically changing temperature at this stage of the analyzed rolling process (rapid overheating after accelerated cooling between the NTM and RSM blocks of a rolling mill and the related inertia of the heat conduction phenomenon) and the geometry of the induction coil used in the STD 812 torsion plastometer.

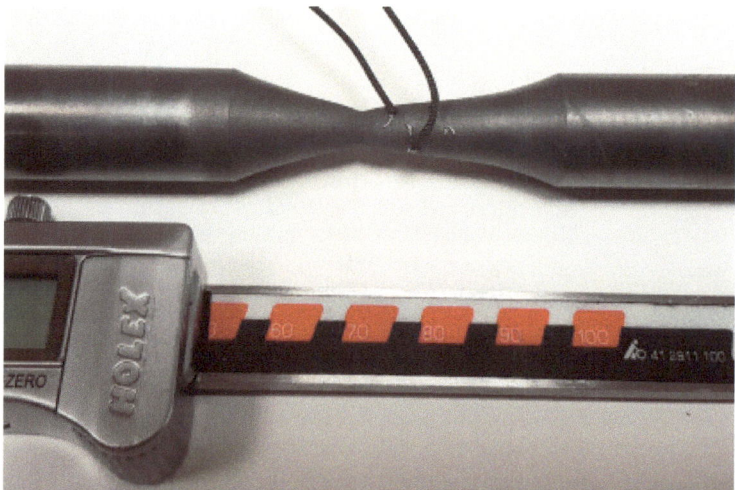

Figure 18. The view of the 20MnB4 steel sample after the deformation process in a complex deformation state (simultaneous torsion and tension) and location of the deformation outside the test area.

In addition, we have planned to carry out tests of the analyzed process with alternating torsion and alternating compression and tensile tests.

5. Summary and Conclusions

The speed of implementing the results of theoretical calculations and laboratory scale tests in industry determines the development and dissemination of new technologies. Industrial research constitutes the final and very costly part of the implementation process, costly due to involving significant levels of production, amounts of manpower and materials, and the consumption of utilities. The costs of implementing new technologies can be significantly reduced and the process itself can be simplified and accelerated with the use of modern research methods. Physical modelling of the dynamic wire rod rolling in modern rolling mills is a complicated issue. This is due to, among others, high deformation speeds in individual rolling stands and short intervals between successive deformations. Therefore, during physical modelling of this process, it is necessary to take advantage of certain simplifications, for example, by replacing the sequence of several deformations with a single deformation. In the case of the analyzed, dynamic plastic treatment processes, such a solution is acceptable, as evidenced by the research results published, among others, in works [17,40].

Results of the parameters of the deformation process from physical modelling that are similar to those occurring in actual processes ensures the high accuracy of the obtained results, and they affect the shape and nature of changes in the plasticizing stress of the tested material and consequently the microstructure and properties of the finished product. In this paper, a similar value of plasticizing stress to that occurring in the actual rolling process [17] was achieved by changing the deformation pattern during the final stage of physical modelling from non-free torsion to the tension (variant V3). This is crucial for activating the microstructure reconstruction processes during the controlled cooling process immediately after the deformation is completed.

On the basis of the results presented in the paper, the following conclusions can be drawn:

1. The best variant is the V3 variant, which consists of physically replicating the rolling process in a bar rolling mill as multi-sequence non-free torsion; the rolling process in an NTM block (No-Twist Mill) as non-free continuous torsion, with the total deformation equal to the actual deformation occurring during this stage of the technological process; and the rolling process in an RSM block (reducing and sizing mill) as a tension, while maintaining the total deformation value in this block;
2. The difference between the ferrite grain size measured in the cross-section of a wire rod and obtained as a result of physical modelling according to the V3 technological variant was 4.3%, whereas the error between the ferrite grain sizes measured in the longitudinal cross-section of a wire rod and obtained as a result of physical modelling for this technological variant was 11.2%. On this basis, it can be concluded that the average ferrite grain size in a wire rod obtained under industrial conditions is similar to the grain size obtained in samples after physical modelling of the rolling process according to variant V3;
3. The differences between the most important mechanical parameters determined during a static tensile test of a wire rod under industrial conditions and material after physical modelling were 1.5% for the yield strength, approximately 6.1% for the tensile strength, and approximately 4.1% for the constriction and plasticity reserve;
4. The developed research methodology (variant V3) allows for replicating the entire deformation cycle while precisely preserving the temperature value of the deformed material during individual stages of the analyzed technological process and allows for quickly and accurately determining the most important mechanical properties during a static tensile test;
5. Using a variable deformation scheme increases the research possibilities of modern torsion plastometers in terms of physical modelling of dynamic thermal and plastic treatment processes.

Funding: This research received no external funding.

Institutional Review Board Statement: Not applicable.

Informed Consent Statement: Not applicable.

Data Availability Statement: The data presented in this study are available on request from the corresponding author.

Conflicts of Interest: The author declares no conflict of interest.

References

1. Grosman, F.; Woźniak, D. Nowoczesne walcownie walcówki (Modern wire rod mills). *Hut.-Wiadomości Hut. (Met.-Lurgist-Metall. News)* **2001**, *3*, 97–104.
2. Laber, K. Problemy fizycznego modelowania procesów walcowania walcówki z dużymi prędkościami (Problems of physical modelling of wire rod rolling processes at high speeds). *Obróbka Plast. Met. (Met. Form.)* **2016**, *XXVII*, 119–132.
3. Kuziak, R. Modelowanie zmian struktury i przemian fazowych zachodzących w procesach obróbki cieplno-plastycznej stali. In *Modeling of Structure Changes and Phase Transformations Occurring in the Processes of Thermo-Plastic Treatment of Steel*; Instytut Metalurgii Żelaza (Institute for Ferrous Metallurgy): Gliwice, Poland, 2005.

4. Kajzer, S.; Kozik, R.; Wusatowski, R. Wybrane zagadnienia z procesów obróbki plastycznej metali. In *Projektowanie Technologii (Selected Problems from Metal Forming Processes. Technology Design)*. Wyd. Politechniki Śląskiej; Publishing House of the Silesian University of Technology: Gliwice, Poland, 1997.
5. Zhauyt, A.; Bukayeva, A.; Tursymbekova, Z.; Bulekbayeva, G.; Begendikova, Z.; Mambetaliyeva, G.; Chazhabayeva, M. Development of hot rolling technology using the method of physical modeling. *Metalurgija* **2021**, *60*, 387–390.
6. Erisov, Y.A.; Grechnikov, F.V. Physical modelling of hot rolling for low-density alloy of the Al-Mg-Li-Zr-Zn-Sc System. *Metallurgist* **2018**, *61*, 822. [CrossRef]
7. Liu, C.; Mapelli, C.; Peng, Y.; Barella, S.; Liang, S.; Gruttadauria, A.; Belfi, M. Dynamic recrystallization behavior of low-carbon steel during the flexible rolling process: Modeling and characterization. *Steel Res. Int.* **2021**, *93*, 1–15. [CrossRef]
8. Kuziak, R. Physical simulation of thermomechanical treatment employing gleeble 3800 simulator. In Proceedings of the 15th International Met-Allurgical & Material Conference—METAL 2006, Červený Zámek, Hradec nad Moravicí, Czech Republic, 23–25 June 2006.
9. Dziedzic, M.; Turczyn, S.; Kuźmiski, Z. Multistage compression modelling during hot rolling of steel products. In Proceedings of the 25th Anniversary International Conference on Metallurgy and Materials—METAL 2016, Brno, Czech Republic, 25–27 May 2016; pp. 279–284.
10. Siciliano, F.; Allen, B.J.; Rodrigues, S.F.; Jonas, J.J. Physical simulation methods applied to hot rolling of linepipe steels. *Mater. Sci. Forum* **2018**, *941*, 438–442. [CrossRef]
11. Siciliano, F.; Allen, B.J.; Ferguson, D. Hot torsion tests—A reliable rolling simulation method for C-Mn steels. *Mater. Sci. Forum* **2016**, *879*, 1783–1787. [CrossRef]
12. Ferreira, J.C.; Machado, F.R.d.S.; Aranas, C.; Siciliano, F.; Pasco, J.; Reis, G.S.; De Miranda, E.J.P.; Paiva, A.E.M.; Rodrigues, S.F. Physical simulation based on dynamic transformation under hot plate rolling of a Nb-microalloyed steel. *Front. Mater.* **2021**, *8*, 716967. [CrossRef]
13. Berdjane, D.; Fares, M.; Baccouche, M.; Lemmoui, A.; Fares, L. Deformation behavior of a Nb-Ti-V microalloyed steel to achieve the HSLA X80 grade by simulation with a torsion test and pilot hot rolling mill. *Metall. Res. Technol.* **2012**, *109*, 465–475. [CrossRef]
14. Felker, C.A.; Speer, J.G.; De Moor, E.; Findley, K.O. Hot strip mill processing simulations on a Ti-Mo microalloyed steel using hot torsion testing. *Metals* **2020**, *10*, 334. [CrossRef]
15. Kuziak, R.; Pietrzyk, M. Zastosowanie nowoczesnych metod modelowania matematycznego do optymalizacji procesu walcowania blach na gorąco (Application of modern methods of mathematical modeling to optimize the process of hot rolling of sheets). Materiały 2 konferencji Walcowanie i przetwórstwo blach i taśm. In Proceedings of the 2nd Conference Rolling and Processing of Sheets and Strips, Poraj, Poland, 18–20 October 1995; Wydawnictwo Politechniki Częstochowskiej (Publishing House of Czestochowa University of Technology: Czestochowa, Poland, 1995; p. 40.
16. Laber, K.; Milenin, A.; Markowski, J. Metodyka fizycznego modelowania zjawisk zachodzących w materiale podczas procesu regulowanego walcowania prętów okrągłych (Methodology of physical modelling of phenomena occurring in the material during the process of controlled rolling of round bars). In *Polska Metalurgia w Latach 2002–2006 (Polish Metallurgy in 2002–2006)*; Świątkowski, K.R., Ed.; Wydawnictwo Naukowe AKAPIT (AKAPIT Scientific Publishing House): Kraków, Poland, 2006; pp. 519–526. ISBN 83-910159-4-7.
17. Laber, K. *Nowe Aspekty Wytwarzania Walcówki Ze Stali Do Spęczania Na Zimno (New Aspects of Wire Rod Production from Steel for Cold Heading)*; Seria: Monografie nr 79 (Series: Monograph No. 79); Czestochowa University of Technology, Faculty of Production Engineering and Materials Technology Publishing House: Częstochowa, Poland, 2018; ISBN 978-83-63989-64-4. ISSN 2391-632X.
18. Knapiński, M.; Markowski, J. The physical modelling of a normalizing rolling of plates of S460NL1 steel grade. *Arch. Mater. Sci. Eng.* **2007**, *28*, 373–376.
19. Knapiński, M.; Dyja, H.; Kwapisz, M.; Frączek, T. Analysis of the finish rolling temperature influence on super fine-grained constructional steel plate microstructure. *Hut. Wiadomości Hut.* **2009**, *76*, 330–335.
20. Laber, K. Modelowanie i optymalizacja procesów regulowanego walcowania i kontrolowanego chłodzenia wyrobów walcowni bruzdowych (Modelling and optimization of the processes of controlled rolling and controlled cooling of products of groove rolling mills). Metalurgia. In *Nowe Technologie i Osiągnięcia (Metallurgy. New Technologies and Achievements)*; Dyja, H., Ed.; Wydawnictwo Wydziału Inżynierii Procesowej, Materiałowej i Fizyki Stosowanej Politechniki Często-chowskiej (Publishing house of the Faculty of Process Engineering, Materials Engineering and Applied Physics of the Czestochowa University of Technology): Czestochowa, Poland, 2009; pp. 99–122. ISSN 2080-2072. ISBN 978-83-87745-13-4.
21. Laber, K.; Knapiński, M. Determining conditions for thermoplastic processing guaranteeing receipt of high-quality wire rod for cold upsetting using numerical and physical modelling methods. *Materials* **2020**, *13*, 711. [CrossRef] [PubMed]
22. Laber, K.; Kułakowska, A.; Dyja, H. Physical modelling of the process of rolling $AlZn_{5.5}MgCu$ aluminum alloy bars on the RSP14/40 three-high reeling mill. In Proceedings of the 27th International Conference on Metallurgy and Materials—METAL 2018, Brno, Czech Republic, 23–25 May 2018; pp. 1599–1604, ISBN 978-80-87294-84-0.
23. Gryc, A.; Bajor, T.; Dyja, H.; Sawicki, S.; Laber, K. Physical modelling of plastic deformation conditions for the rolling proces of AZ31 bars in a three high skew rolling mill. *Metalurgija* **2014**, *53*, 489–492.
24. Wong, S.; Hodgson, P.; Thomson, P. Comparison of torsion and plane-strain compression for predicting mean yield strength in single-and multiple-pass flat rolling using lead to model hot steel. *J. Mater. Process. Technol.* **1995**, *53*, 601–616. [CrossRef]

25. Laber, K. Wpływ historii odkształcenia, stanu odkształcenia oraz prędkości odkształcenia na naprężenie uplastyczniające, mikrostrukturę i własności mechaniczne stali 30MnB4 podczas fizycznego modelowania procesu walcowania prętów. Influence of the strain history, strain state and strain rate on the flow stress, microstructure and mechanical properties of the 30MnB4 steel grade during physical modeling of the bar rolling process. *Hut.-Wiadomości Hut. (Metall.-Metall. News)* **2016**, *83*, 232–237. [CrossRef]
26. Danno, A.; Tanaka, T. Hot forming of stepped steel shafts by wedge rolling with three rolls. *J. Mech. Work. Technol.* **1984**, *9*, 21–35. [CrossRef]
27. Laber, K.; Leszczyńska-Madej, B. Theoretical and experimental analysis of the hot torsion process of the hardly deformable 5XXX series aluminium alloy. *Materials* **2021**, *14*, 3508. [CrossRef] [PubMed]
28. Hodgson, P.D.; Gibbs, R.K. A Mathematical model to predict the mechanical properties of hot rolled C-Mn and microalloyed steels. *ISIJ Int.* **1992**, *32*, 1329–1338. [CrossRef]
29. Sawada, Y.; Foley, R.P.; Thompson, S.W.; Krauss, G. *35th MWSP Conference Proceedings*; ISS-AIME: Pitsburgh, PA, USA, 1994; p. 263.
30. PN-EN 10263-4:2004; Stal-Walcówka, pręty i drut do spęczania i wyciskania na zimno. Część 4: Warunki techniczne dostawy stali do ulepszania cieplnego (Wire rod, rods and wire for upsetting and cold extrusion. Part 4: Technical delivery conditions for steel for heat treatment); Polska Norma (Polish Standard). Polski Komitet Normalizacyjny (Polish Committee for Standardi-zation): Warszawa, Poland, 2004.
31. Dobrzański, L.A. *Metaloznawstwo Z Podstawami Nauki O Materiałach (Metal Science with the Basics of Materials Science)*; Wydawnictwa Naukowo-Techniczne (Scientific and Technical Publishers): Warszawa, Poland, 1996.
32. Gorbanev, A.A.; Zhuchkov, S.M.; Filippov, V.V.; Timoshpolskij, V.I.; Steblov, A.B.; Junakov, A.M.; Tishhenko, V.A. *Teoreticheskie I Tekhnologicheskie Osnovy Vysokoskorostnoj Prokatki Katanki (Theoretical and Technological Basis of High Speed Wire Rod Production)*; Izdatelstvo Vyshehjshaja Shkola (College Publishing House): Minsk, Ukraine, 2003.
33. Kajzer, S.; Kozik, R.; Wusatowski, R. *Walcowanie Wyrobów Długich. Technologie Walcownicze (Rolling of Long Products. Rolling Technologies)*; Wyd. Politechniki Śląskiej (Publishing house of the Silesian University of Technology): Gliwice, Poland, 2004.
34. PN-EN 10025; Wyroby Walcowane Na Gorąco Z Niestopowych Stali Konstrukcyjnych. Warunki Techniczne Dostawy (Hot Rrolled Products of Non-Alloy Constructional Steels. Technical Conditions of Delivery); Polska Norma (Polish Standard). Polski Komitet Normalizacyjny (Polish Committee for Standardization): Warszawa, Poland, 2002.
35. Paduch, J.; Szulc, W. (Eds.) Kształtowanie nowych jakości oraz racjonalizacja kosztów wytwarzania wyrobów stalowych dostosowanych do wymagań konkurencyjnych rynku. In *CZ. 3: Dostosowanie Technologii Hutniczych do Aplikacyjnych i Jakościowych Potrzeb Rynku (Shaping New Qualities and Rationalizing of the Steel Products Production Costs Adapted to the Market's Competitive Requirements. Part 3: Adaptation of Metallurgical Technologies to the Application and Quality Needs of the Market)*; Works of the Iron Metallurgy Institute: Gliwice, Poland, 2000; Volume 52, pp. 17, 19, 24–25.
36. Grosman, F.; Hadasik, E. *Technologiczna Plastyczność Metali, Badania Plastometryczne (The Technological Plasticity of Metals, Plastometric Testing)*; Publishing house of the Silesian University of Technology: Gliwice, Poland, 2005; pp. 11–12. ISBN 83-7335-204-X.
37. Horsinka, J.; Drozd, K.; Kliber, J.; Ostroushko, D.; Černý, M.; Mamuzic, I. Strain and strain rate in torsion test. In Proceedings of the 20th International Conference on Metallurgy and Materials—METAL 2011, Brno, Czech Republic, 18–20 May 2011; Publisher TANGER Ltd.: Ostrava, Czech Republic, 2011; pp. 383–389, ISBN 978-80-87294-24-6. ISSN 2694-9296.
38. Dyja, H.; Krakowiak, M. *Kronika 60-Lecia—Od Wydziału Metalurgicznego do Wydziału Inżynierii Procesowej, Materiałowej i Fizyki Stosowanej (The 60th Anniversary Chronicle—From the Faculty of Metallurgy to the Faculty of Process, Materials Engineering and Applied Physics)*; Wydawnictwo Wydziału Inżynierii Procesowej, Materiałowej i Fizyki Stosowanej Politechniki Częstochowskiej (Czestochowa University of Technology, Faculty of Process, Materials Engineering and Applied Physics Publishing House): Częstochowa, Poland, 2010.
39. PN-84/H-04507/01; Metale—Metalograficzne badania wielkości ziarna. Mikroskopowe metody określenia wielkości ziarna (Metals—Metallographic tests of grain size. Microscopic methods of determining grain size); Polski Komitet Normalizacji, Miar i Jakości (Polish Committee for Standardization, Measurement and Quality). Polska Norma (Polish Standard), Wydawnictwa Normalizacyjne Alfa (Alfa Standardization Publishing House): Warszawa, Poland, 1985.
40. Laber, K.; Dyja, H.; Koczurkiewicz, B.; Sawicki, S. Fizyczne modelowanie procesu walcowania walcówki ze stali 20MnB4 (Physical modeling of the wire rod rolling process of 20MnB4 steel). In Proceedings of the VI Konferencja Naukowa WALCOWNICTWO 2014. Procesy-Narzędzia-Materiały (VI Scientific Conference Rolling Mill Practice 2014. Process-es-Tools-Materials), Ustroń, Poland, 20–22 October 2014; pp. 37–42.
41. Dyja, H.; Gałkin, A.; Knapiński, M. *Reologia Metali Odkształcanych Plastycznie (Rheology of Plastically Deformed Metals)*; Seria: Monografie nr 190 (series: Monograph No. 190); Publishing house of the Czestochowa University of Technology: Częstochowa, Poland, 2010; pp. 160–165, 203, 217–222. ISBN 978-83-7193-471-1. ISSN 0860-5017.

Disclaimer/Publisher's Note: The statements, opinions and data contained in all publications are solely those of the individual author(s) and contributor(s) and not of MDPI and/or the editor(s). MDPI and/or the editor(s) disclaim responsibility for any injury to people or property resulting from any ideas, methods, instructions or products referred to in the content.

Article

FEM Numerical and Experimental Study on Dimensional Accuracy of Tubes Extruded from 6082 and 7021 Aluminium Alloys

Dariusz Leśniak [1,*], Józef Zasadziński [1], Wojciech Libura [1], Krzysztof Żaba [1], Sandra Puchlerska [1], Jacek Madura [1], Maciej Balcerzak [1], Bartłomiej Płonka [2,*] and Henryk Jurczak [3]

1. Faculty of Non-Ferrous Metals, AGH University of Science and Technology, 30-059 Kraków, Poland
2. Łukasiewicz Research Network—Institute of Non-Ferrous Metals, 44-100 Gliwice, Poland
3. Albatros Aluminium Corporation, 78-600 Wałcz, Poland
* Correspondence: dlesniak@agh.edu.pl (D.L.); bartlomiej.plonka@imn.lukasiewicz.gov.pl (B.P.); Tel.: +48-12-617-31-96 (D.L.); +48-12-276-40-88 (B.P.)

Citation: Leśniak, D.; Zasadziński, J.; Libura, W.; Żaba, K.; Puchlerska, S.; Madura, J.; Balcerzak, M.; Płonka, B.; Jurczak, H. FEM Numerical and Experimental Study on Dimensional Accuracy of Tubes Extruded from 6082 and 7021 Aluminium Alloys. *Materials* **2023**, *16*, 556. https://doi.org/10.3390/ma16020556

Academic Editor: Francesco Iacoviello

Received: 7 December 2022
Revised: 23 December 2022
Accepted: 26 December 2022
Published: 6 January 2023

Copyright: © 2023 by the authors. Licensee MDPI, Basel, Switzerland. This article is an open access article distributed under the terms and conditions of the Creative Commons Attribution (CC BY) license (https://creativecommons.org/licenses/by/4.0/).

Abstract: The extrusion of hollow profiles from hard-deformable AlZnMg alloys by using porthole dies encounters great technological difficulties in practice. High extrusion force accompanies the technological process, which is caused by high deformation resistance and high friction resistance in extrusion conditions. As a result of high thermo-mechanical loads affecting the die, a significant loss of dimensional accuracy of extruded profiles can be observed. The different projects of porthole dies for the extrusion of Ø50 × 2 mm tubes from the 7021 alloy were numerically calculated and then tested in industrial conditions by using a press of 25 MN capacity equipped with a container with a diameter of 7 inches (for 7021 alloy and 6082 alloy for comparison). New extrusion die 3 with modified bridge and mandrel geometry and a special radial–convex entry to the die opening was proposed. FEM was applied to analyse the metal flow during extrusion, geometrical stability of extruded tubes and the die deflection. The photogrammetric measuring method was used to evaluate dimensional accuracy of tubes extruded in different conditions and geometrical deviations in porthole dies elements, especially the bridges and the mandrels. Research revealed a high dimensional accuracy of tubes extruded from the 6082 alloy and from the 7021 alloy by using original extrusion die 3, while much higher dimensional deviations were noted for tubes extruded from the 7021 alloy by using extrusion dies 1 and 2, particularly in relation to the circularity, centricity and wall thickness.

Keywords: AlZnMg alloys; extrusion; porthole dies; metal flow; die deflection; extrudates dimensional accuracy

1. Introduction

The hollow extruded profiles from 7000 series aluminium alloys are currently used in a wide application of the construction of various means of transportation, including important automotive systems. Therefore, the profiles must fulfil high quality expectations, in particular, concerning dimensional accuracy. The extrusion of aluminium alloys becomes more difficult as its strength is increased. The 7000 series alloys not containing copper are easier to extrude compared to, e.g., high strength 7075 alloys, but the extrusion force is high, which is what implicates various dimensional defects in the extruded profiles. Additional difficulties are connected with a high friction at the tool surface, which produces non uniform metal flow from the die cavity. The extrusion parameters and the porthole die design can also have an impact on the extrudate geometry. The complex set of reasons of the geometrical defects in extrudate comprises the result from the unsuitable design of the porthole die and parameters of the whole production line, including extrusion process, heat treatment on the run-out table and stretching. In turn, the production parameters cover types of alloy, extrusion force, billet temperature, extrusion speed, rate of cooling in the quenching tunnel and force of stretching.

The most important tolerance measures are as follows: wall thickness, circularity and eccentricity, convexity and concavity of the walls, angularity, straightness, bow and twist along the profile. The direct reasons for the loss of profile dimensional accuracy are as follows: elastic deflection of die, including deflection of mandrels; non-uniform metal flow within the die cavity; imperfect design of the welding chamber and bearings; die wear; non-uniform temperature distribution within the billet; too high a cooling rate within the quenching tunnel; and sometimes excessive plastic deformation during stretching. When designing the extrusion process, one should take account of the fact that tighter tolerances will push up the cost and make the production longer.

When a new profile is produced, trials and corrections of the die are necessary in order to fulfil required geometrical tolerances, as well as to achieve the compromise between effectiveness, surface quality and die life. Therefore, that finite element simulation (FEM) is a useful tool for analysing the relation between the process parameters, the tool design and the profile quality.

Many researchers analysed the influence of the porthole die geometry on die deflection and the resulting geometrical quality of the extruded aluminium profiles [1–5]. Pinter et al. [1] analysed the influence of process parameters (ram speed, billet length and alloy) on the achieved die deformation after a determined extruded amount in the case of 6005A alloys. The authors have found that the mandrel deflection, which influences profile geometry, rises as the amount of extruded alloys increases. In the study [2], the mandrel fracture behaviour was examined through investigating the elastic deformation of the mandrel during the extrusion using FE analysis. After optimisation with the use of HyperXtrude 13.0 software, the desired die geometry giving proper material flow at the die exit and small mandrel deflection were obtained [3]. Xue et al. [4] applied different modifications of the porthole die, including two-step welding chamber to obtain uniform metal flow. An original system for the monitoring of the tool deflection, of the tool temperature and at the die bearings, as well as the pressure in the die opening, were developed and tested within [5]. The numerical simulation was applied to find the optimal die design and a proper metal flow, guaranteeing the good geometry of the product [6–8]. Guan et al. [9] analysed the design of a multihole extrusion die and investigated the effects of the layout of holes on the extrusion process. HyperXtrude program was used to process simulation and the die with three portholes was the optimum one, where the uniform velocity distribution, maximum welding pressure, minimum required extrusion load, and minimum die stress were obtained.

The influence of die deformation on the speed, temperature distribution and distortion of the two profiles from AA6082 alloy is reported and analysed in [10]. As a consequence of the die deflection, a bended profile with a large curvature radius was produced.

Refs. [11,12] reported that to obtain a uniform metal flow at the die exit, the porthole die was optimized by adding baffle plates on the die insert. After optimization, the concave defects on the profile were remarkably limited. Various modifications in the porthole die design were proposed by Xue et al. [4] to improve the homogeneity of metal flow from the die, giving a relative high product accuracy for a complex section from the 6060 alloy.

Hsu et al. [13] studied the varying of the welding chamber geometry and bearing length to obtain a uniform material flow. Finite element analysis and Taguchi method were used to obtain a better porthole die design for the extrusion of the 7075 alloy. Different welding chamber geometry and different length of die bearing caused metal flow to be more uniform.

Incidentally, it is worth noting that the AlZnMgCu alloys are very difficult as far as the dimensional accuracy is considered because of both the high extrusion force and deformation resistance leading to large die deflection and non-uniform metal flow. The large distortion of the rectangular hollow profile from the 6061 alloy was observed as a result of the non-uniform exit velocity [14]. To decrease this phenomenon, the second welding-chamber was applied and the bearing length and baffle block were modified in profile extrusion. In [15], the extrusion process of a large, aluminium profile for high-speed train was simulated to obtain uniform metal flow. Modifications to the porthole die, including shape of mandrels and length of the bearings, were proposed, which influenced the metal flow effectively. The baffle plates were applied in the welding chamber to balance the material flow velocity in the die cavity during the extrusion of the profile [16]. Through a series of modifications, the velocity difference in the cross-section of the profile decreased significantly. The exit velocity distribution of the profile was investigated by Chen et al. [17] by using the effects of eccentricity, the shape of the welding chamber, and uneven bearing length.

The study [18] presented 3D FEM simulations aimed at die design and process optimisation for the 7075 alloy with large differences in wall thickness. The influence of bearing length and extrusion speed on profile temperature and extrusion pressure were analysed. A longer bearing leads to a greater dimensional accuracy of the profile. In the work [19], the porthole die for the extrusion of a solid heatsink profile with high wall thickness variation was designed using finite element (FE) simulations. The structure of the die elements effectively influenced the flow behaviour of the metal. In consequence, the proposed solution can be implemented in industrial practice. The paper [20] presented the 3D FE simulation of the extrusion processes of the 7003 alloy through the porthole dies in different process variables, including billet temperature, bearing length, tube thickness and extrusion ratio. The products surface was also examined. In the work [21], the effects of the length and geometry of die land on curved profiles produced by a novel process, differential velocity sideways extrusion (DVSE), were studied through physical experiments and FEM.

In this work, the influence of different porthole die design on the metal flow during extrusion and on the dimensional accuracy of round tubes of Ø50 × 2 mm from the 6082 and 7021 alloys was investigated. The die deflection was experimentally measured for different design solutions to investigate the effect of the die deflection on the material flow. FEM simulation was applied in the design of the porthole dies and to predict the metal flow and the die deflection. The photogrammetric measuring method was used to evaluate the dimensional accuracy of tubes extruded in different conditions and the geometrical deviation of porthole dies, especially the bridges and the mandrels. The statistical correlation method was used to determine the influence of the technological parameters of the extrusion process on the dimensional accuracy of profiles. The ANOVA analysis was applied for extrusion dies.

2. Materials and Methods

2.1. Characteristics of AlMgSi and AlZnMg Alloys

The billets of 178 mm in diameter, of which chemical compositions are presented in Table 1 (6082 alloy) and Table 2 (7021 alloy), were DC cast in semi-industrial conditions. In the case of 7021 alloy, the low-melting microstructure components were dissolved during homogenization to a degree that was sufficient in practice—no incipient melting peaks on the DSC curves were observed (Figure 1). As a result, the significant increase in solidus temperature was obtained for 7021 alloy after homogenisation—572.1 °C with regard to the temperature of as-cast alloy at the level of 478.1 °C (Table 3).

Table 1. Chemical composition of 6082 alloy in mass percentage.

Alloy Denotation	Si	Fe	Cu	Mg	Cr	Zn	Ti	Zr
6082	1.3	0.21	0.03	0.59	0.00	0.4	0.2	0.1

Table 2. Chemical composition of 7021 alloy in mass percentage.

Alloy Denotation	Si	Fe	Cu	Mg	Cr	Zn	Ti	Zr
7021	0.08	0.21	0.00	2.12	0.00	5.47	0.01	0.15

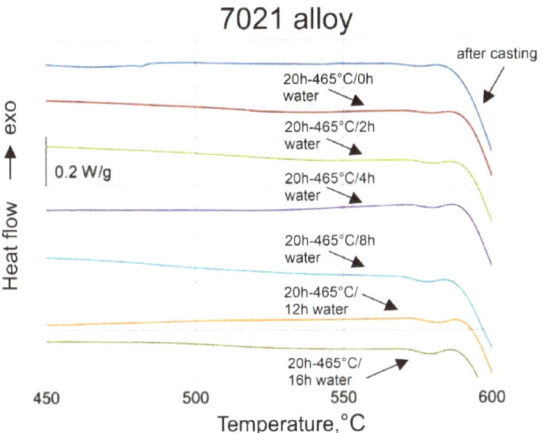

Figure 1. DSC curves for 7021 alloy after homogenization in different conditions.

Table 3. DSC test results of homogenized 7021 alloy [22].

Alloy	Solidus Temperature, °C	Incipient Melting Heat, J/g
7021 alloy	572.1	0.29

2.2. FEM Numerical Modeling of Extrusion Process

The QForm-Extrusion software and specially prepared material model were used to analyse the extrusion process of tubes of Ø50 × 2 mm from aluminium alloy grade 7021 (AlZnMg) using porthole dies of various geometry. Combined Lagrangian–Eulerian approach is the method that is used in QForm-Extrusion in order to describe the deformed material motion. This approach combines two basic formulations, taking advantages of both of them—from the Eulerian: adapted stationary mesh that allows improving the accuracy of metal flow prediction significantly and reduces simulation time, and from the Lagrangian: dynamically movable mesh to animate the profile flow after the bearing exit. These features, consolidated by coupled thermal and mechanical tasks available in the software, allow the obtaining of a precise distribution of the profile velocities that, in turn, means the accurate prediction of parameters defining dimensional accuracy of the extrudates.

The first step in the numerical modelling FEM was preparation of proper 3D models of tools and billet using CAD Solid Works program. The geometry of the porthole die has a crucial importance from the point of view of minimum deformation resistance in the extrusion process, which, in turn, relates to minimisation of extrusion force, elastic deflection of the die, improvement in metal flow, dimensional accuracy of the profile and maximisation of the exit speed.

Three different solutions of 2-hole die geometry were taken to calculation—the die with maximum opening of the inlet channels (die 1), conventional porthole die for 6xxx alloys based on the local 3-armed bridges (die 2) and die 3, which is a modified version of die 2. The geometry of die 2 was taken from extrusion industry practice for 6082 alloy. The final design of extrusion die 3 was developed based on gradual optimization of different die geometry using FEM numerical simulations based on three assumed criteria. The first criterion defined a maximum force not exceeding 25 MN. The second optimization factor minimized the elastic deflection of the die during the process, not exceeding 0.5 mm. The third criterion included ensuring an even outflow of metal from the die bearing, so the velocity deviation did not exceed ±20% of the average value. In total, almost 100 numerical

FEM simulations were carried out analysing the influence of portholes, webs, cores, welding chambers, pockets and bearing geometry to determine the final specific die geometry.

The massive bridges were assumed for die 1 and adapted for hard 7xxx alloys; the thickness of bridges for die 3 was increased by 4 mm, whereas their length was increased by 40 mm in relation to die 2. In addition, a radial and convex entry to cavity of die 3 was added. The thickness of bridges was equal to 28, 16 and 20 mm for die 1, die 2 and die 3, respectively; length of bridges was 70, 55 and 95 mm, whereas height of the welding chambers was 25, 30 and 27 mm. The maximal broad inlet channels, shaped pockets, bearings of varied length and proper geometry of the central baffle and mandrels were adapted for die 3. The 3D models of the porthole dies discussed and the view from the bottom are presented in Figure 2a,b.

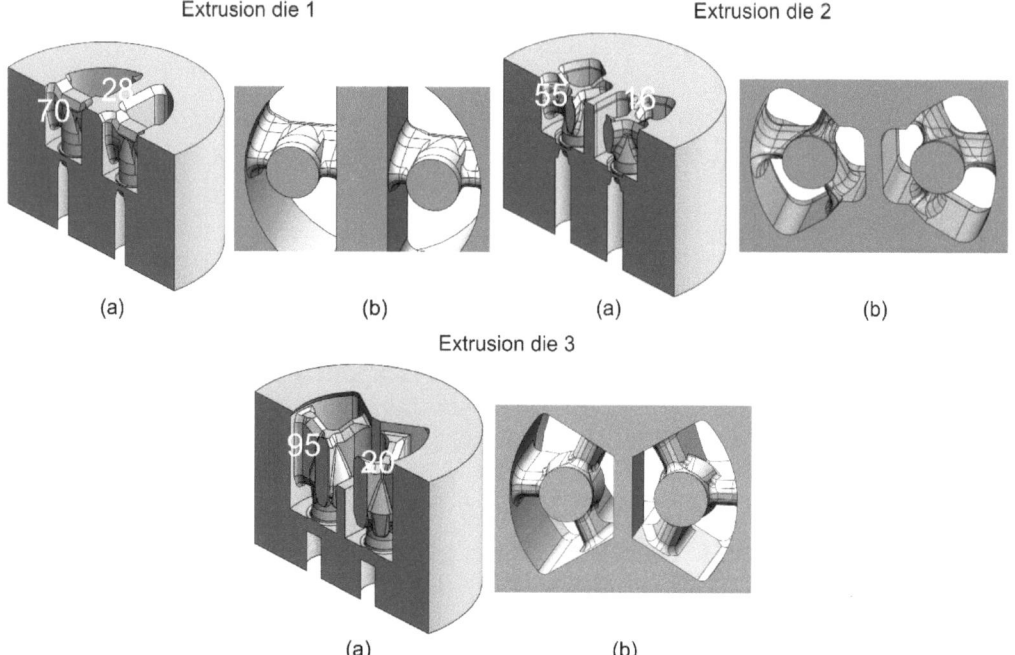

Figure 2. The 3D CAD models of different porthole dies for extrusion of tubes of Ø50 × 2 mm from 7021 alloy: (**a**) in the cross-sectional view and (**b**) in the bottom view—with the bridge dimensions marked (height and thickness).

The second step of the FEM modelling was preparation and implementation of the material model for the alloy 7021, which after approximation was expressed by the constitutive equation of Hansel–Spittel (Equation (1)). The starting point was the plastometric compression tests with the use of Gleeble 3800 simulator (Dynamic Systems Inc., Poestenkill, NY, USA), which allowed for determining stress–true strain equation, dependent on temperature and strain rate (Figure 3).

$$\sigma = 1090 e^{-0.0675T} \cdot T^{0.056} \cdot \varepsilon^{-0.055} \cdot e^{\frac{-0.015}{\varepsilon}} \cdot (1+\varepsilon)^{0.019T} \cdot e^{-0.026\varepsilon} \cdot \dot{\varepsilon}^{0.21} \cdot \dot{\varepsilon}^{-0.00019T} \quad (1)$$

where σ—plastic stress, ε—plastic strain, $\dot{\varepsilon}$—strain rate, T—temperature of deformation.

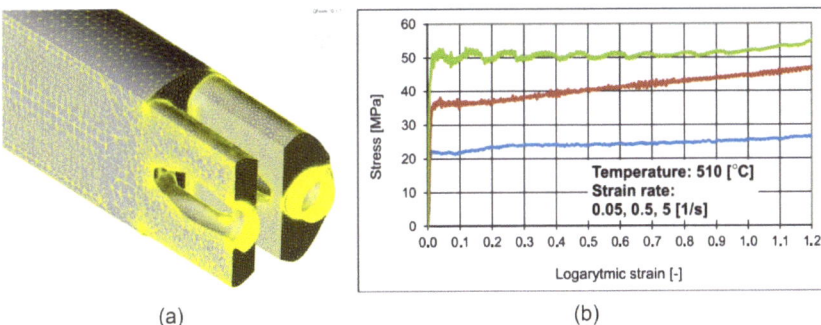

Figure 3. The geometrical model of extruded material with mesh elements (**a**) and determined material model for FEM calculations—dependence of flow stress vs. logarithmic strain for different strain rates for alloy 7021 (**b**): blue colour for 0.05 1/s, red colour for 0.5 1/s and green colour for 5 1/s.

The third step of the FEM modelling was defining of boundary conditions. Friction model used in QForm allowed for taking into account the adhesion effect between aluminium and steel. The constant shear friction law is adapted everywhere between workpiece and tool materials, except in bearings where the contact pressure is relatively low. The friction on the bearings was calculated based on the phenomenological model that includes a number of parameters, such as resulting bearings angle, pressure, etc. This approach allowed us to obtain the realistic behaviour of the metal in simulation with three possible zones: sticking, sliding, and separation zone. All the defined extrusion process parameters are presented in Table 4.

Table 4. All the defined parameters of the FEM modelled extrusion process of tubes of Ø50 × 2 mm from 7021 alloy.

Alloy	7021
Billet dimensions	Ø178 × 800 mm
Billet temperature	480, 500, 510, 520 °C
Container/Die temperature	450 °C
Extrusion ratio	42
Stem velocity	1.5–4 mm/s
Metal exit speed	2.5–10 m/min
Friction coefficient	1

2.3. Extrusion Trials of Round Tubes

The extrusion trials for the round tube of Ø50 × 2 mm from the 7021 alloy with the use of the discussed porthole dies were performed on the hydraulic direct press of 25 MN capacity and equipped with container of 7″ in diameter.

The conditions of the trials were identical to those in the numerical simulation for different billet heating temperature and extrusion speed. In the case of extrusion through die 3, the trials were performed for billet temperature of 480 °C and ram speed of 1.5 mm/s as a result of the experiences gained during previous extrusion with dies 1 and 2, which were also tested with higher extrusion speeds and higher billet temperatures. The run-out table of the press with the cooling system and the tested porthole dies of two holes are shown in Figure 4. Extruded tubes were next submitted to stretching with the true strain equal to 0.5%.

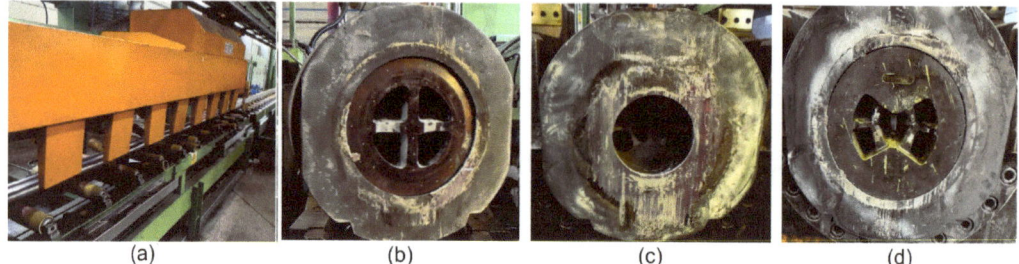

Figure 4. The 25 MN 7 inch extrusion press run-out table (**a**) and different extrusion porthole dies: extrusion die 1 (**b**), extrusion die 2 (**c**) and extrusion die 3 (**d**)—designed for 2-hole extrusion tests of tubes of Ø50 × 2 mm from 7021 alloy in industrial conditions.

During the trials, the following process parameters were recorded: ram speed, exit speed, extrusion pressure and profile temperature with the use of the data acquisition system.

The surface quality was inspected on-line with respect of cracking or tarnish. The maximal extrusion speed was recorded when cracking on profile surface began. The samples taken from the tubes extruded with different process conditions were submitted to optical scanning to check their dimensional accuracy. Similar optical scanning was performed on the used dies and their dimensions were compared with these for new ones.

2.4. 3D Optical Scanning of Extruded Tubes and Dies

The scanner GOM Atos Core 200 (Lenso, Poznań, Poland) (Figure 5a) was used for scanning of the extruded elements on the inner and outer surface of the samples to measure diameters of the tubes, deviations from circularity and wall thickness. The surface was cleaned off from the impurities before scanning. The coloured map of deviations was also obtained on the basis of the CAD model and scanned element.

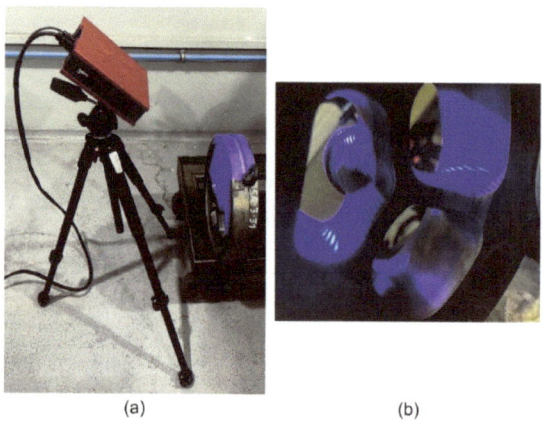

Figure 5. The scanner GOM Atos Core 200 for optical measuring the geometry of extruded tubes and dies (**a**) and the scanned extrusion porthole die (**b**).

The first step of the dies measurements included gathering coordinates of all the selected points on the samples' surface. Next, the detailed scanning from the different perspectives was realized to digitalize the whole available surface. Due to complicated shape of the dies, the process needed about 75 scans from each side. The surfaces of bridges, mandrels and bearing lands were inspected, in particular. The coloured maps of deviations allowed indications of the regions evidently subjected to wearing.

2.5. Statistical Analysis of Results for Extruded Tubes and Extrusion Dies

The analysis of the results was performed with the use of Gom Inspect program, and the deviations from the nominal geometry were generated. The deviations concerned diameters, ovality, eccentricity, wall thickness and surface quality of the extruded tubes. The results were analysed and the statistical correlations was obtained using Origin Lab code. The correlation determined interactions between selected variables but did not explain the reasons or the method of their origination. A positive correlation (correlation coefficient from 0 to 1) suggested that as one feature rose, the mean values of the second increased too. A negative correlation (correlation coefficient from -1 to 0) indicated the opposite dependence. The power of correlation is defined in the following way: 0.2 poor correlation; $0.2 \div 0.4$ low correlation (distinct dependence); $0.4 \div 0.6$ moderate correlation (key dependence); $0.6 \div 0.8$ high correlation (significant dependence); $0.8 \div 0.9$ very high correlation (very high dependence); $0.9 \div 1.0$ full dependence. In the case of dies, the results were converted to ANOVA statistics.

3. Results
3.1. FEM Numerical Calculations

Figure 6 presents the distributions of metal particles velocity during the extrusion of tubes. The metal flows from the inlet channels of the porthole die via welding chambers up to die opening—for design of die 1 (Figure 6a, on the left), die 2 (Figure 6b in the middle) and die 3 (Figure 6c, on the right). Some non-uniformity of metal flow while extruding through die 1 (Figure 6a) can be observed, which manifests itself by the diversification of particles velocity in the die cavity on the tube perimeter. In the outside regions of the die opening, the particles velocity is as high as 35.6 mm/s and is lower by about 31% in relation to the mean value, whereas in the inside regions of the die cavity (and simultaneously close to the central axis of extrusion) it is 53.8 mm/s and is higher by about 16% in relation to the mean value. Some lower values of the particles velocity were observed for die 2, equipped with two local three-dorsal bridges; however, such die design, based on three points feeding of the die opening, provided more uniform distribution of metal particles on the perimeter of the extruded tube in comparison to die 1 (Figure 6b). In the outside region of the die cavity, the particles velocity is as high as 45.1 mm/s (in case of die 1 was 35.6 mm/s) and is only around 5% lower than the mean values, whereas in the inside regions of the die cavity it is 47.5 mm/s and is higher by about 4% in relation to the mean value. The highest and the most uniform metal exit speed was observed in the case of die 3 (Figure 6c). The average metal exit speed for die 3 is about 25% higher compared to that of die 2, as a result of applying the radial and convex entry to the die cavity and original geometry of a central baffle wall regulating the metal flow.

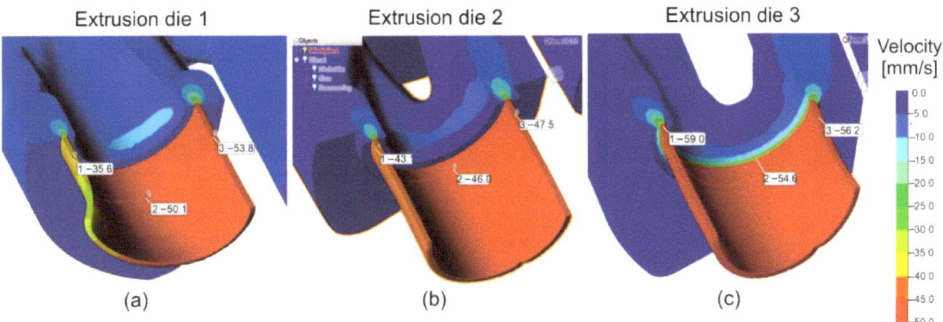

Figure 6. Distribution of metal velocity during extrusion of tubes of Ø50 × 2 mm from 7021 aluminium alloy through porthole dies of different geometry—die 1 (**a**), die 2 (**b**) and die 3 (**c**).

Figure 7 presents the distribution of temperature of the metal during extrusion for the design 1 of die (Figure 7a—on the left), design 2 (Figure 7b—in the middle) and design 3 (Figure 7c—on the right). The higher values of metal temperature of about 9–10 °C were predicted in the case of die 1 and die 3 (516.1 °C and 514.1 °C, in relation to 505.2 °C). The reason for this can be because of the existence of some higher mean velocities of particles and the higher extent of plastic deformation and its conversion into heat.

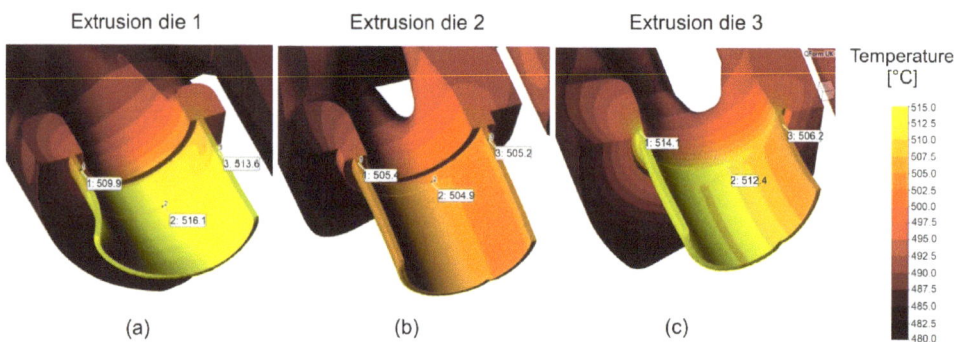

Figure 7. Distribution of metal temperature during extrusion of tubes of Ø50 × 2 mm from 7021 aluminium alloy through porthole dies of different geometry—die 1 (**a**), die 2 (**b**) and die 3 (**c**).

Figure 8 presents the distribution of means for the design of die 1 (Figure 8a—on the left), die 2 (Figure 8b—in the middle) and die 3 (Figure 8c—on the right). More beneficial states of stress occur for die 2 and die 3; whereas on the whole perimeter of the die land, the compression pressures are observed, favourable for the deformability of the material and the averages. For die 2, compressive stresses are higher in comparison to die 1 (maximum value of the compression pressure is 40.9 MPa in relation to 25.1 MPa). The most uniform compressive stresses in the die orifice on the extrudate perimeter are observed for die 3. In the case of die 2, within the inside region of the tube (die land of the die opening), positive values of pressure appear at the level of 12.1 MPa, which, in practice, can limit the exit velocity from the die opening.

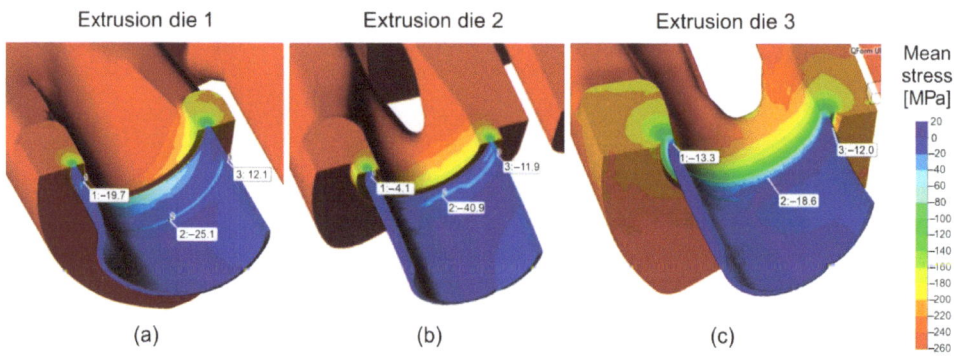

Figure 8. Distribution of mean stress during extrusion of tubes of Ø50 × 2 mm from 7021 aluminium alloy through porthole dies of different geometry—die 1 (**a**), die 2 (**b**) and die 3 (**c**).

Figure 9 presents the distribution of elastic deflection of the porthole die in the Z direction, particularly within the regions of the mandrel and the die land for all the die geometrical solutions. Coloured maps indicate that the highest dimensional deviations of the mandrel were made in the case of die 2, where an elastic deflection of 0.66 ÷ 0.80 mm was predicted within the area of the die land. In the corresponding area for die 1, the mandrel deflection was much lower and kept itself within the range 0.51 ÷ 0.55 mm. The lowest mandrel deflection of 0.37 ÷ 0.48 mm was predicted for die 3, depending on the mandrel position.

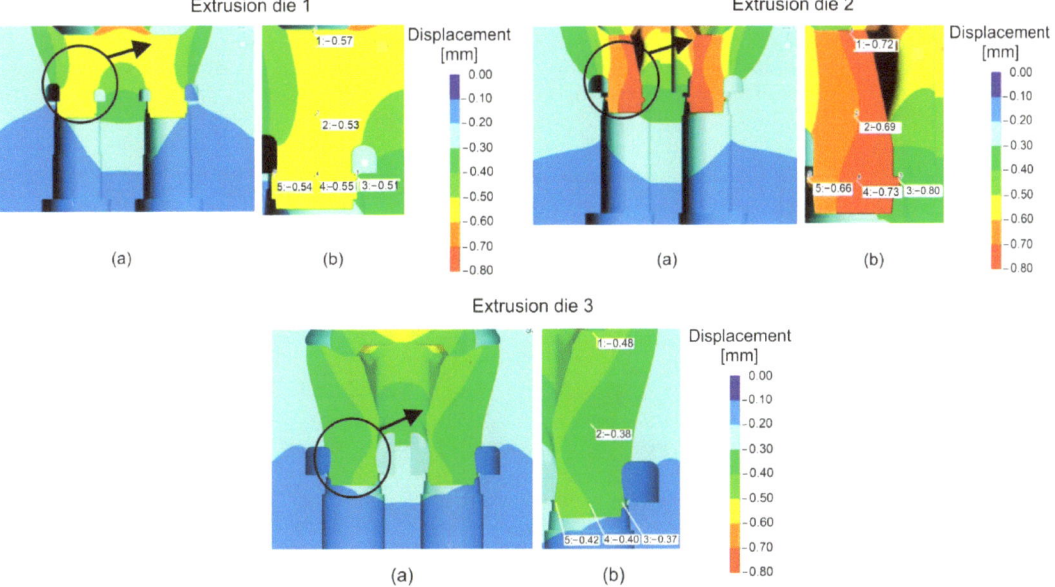

Figure 9. Distribution of die deflection during extrusion of tubes of Ø50 × 2 mm from 7021 aluminium alloy through porthole dies of different geometry—global view of mandrels (**a**) and local view of chosen part of mandrel (**b**).

The detailed data concerning the metal exit speed in the die cavity on the profile perimeter and elastic deflection in different regions of the mandrel for the all analysed dies are presented in Figure 10. Based on the calculations, it can be concluded that porthole die 3 is the most advantageous and guarantees the highest and most uniform metal exit speed and the lowest elastic deformation of the mandrels in the extrusion direction. This allows the forecasting of the smallest dimensional deviations of extruded tubes for die 3.

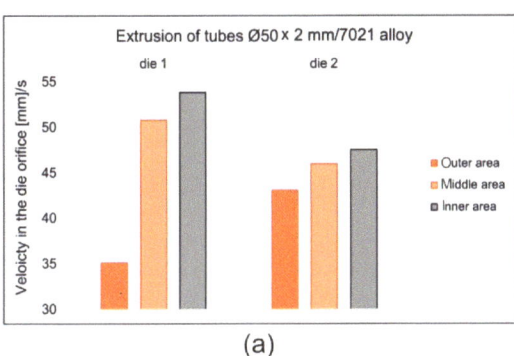
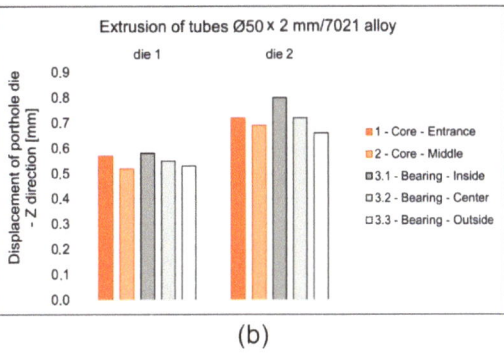

Figure 10. Diagrams of the metal velocity in the die cavity during extrusion of tubes of Ø50 × 2 mm from 7021 aluminium alloy through porthole dies of different geometry (**a**) and porthole die deflection in Z direction in different areas of mandrel for different die geometry (**b**).

Figure 11 presents the distributions of predicted indicator of the strength status for the porthole dies of different geometry. The indicator of the strength status—red area of the scale—means a high risk of cracking or even mechanical tearing of the die. The maximal values of the indicator for die 1 and die 2 occur in the region of mandrel tops. The significant improvement in the strength conditions in case of die 3 is observed as a result of the elimination of dangerous areas of the bridge arms and the mandrels. For die 3, the indicator of strength status is placed within the range of 0.92–0.95. These benefits result from appropriate changes in the design of the porthole die, i.e., the geometry of the inlet channels, bridges and mandrels.

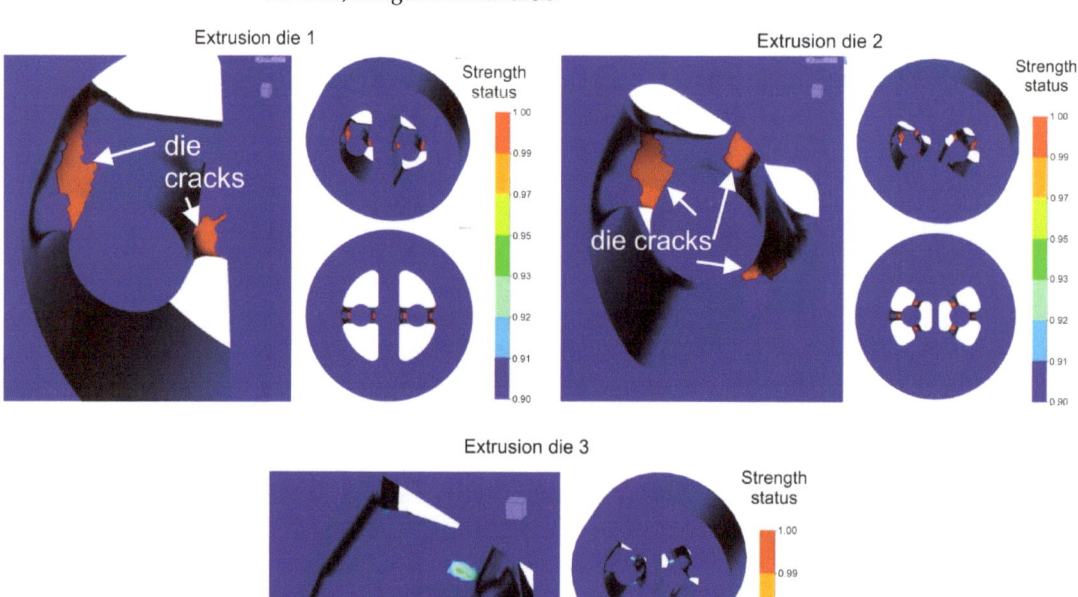

Figure 11. Distribution of strength status for the mandrel tops for different die geometry during extrusion of tubes of Ø50 × 2 mm from 7021 aluminium alloy.

3.2. Extrusion Trials

The extrusion of the examined tubes was performed for die 1, die 2 and die 3 for variable heating temperatures of billet: 520 °C, 510 °C, 500 °C and 480 °C and for variable extrusion speed: 1.5 ÷ 4 mm/s. In the case of die 3, the billet temperature was of 480 °C and the extrusion speed was of 1.5 mm/s. Figure 12 presents the tubes on the run-out of the press extruded by using different porthole dies. In general, for the all tested dies, high geometrical stability of the tubes and high quality of the tubes surface, free of tarnish and cracks, can be noticed. The extrusion pressure of 80–90% of the press capacity and exit speed from dies of 4.5 m/min (for die 1), 3.5 m/min (for die 2) and 4.5 m/min (for die 3) were recorded using a data acquisition system. The tubes of Ø50 × 2 mm, extruded from 7021 alloy in cross-section by using different dies, are presented in Figure 13.

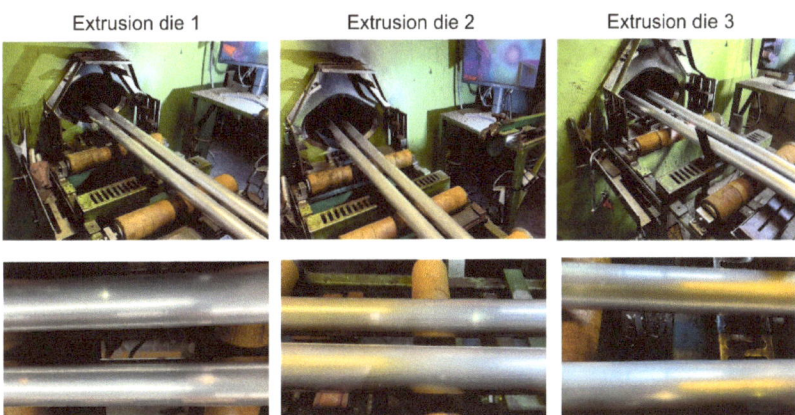

Figure 12. Extruded tubes of Ø50 × 2 mm from 7021 alloy on the press run-out table for dies of different geometry: die 1, T_0 = 480 °C, V_1 = 4.5 m/min; die 2, T_0 = 480 °C, V_1 = 3.5 m/min; die 3, T_0 = 480 °C, V_1 = 4.5 m/min.

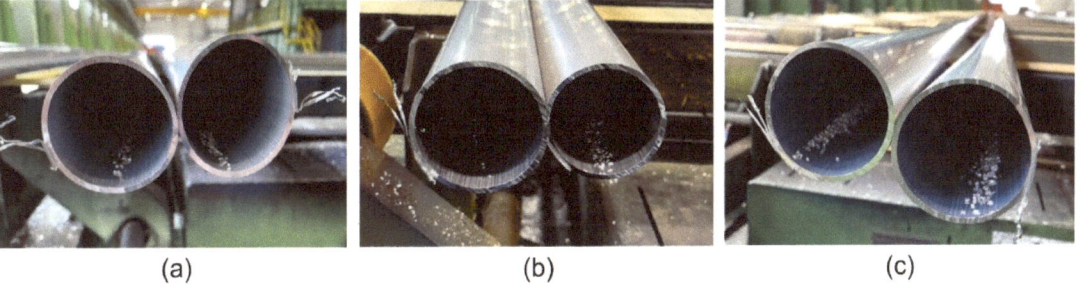

Figure 13. Tubes of Ø50 × 2 mm extruded from 7021 alloy in cross-section by using different dies—die 1 (**a**), die 2 (**b**) and die 3 (**c**).

Exemplary technological parameters of extrusion process for die 3 were shown in Figure 14. Experimental trials confirmed the possibility of obtaining a relatively high metal exit speed with good surface quality of profiles for die 3 with radial and convex entry to the die cavity.

During the tests for die 1 and 2 and for the lowest billet temperature, the cracking of a mandrel occurred in both the tested dies and the mandrels were damaged (Figure 15 on the left). In the case of die 3, bridges and mandrels were free from cracks after extrusion (Figure 15 on the right), which confirmed the earlier FEM predictions. The qualitative assessment of die 3 indicated minimal deflection of the bridges. The dies were submitted to quantitative analysis—3D optical scanning.

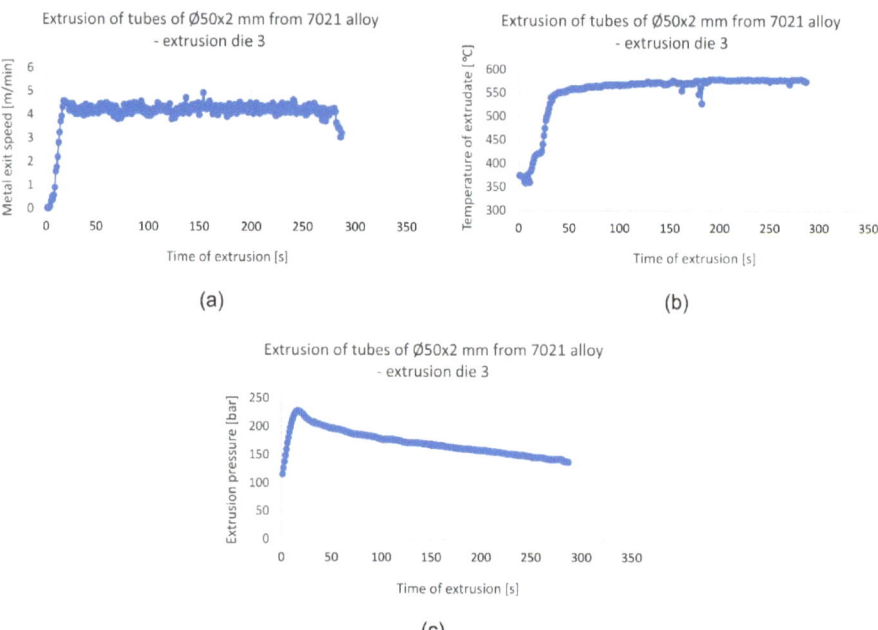

Figure 14. Recorded technological parameters of extrusion of tubes of Ø50 × 2 mm from 7021 alloy for extrusion die 3: metal exit speed (**a**), extrudates temperature (**b**) and extrusion pressure (**c**).

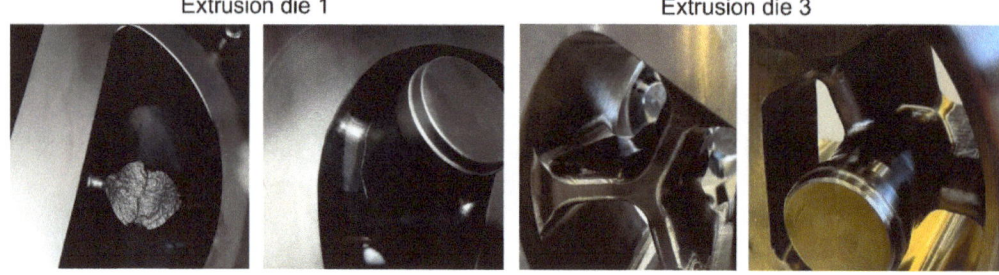

Figure 15. Porthole dies after extrusion of tubes of Ø50 × 2 mm from 7021 alloy; broken mandrel view for extrusion die 1 (on the **left**), bridge and mandrel view for extrusion die 3, free from cracking (on the **right**).

3.3. 3D Optical Scanning of Extruded Tubes

Figures 16 and 17 present the results of the optical scanning with the use of a ATOS Core 200 system, purchased from the GOM Company, for tubes of Ø50 × 2 mm extruded from 7021 alloy through the porthole dies: 1, 2 and 3 and, for comparison, through the tubes of identical geometry from 6082 alloy extruded through the classic porthole die for 6xxx series alloys. Figure 16 refers to the measurements of wall thickness in 16 points along the perimeter of the tube, whereas Figure 16 shows the measurement results of the inner diameter of the tube and deviations from the circularity in six points along the perimeter of the tube. The figures also present results of the inspection of the dimensional accuracy of the wall thickness, inner diameter and circularity of the extruded tubes in comparison with permissible deviations from the nominal dimensions described in the standard. The green colour means that a given dimension fits permissible limits in the standard, the yellow colour means that a dimension is placed on the edge of the standard and the red colour means that the dimension is out of the standard.

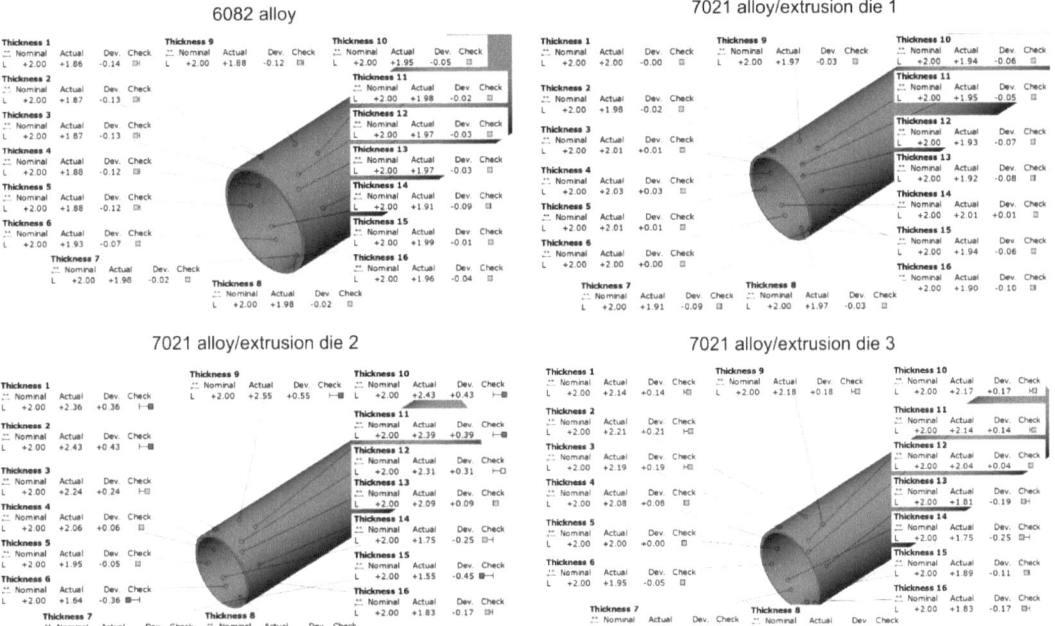

Figure 16. Results of optical scanning of the tubes of Ø50 × 2 mm: measured wall thickness of tubes extruded from 6082 alloy by using 4-hole die (on the **top left**) and extruded from 7021 alloy by using 2-hole die 1 (on the **top right**), 2-hole die 2 (on the **bottom left**), as well as extruded from 7021 alloy by using 2-hole die 3 (at the **bottom right**).

On the basis of the measurements, a high mapping result was obtained for the wall thickness, inner diameter and circularity for tubes extruded from the 6082 alloy at all the measurement points in which the dimensional deviations were acceptable (Figure 16 on the left and Figure 17 on the left). In the case of tubes from the 7021 alloy extruded through die 1 and 2, the relatively high dimensional deviations of wall thickness and circularity were observed, with the simultaneous preservation of narrow deviations of outer and inner diameters. The maximal deviations of the wall thickness reached 0.48 mm (Figure 16, die 2), which, together with concurrent positive deviation at the level of +0.43 mm, leads to a strong tube eccentricity phenomenon. However, maximal circularity deviations reached as much as 1.6 mm (Figure 16, die 1), which considerably exceeds an acceptable deviation of circularity, defined at the level of 0.25 mm. At the same time, for die 1 the narrow dimensional tolerances of the wall thickness for the tube from the 7021 alloy occurred and were better in relation to the tube from the 6082 alloy, maximal measured deviations did not exceeded 0.1 mm and permissible deviation was at the level of ±0.25 mm. At the same time, the use of die 3 is associated with a radical improvement in the dimensional tolerances of the extruded tubes—the average deviation of wall thickness was 0.14 mm, the average deviation of circularity was already acceptable at the level of 0.27 mm and the average deviation of centricity was 0.25 mm.

The results described above are illustrated in Figures 18 and 19 in the form of coloured maps of deviations, as well as in Figure 20, where the deviation of circularity, eccentricity and wall thickness for both the analysed dies are jointly shown in comparison to the standard (dashed line). In addition, an influence of the extrusion speed (within the range of

1.5 ÷ 4 mm/s) on dimensional deviations of tube for dies 1 and 2 in Figure 20 is presented. In particular, a strong influence of the extrusion speed on the circularity deviation for die 1 can be observed. This is connected with the occurring of the considerable ovality of the tube at relatively high extrusion speeds for the 7021 alloy.

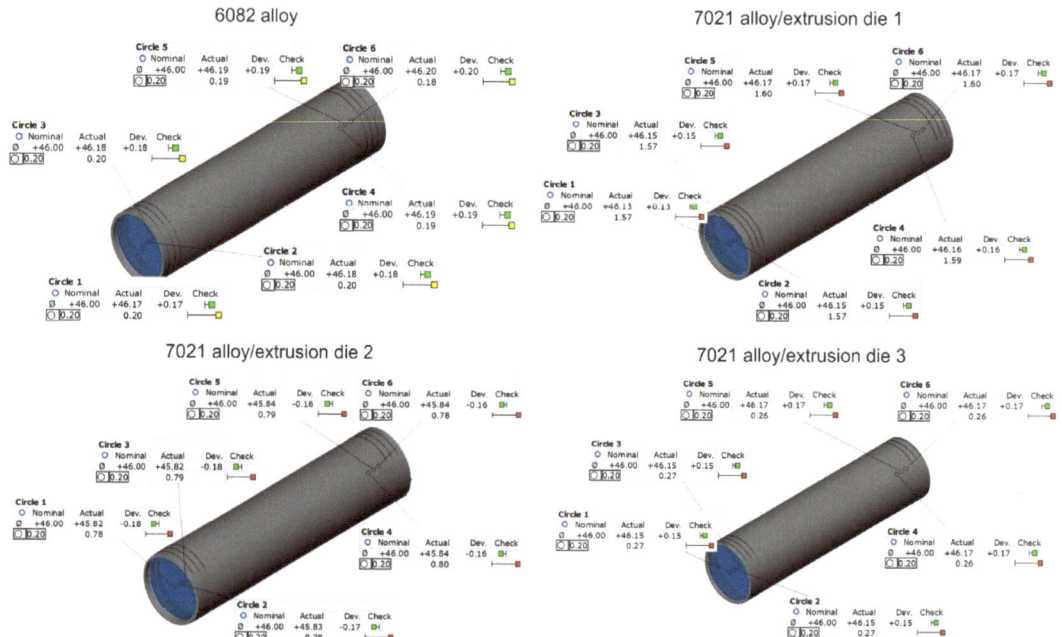

Figure 17. Results of optical scanning of the tubes of Ø50 × 2 mm: measured inner diameter and roundness deviations of tubes extruded from 6082 alloy by using 4-hole die (on the **top left**) and extruded from 7021 alloy by using 2-hole die 1 (on the **top right**), 2-hole die 2 (on the **bottom left**), as well as extruded from 7021 alloy by using 2-hole die 3 (on the **bottom right**).

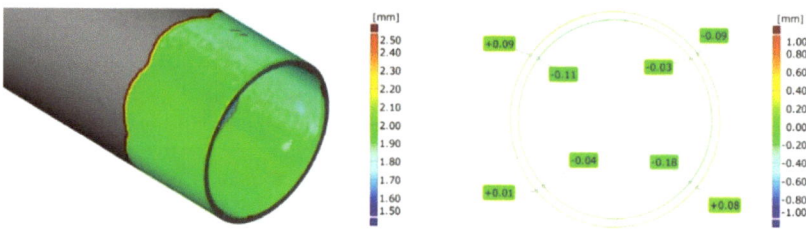

Figure 18. Results of optical scanning of the tubes of Ø50 × 2 mm from 6082 alloy: coloured maps of deviations for 6082 alloy using a 4-hole die (wall thickness on the **left** and roundness/circularity deviations on the **right**).

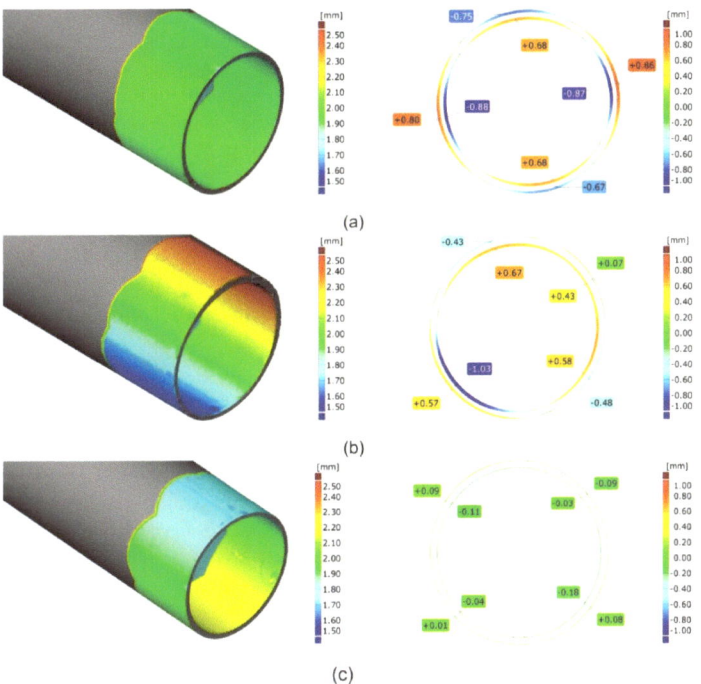

Figure 19. Results of optical scanning of the tubes of Ø50 × 2 mm: coloured maps of deviations for tubes extruded from 7021 alloy with using 2-hole die 1 (**a**), 2-hole die 2 (**b**) and 2-hole die 3 (**c**) (wall thickness on the left and roundness deviations on the right).

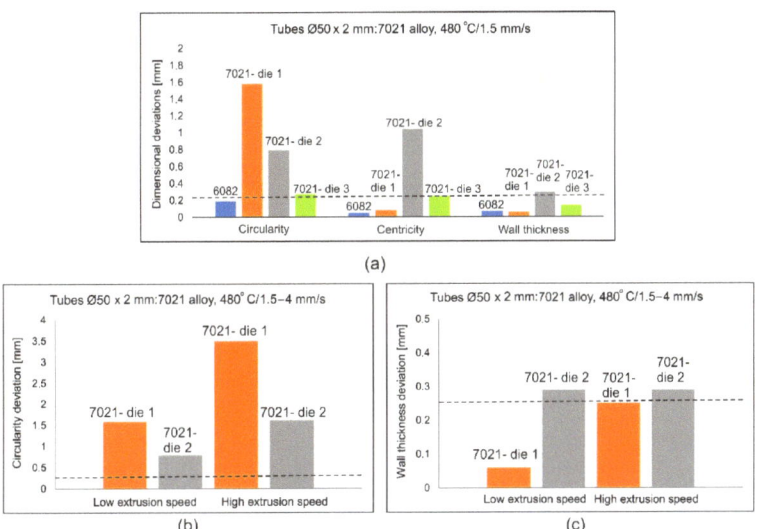

Figure 20. Results of optical scanning of the tubes of Ø50 × 2 mm extruded from 6082 alloy through 4-hole die and from 7021 alloy extruded through die 1, die 2 and die 3, (**a**)—comparison of deviations for tubes extruded from 6082 and 7021 alloys, (**b**)—circularity deviations for 7021 alloy, (**c**)—wall thickness deviations for tubes from 7021.

The influence of extrusion speed on the level of circularity deviation for die 2, which rises from 0.8 mm up to nearly 1.5 mm (as extrusion speed changes within the range of 1.5 ÷ 4 mm/s), is also significant. Regarding the eccentricity of the tube in the case of die 1, a low acceptable value is observed even for high extrusion speeds. However, a large value of eccentricity for the tube of die 2 is observed (over 1 mm) for a low extrusion speed of 1 mm/s, which unexpectedly drops twice for high extrusion, with a speed of 4 mm/s. As reported before, a small deviation of the wall thickness below 0.1 mm for die 1 and a low extrusion speed of 1.5 mm/s occurred, whereas for a high extrusion speed of 4 mm/s the deviation slightly exceeds the permissible value of ±0.25 mm. In the case of die 2, a large deviation of the wall thickness, exceeding the acceptable limit is observed, independently of the extrusion speed (Figure 20). Die 3 provided the lowest and most acceptable geometrical deviations of extruded tubes from the 7021 alloy.

3.4. Corelation Statistical Analysis of Extruded Tubes

Based on the analysis results, a correlation coefficient was established. In particular, a coefficient from 0.6 to 1.0 (high, very high and full correlation) was taken into consideration. The significant coefficients of correlation were indicated in red in Figure 21. A strong correlation of variables, such as extrusion speed, circularity, wall thickness, inner diameter, and outer diameter of a tube, and a moderate correlation of variables, such as billet temperature and centricity, was found. In particular, for die 1 a high correlation of extrusion speed and centricity of the tube (coefficient of 0.74) was obtained, which means that the rise in extrusion speed within the range of 1–4 mm/s leads to high increase in the tube centricity. In turn, a very high negative correlation between extrusion speed and wall thickness (−0.90), as well as between outer diameter (−0.62), indicates that the increase in the extrusion speed leads to significant decreasing of both wall thickness and outer diameter of the tube. The rise in temperature moderately influences the decrease in centricity of the tube (correlation coefficient of −0.43).

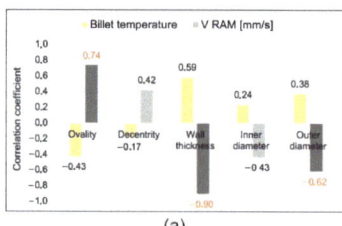
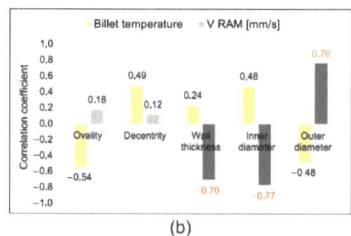

Figure 21. Results of statistical correlation analysis for optically scanned tubes of Ø50 × 2 mm extruded from 7021 alloy through porthole dies for different billet temperatures and different ram speeds, (a) die 1, (b) die 2.

In the case of die 2, a negative and high correlation of the extrusion speed and wall thickness is observed (correlation coefficient of—0.70), as well as the extrusion speed—inner diameter (correlation coefficient of—0.77), whereas a high correlation between extrusion speed and the outer diameter (coefficient of correlation 0.76) occurred. This means, in practice, a high decrease in the wall thickness and inner diameter, as well as an increase in the outer diameter, will cause a rise in extrusion speed. Moderate decreases in the circularity of the tube with an increase of billet temperature (correlation coefficient of—0.54) is also visible.

Figure 22 presents the scatter diagrams of results for the tube extrusion of Ø50 × 2 mm from 7021 alloy through die 1 (Figure 22a,b; correlation of extrusion speed and ovality, as well as wall thickness) and through die 2 (Figure 22c,d; correlation of extrusion speed and inner diameter, as well as thickness of the tube).

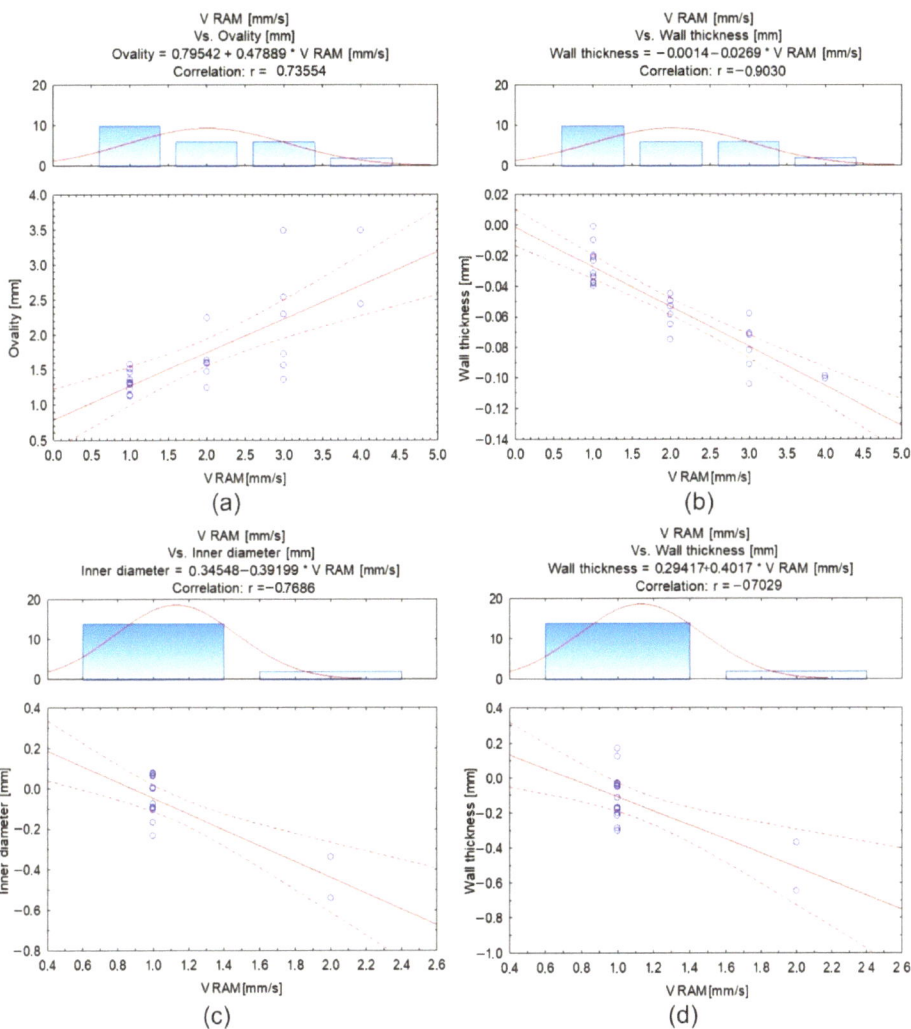

Figure 22. Scatterplots for optically scanned tubes of Ø50 × 2 mm extruded from 7021 alloy through porthole dies for different billet temperatures (480 °C, 500 °C, 510 °C and 520 °C) and different ram speeds (1.5, 2, 3 and 4 mm/s): die 1—ovality and wall thickness deviations (**a,b**), and die 2—inner diameter and wall thickness deviations (**c,d**).

3.5. 3D Optical Scanning of Extrusion Porthole Dies

Figure 23 shows the maps of dimensional deviations of the porthole die after extrusion of tubes of Ø50 × 2 mm from the 6082 alloy (bridges view—on the left, mandrel view—on the right). It should be noted that there are minimal dimensional deviations of bridges, both in the central part and on the bridge arms, not exceeding 0.03 mm, indicating a small elastic deflection in the extrusion direction and finally, on a small permanent deformation during the extrusion process. A similar situation exists for mandrels, particularly in the region of bearing length, where dimensional deviations are small at the level between -0.01 mm and 0.03 mm. Analogical maps for porthole dies after the extrusion of tubes from the 7021 alloy are presented in Figure 24.

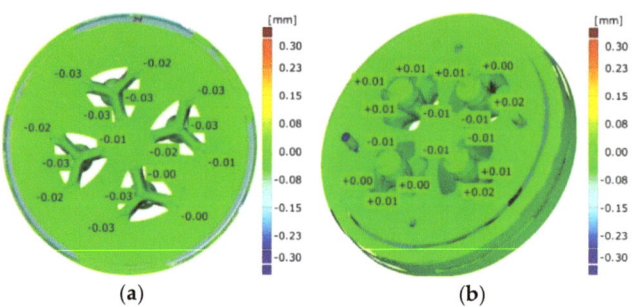

Figure 23. Results of optical scanning of the porthole dies for extrusion of tubes of Ø50 × 2 mm from 6082 alloy: die deflection—bridges view (**a**) and the mandrels (**b**).

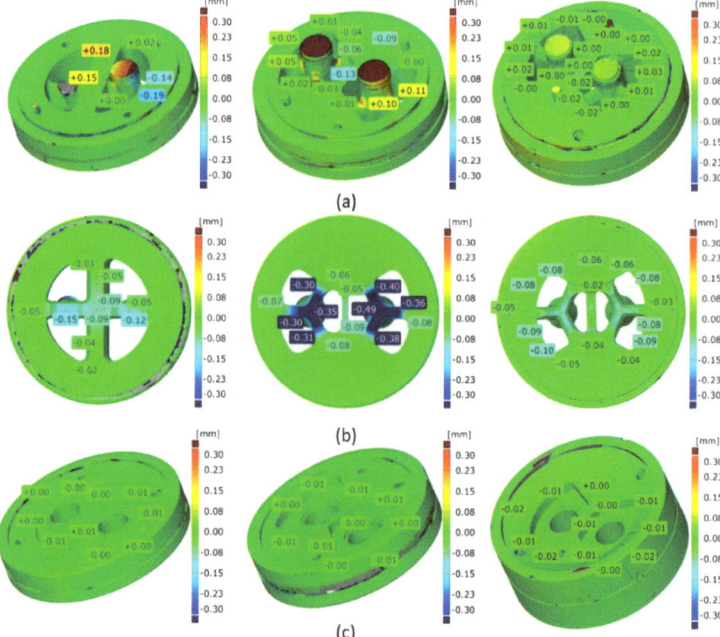

Figure 24. Results of optical scanning of the porthole dies for extrusion of tubes of Ø50 × 2 mm from 7021 alloy: die deflection for mandrels (**a**), bridges (**b**) and inserts (**c**)—die 1 (on the left), die 2 (in the middle) and die 3 (on the right).

Figure 24a shows mandrels, Figure 24b—bridges and Figure 24c—die inserts (die 1 on the left, die 2 in the middle, and die 3 on the right). The significant dimensional deviations of mandrels for die 1 and die 2 are visible (Figure 24a on the left and in the middle). Slightly higher deviation for die 1 is observed from −0.19 mm at the outer part of the mandrel up to +0.18 mm at the inner part of the mandrel. This suggests permanent deviation of the mandrel to the centre of the die; the tearing of one mandrel region is also visible. The situation for die 2 looks in reverse, i.e., there is a permanent deviation of mandrels outside the die, whereby the dimensional deviations are smaller compared to die 1. Very small dimensional deviations of the mandrels, not exceeding 0.3 mm, were stated for die 3 (Figure 24a on the right). A moderate deflection of bridges in the extrusion direction for die 1 is observed (Figure 24b on the left)—deviation is at the level of −0.12 mm down to −0.15 mm, whereas the significant deflection of the bridges for die 2 was stated at the level between −0.35 and −0,49 (Figure 24b in the middle). The minor bending of the bridges in the extrusion direction, not exceeding −0.1 mm, was stated for die 3 (Figure 24b on the right). In practice, zero dimensional deviations were stated for regions of die lands of all the porthole dies (Figure 24c).

The close-up of a chosen bridge (on the left) and a chosen mandrel (on the right) for the discussed tube from the 6082 alloy in Figure 25 is presented. Figure 26 presents the close-up of the chosen bridges (on the left) and the chosen mandrels (on the right) for the extrusion of discussed tubes from 7021 alloy—for die 1 (Figure 26a), die 2 (Figure 26b) and die 3 (Figure 26c).

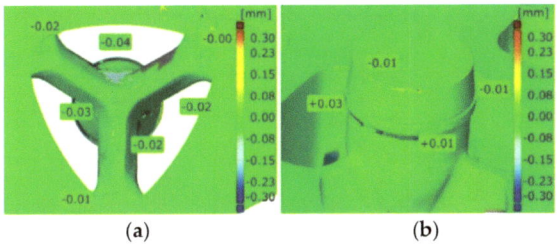

Figure 25. Results of optical scanning of the porthole dies for extrusion of tubes of Ø50 × 2 mm from 6082 alloy: die deflection—close-up of the bridge view (**a**) and the mandrel view (**b**).

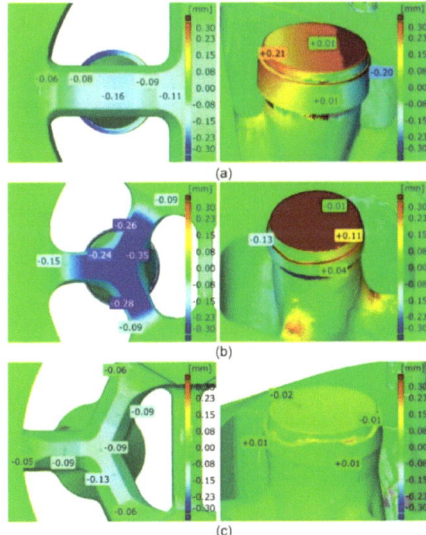

Figure 26. Results of optical scanning of the porthole dies for extrusion of tubes of Ø50 × 2 mm from 7021 alloy: die deflection—close-up of view for die 1 (**a**), die 2 (**b**) and die 3 (**c**).

3.6. ANOVA Statistical Analysis of Extrusion Porthole Dies

The results from the measurements with the use of optical scanning 3D were utilized in a statistical analysis. Surfaces of the bridge part and insert die were divided between the regions in which the dimensional deviations were determined, which was the result of the matching CAD models of the dies and real scans.

In the case of the bridge-part of die 1 (H90259), three crucial areas were analysed. Area 1 relates to the frontal part of the bridge. It was considered as a whole without demarking between particular apertures because of the die design, areas 2–5 concern the side surface of the channel and areas 6–7 indicate the surface of the mandrel/core land (Figure 27a).

The lack of area 7, which marks the surface of the calibrating land on the core, is caused by its damage. In the case of this element, the largest differences, at the level of −0.3 mm, are visible for areas 1 and 2. This is due to the significant deformation of this part of the porthole die during use. The deformation of area 1 results from the bending of the die bridges and area 2 from the excessive wear of the side surface of the channel.

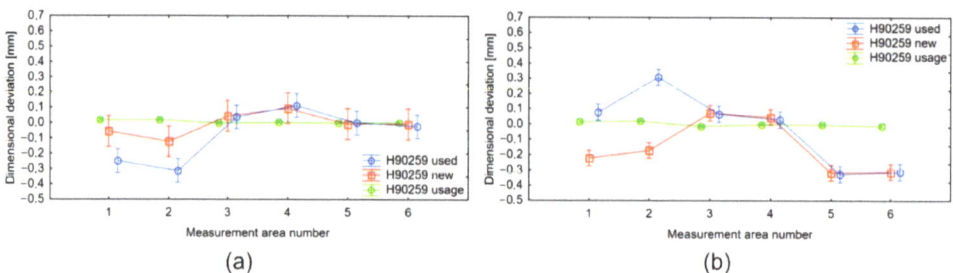

Figure 27. Results of ANOVA analysis for the porthole die 1: die core/bridge part (**a**) and die disc/insert (**b**).

In the case of the insert-part of die 1 (H90259), three crucial areas were analysed. Areas 1–2 indicate the frontal part of the channel, 3–4 the side surface of the channel and areas 5–6 relate to the die land. In the case of this element, the highest deviation values, at a maximum level of −0.3 mm, occur for the areas 5–6 (Figure 27b). The obtained values result from insufficient precision in the workmanship of the element, since the diagram representing wear indicates small deviations from the nominal value.

Three regions for each die aperture were marked in the case of the bridge-part of die 2 (H90227). The area marked with 1–4 means the dimensional deviations of the channel surface, 5–8 refers to deviations from the surface of the core land and 9–12 are related to the frontal surface of the bridge (Figure 28a). The highest deviations, at the level of −0.45 mm for the new die and −0.40 mm for the used die, occur for the area 9–12, which means that these regions seem most exposed to wear.

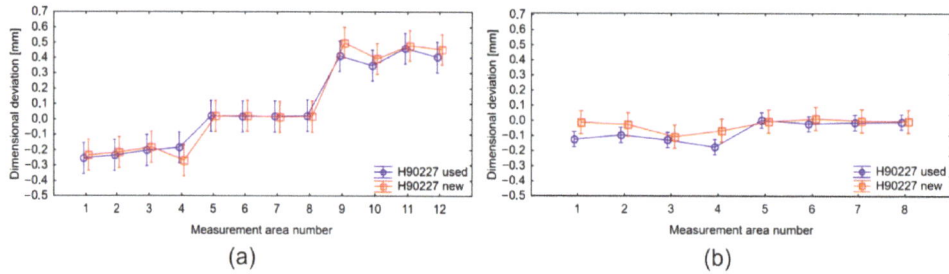

Figure 28. Results of ANOVA analysis for the porthole die 2: die core/bridge part (**a**) and die disc/insert (**b**).

In the case of the insert-part of die 2 (H90227), two areas were marked and analysed. Areas 1–4 indicate deviations obtained from the side surface of channels in the die insert, whereas areas 5–8 relate to quantities obtained from the die land surface (Figure 28b). The results indicate small dimensional deviations at a level of 0.07 mm, whereas the largest differences occur in the case of the side surface of the channel in comparison to the die land values.

4. Discussion

The distribution of the metal velocity during the extrusion of the tubes of Ø50 × 2 mm from a hard 7021 alloy, with the use of die 1 (Figure 6a), indicates a significant difficulty in metal supply to the outer regions of the die cavity in the sub-bridge area. In this area, the metal flows over the bridge, then it flows across the welding chamber, before it reaches the die cavity. The fastest flowing was obtained for the inner regions of the die cavity, as well as for the sub-bridge region, but is located close to the central extrusion axis, where, by nature, the metal flows faster. The high diversity in the metal velocity within the die cavity on the perimeter of the tube may cause the formation of the tensile stresses at the die land (Figure 8a). This, in turn, can lead to the surface tearing of the product and, in consequence, to a decrease in the exit speed from the die, which reduces the effectiveness of the process. Moreover, it may result in a decrease in the dimensional accuracy of the tube and, in particular, in the high deviations from the circularity.

It is worth noting that the improvement of the die 1 design, aimed at the acceleration of the metal flow within outer regions of the die aperture in the sub-bridge area and connected with the extending of the welding chambers and lowering of the die lands, has proved to be unsuccessful. On the one hand, the maximum opening of the inlet channels in die 1, along with an application of the central two-ridge bridge, offered a chance for a decrease in the metal flow resistance during extrusion and to an increase in the average exit speed. On the other hand, it resulted in a non-uniform metal flow at the perimeter of the tube and, finally, impeded a loss of geometrical stability of the extruded tubes (Figures 16–20).

As shown in the subsequent photogrammetric investigations, a correlation can be found between the tendency to lose roundness when using die 1 and high uniformity of the metal flow within the die aperture. In the case of die 2, where a uniform metal flow was observed (Figure 6b), deviations from the tube ovality were significantly lower (Figures 16–20); however, they also exceeded the values permissible by proper standard. The permanent deviation of mandrels from the extrusion axis in the transverse direction is a main reason of the ovality of tubes (Figures 24 and 26). However, there is also a correlation between deviations of the wall thickness—combined with the eccentricity of the tube—an elastic deflection of mandrels in the extrusion direction (Z direction), and the further permanent deviation of mandrels, as well as bridge bending in extrusion direction (Figures 24 and 26). The differences in elastic deflection of the mandrels in the porthole die results from a diverse mandrel design, particularly its thickness and length.

Lower values of the elastic deflection of the mandrel were obtained for die 1, with bridges of 28 mm in thickness and 70 mm in length, whereas significantly higher elastic deflections occurred for die 2 of 16 mm bridge in thickness and of 55 mm in length. In addition, higher permanent deflections of bridges were measured for die 2. Consistently, the higher dimensional deviations of the wall thickness and eccentricity phenomenon of die 2 were confirmed. However, a large ovality of extruded tubes was stated for the extrusion of die 1 as a result of non-uniform metal outflow from the die opening and back bending of mandrels.

The above observations suggest the inadequate porthole die design of die 1 and die 2, which were modified in terms of better metal flow control and higher strength and stiffness. The following corrections were performed to design the extrusion of die 3: slightly lowered/rounded entry to the inlet channels, the original geometry of central die compartment, thicker bridges of up to 20 mm, elongated bridges of up to 95 mm, greater height of the welding chambers of 21–25 mm, elongated mandrels of modified geometry, and radial-convex entry to the die cavity.

The FEM calculations indicated lower elastic deflections of both bridges and mandrels, as well as more uniform supply of the metal to the die aperture, while also eliminating regions where the material passes from the elastic to plastic state (the potential places of the mandrel cracking). The industrial trials of the extrusion of the discussed tubes using modified die 3 and the subsequent 3D optical scanning of extruded tubes and used dies positively validated the established assumptions. A high compliance of extrusion force was obtained for FEM numerical calculations and experimental extrusion tests of the analysed tubes (Figure 29).

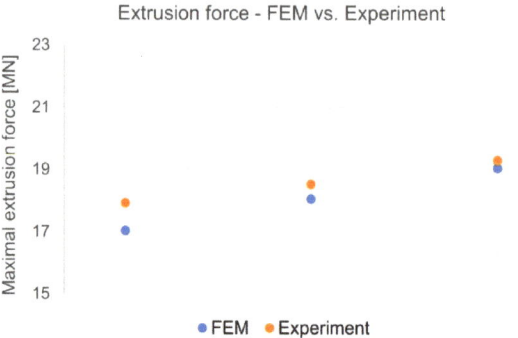

Figure 29. Comparison of extrusion force for the analysed tube from 7021 alloy—FEM predicted and experimentally validated.

5. Conclusions

Based on the obtained results, the following conclusions can be formulated:

1. The extrusion of tubes of Ø50 × 2 mm from the 7021 alloy with the use of the porthole dies is connected with high deformation resistance and high frictional resistance in comparison to the extrusion of analogical tubes from the 6082 alloy, which translates to higher extrusion pressures and lower metal exit speeds.
2. High unit pressures during extrusion tubes of Ø50 × 2 leads to high thermo-mechanical loads of dies connected with permanent deflection of bridges and mandrels, which leads to significant dimensional deviations of circularity and eccentricity, exceeding standard's limits.
3. Strong plus correlation between extrusion speed and tube circularity, strong minus correlation of extrusion speed and the wall thickness, inner and outer diameter of tubes, and moderate minus correlations between billet heating temperature and circularity of tubes exist. This means that maximal permissible extrusion speed cannot be applied, but high billet heating temperatures should possibly be advised.
4. The application of the modified die 3 resulted in successful control of the metal flow, which consisted in the effective facilitating of the inflow of metal to the sub-bridge areas, thus minimizing the side deviation of mandrels.
5. The design of the porthole dies for the extrusion of the 7xxxx series aluminium alloys needs new assumptions in relation to classic dies for the extrusion of the 6xxxx series alloys. In the extrusion of tubes from a 7021 alloy, we recommend: slightly thicker and considerably elongated bridges, special central die compartment, higher welding chambers, shaped pocket dies, geometrically modified mandrels and smooth entries to the die cavity.

Author Contributions: Conceptualization, J.Z. and W.L.; Methodology, J.M., K.Ż. and S.P.; Software, J.M.; Validation, D.L., Formal analysis, D.L.; Investigation, M.B., J.M., D.L., B.P. and H.J.; Writing—original draft preparation, D.L.; Writing—review and editing, D.L., B.P. and K.Ż.; Visualization, J.M and M.B.; Supervision, D.L.; Project administration, D.L. All authors have read and agreed to the published version of the manuscript.

Funding: This research was funded by THE NATIONAL CENTRE FOR RESAERCH AND DEVELOPMENT, grant number TECHMATSTRATEG2/406439/10/NCBR/2019 "Extrusion welding of high-strength shapes from aluminium alloys 7xxx series".

Institutional Review Board Statement: Not applicable.

Informed Consent Statement: Not applicable.

Data Availability Statement: Not applicable.

Acknowledgments: The authors are grateful to Albatros Aluminium Company for the opportunity to perform wide extrusion trials in industrial conditions.

Conflicts of Interest: The authors declare no conflict of interest.

References

1. Tomesani, L. Numerical Assessment of the Influence of Process and Geometric Parameters on Extrusion Welds and Die Deformation after Multiple-cycles. *Mater. Today Proc.* **2015**, *2*, 4856–4865.
2. Kang, C.K. Effects of chamber shapes of porthole die on elastic deformation and extrusion process in condenser tube extrusion. *Mater. Des.* **2005**.
3. Liu, Z.; Li, L.; Li, S.; Yi, J.; Wang, G. Simulation analysis of porthole die extrusion process and die structure modifications for an aluminium profile with high length–width ratio and small cavity. *Materials* **2018**, *11*, 1517. [CrossRef]
4. Xue, X.; Vincze, G.; Pereira, A.B.; Pan, J.; Liao, J. Assessment of Metal Flow Balance in Multi-Output Porthole Hot Extrusion of AA6060 Thin-Walled Profile. *Metals* **2018**, *8*, 462. [CrossRef]
5. Selvaggio, A.T.; Kloppenborg, T.; Schwane, M.; Hölker, R.; Jäger, A.; Donati, L.; Tomesani, L.; Tekkaya, A.E. Experimental Analysis of Mandrel Deflection, Local Temperature and Pressure in Extrusion Dies. *Key Eng. Mater.* **2014**, *585*, 13–22. [CrossRef]
6. Biba, N.; Rezwych, R.; Kniazkin, I. Quality prediction and improvement of extruded profiles by means of simulation. *Alum. Extrus. Finish.* **2019**, *2*, 13–17.
7. Biba, N.; Stebunov, S.; Lishny, A. Simulation of material flow coupled with die analysis in complex shape extrusion. *Key Eng. Mater.* **2013**, *585*, 85–92. [CrossRef]
8. Chathuranga, P. Case study of extrusion die optimization using innovative cartridge type die. *Light Met. Age* **2019**, *77*, 20–24.
9. Li, P. Design of a Multihole Porthole Die for Aluminum Tube Extrusion. *Mater. Manuf. Process.* **2012**, *27*, 147–153.
10. Pietzka, D.; Ben Khalifa, N.; Donati, L.; Tomesani, L.; Tekkaya, A.E. Extrusion Benchmark 2009 Experimental analysis of deflection in extrusion dies. *Key Eng. Mater.* **2010**, *424*, 19–26. [CrossRef]
11. Yi, J.; Wang, Z.; Liu, Z.; Zhang, J.; Xin He, X. FE analysis of extrusion defect and optimization of metal flow in porthole die for complex hollow aluminium profile. *Trans. Nonferr. Met. Soc. China* **2018**, *28*, 2094–2101. [CrossRef]
12. Chen, H.; Zhao, G.Q.; Zhang, C.S.; Guan, Y.J.; Liu, H.; Kou, F.J. Numerical simulation of extrusion process and die structure optimization for a complex aluminium multicavity wallboard of high-speed train. *Mater. Manuf. Process.* **2011**, *26*, 1530–1538. [CrossRef]
13. Hsu, Q.C.; Chen, Y.L.; Lee, T.H. Non-symmetric hollow extrusion of high strength 7075 aluminium alloy. *Sci. Direct Procedia Eng.* **2014**, *81*, 622–627. [CrossRef]
14. Lou, S.; Wanga, Y.; Lub, S.; Sua, C. Die structure optimization for hollow aluminium profile. *Sci. Direct Procedia Manuf.* **2018**, *15*, 249–256.
15. Cheng, L. Die structure optimization for a large, multi-cavity aluminium profile using numerical simulation and experiments. *Mater. Des.* **2012**, *36*, 152–160.
16. Lee, X. Virtual try-out and optimization of the extrusion die for an aluminium profile with complex cross-sections. *Int. J. Adv. Manuf. Technol.* **2015**, *78*, 927–937.
17. Chen, L.; Zhao, G.; Yu, J.; Zhang, W.; Wu, T. Analysis and porthole die design for a multi-hole extrusion process of a hollow, thin-walled aluminium profile. *Int. J. Adv. Manuf. Technol.* **2014**, *74*, 383–392. [CrossRef]
18. Esund, J. Extrusion of 7075 aluminium alloy through double-pocket dies to manufacture a complex profile. *J. Mater. Process. Technol.* **2009**, *209*, 3050–3059.
19. Truong, T.T.; Hsu, C.C.; Tong, V.C.; Sheu, J.J. A Design Approach of Porthole Die for Flow Balance in Extrusion of Complex Solid Aluminum Heatsink Profile with Large Variable Wall Thickness. *Metals* **2020**, *10*, 553. [CrossRef]
20. Jo, H.H.; Jeong, S.C.; Lee, S.K.; Kim, B.M. Determination of welding pressure in the non-steady-state porthole die extrusion of improved Al7003 hollow section tubes. *J. Mater. Process. Technol.* **2003**, *139*, 428–433. [CrossRef]
21. Zhou, W.; Yu, J.; Lin, J.; Dean, T.A. Effects of die land length and geometry on curvature and effective strain of profiles produced by a novel sideways extrusion process. *J. Mater. Process. Technol.* **2020**, *282*, 116682. [CrossRef]
22. Woźnicki, A.; Leszczyńska-Madej, B.; Włoch, G.; Grzyb, J.; Madura, J.; Leśniak, D. Homogenization of 7075 and 7049 aluminium alloys intended for extrusion welding. *Metals* **2021**, *11*, 338. [CrossRef]

Disclaimer/Publisher's Note: The statements, opinions and data contained in all publications are solely those of the individual author(s) and contributor(s) and not of MDPI and/or the editor(s). MDPI and/or the editor(s) disclaim responsibility for any injury to people or property resulting from any ideas, methods, instructions or products referred to in the content.

MDPI AG
Grosspeteranlage 5
4052 Basel
Switzerland
Tel.: +41 61 683 77 34

Materials Editorial Office
E-mail: materials@mdpi.com
www.mdpi.com/journal/materials

Disclaimer/Publisher's Note: The title and front matter of this reprint are at the discretion of the Guest Editors. The publisher is not responsible for their content or any associated concerns. The statements, opinions and data contained in all individual articles are solely those of the individual Editors and contributors and not of MDPI. MDPI disclaims responsibility for any injury to people or property resulting from any ideas, methods, instructions or products referred to in the content.